PROCESS PLANTS

A Handbook for Inherently Safer Design

Second Edition

PROCESS PLANTS

A Handbook for
Inherently Safer Design

Second Edition

Trevor Kletz
Paul Amyotte

CRC Press
Taylor & Francis Group
Boca Raton   London   New York

CRC Press is an imprint of the
Taylor & Francis Group, an **informa** business

CRC Press
Taylor & Francis Group
6000 Broken Sound Parkway NW, Suite 300
Boca Raton, FL 33487-2742

© 2010 by Taylor and Francis Group, LLC
CRC Press is an imprint of Taylor & Francis Group, an Informa business

No claim to original U.S. Government works

International Standard Book Number: 978-1-4398-0455-1 (Hardback)

Library of Congress Cataloging-in-Publication Data

Kletz, Trevor A.
    Process plants : a handbook for inherently safer design. -- 2nd ed. / Trevor A. Kletz, Paul Amyotte.
        p. cm.
    Includes bibliographical references and index.
    ISBN 978-1-4398-0455-1 (hardcover : alk. paper)
    1. Chemical plants--Safety measures. I. Amyotte, Paul. II. Title.

TP155.5.K538 2010
660'.2804--dc22   2010009477

Visit the Taylor & Francis Web site at
http://www.taylorandfrancis.com

and the CRC Press Web site at
http://www.crcpress.com

This is all that I have learnt: God made us plain and simple but we have made ourselves very complicated.

—**Ecclesiastes 7:29, *Good News Bible***

"Ay, but how chance this was not done before?"
"Because, my lords, it was not thought upon."

—**Christopher Marlowe, *Edward the Second***

A clever man is the one who finds ways out of an unpleasant situation into which a wise man would not have got himself.

—**Dan Vittorio Segre, *Memoirs of a Fortunate Jew***

It is preferable to expend resources eliminating a problem than to expend resources dealing with its effects.

—**G. R. Ward and A. R. Jeffs**

It is quite impossible for any design to be "the logical outcome of the requirements" simply because, the requirements being in conflict, their logical outcome is an impossibility.

—**David Pye (quoted by Henry Petroski, *To Engineer is Human*)**

It is not for you but for a later age.

—**Beethoven, to a musician who was puzzled by some of his music**

"The time has come," the Walrus said.

—**Lewis Carroll, "The Walrus and the Carpenter"**

Contents

Preface to the First Edition

This book is based on a much shorter one, *Cheaper, Safer Plants or Wealth and Safety at Work—Notes on Inherently Safer and Simpler Plants*, published by the U.K. Institution of Chemical Engineers in 1984 (second edition, 1985). It dealt only with inherently safer and simpler plants. An extended edition that also considered other ways of making plants more user friendly was published by Hemisphere Publishing Corporation (now part of Taylor & Francis) in 1991 with the title *Plant Design for Safety—A User-Friendly Approach*. For this new edition, 70% longer than the last one, the title has again been changed, as the phrase *inherently safer* is now much more familiar to engineers than *user friendly*. The difference between the two terms is described in Chapter 1. This new edition contains many more examples of inherently safer and user-friendly design and an expanded section (in Chapter 10) on procedures that can be followed to make sure that the subject is considered systematically during design.

Thanks are due to all those who have provided ideas for inclusion in the book or commented on the earlier versions—especially those whose employers wish them to remain anonymous. I would also like to thank those companies that have allowed me to describe accidents that have shown the need for inherently safer and friendlier designs.

The original book was based on lecture notes I had used for teaching inherently safer design to adult students attending short courses in loss prevention and to undergraduate and graduate students in college courses. University departments of chemical engineering may, therefore, find the book useful for teaching the elements of friendlier design. I argue in Section 10.2 that this subject should be included in all chemical engineering courses. The examples in the book can be used as subjects for discussion, participants being asked to suggest better designs. However, the book is intended primarily for process and design engineers, production staff, safety advisers, and research chemists in the process industries. I hope it will encourage them to design inherently safer plants and help them to do so. The book will also interest engineers in other industries and everyone interested in innovation.

As many design engineers have told me, inherently safer design needs more than new information; it needs a change in the way design is accomplished and this will not come about until senior managers are convinced of the need. I hope, therefore, that readers will encourage their senior managers to read the book or, if they cannot do that, to watch the video lecture on inherently safer design that has been prepared by the U.K. Institution of Chemical Engineers. It is available from 165 Railway Terrace, Rugby CV21 3HQ, England.

I have been told that the original work on which this book is based, *Cheaper, Safer Plants*, was before its time. If so, its time has now come. As will be seen, interest in inherently safer design is substantial and growing. Since the last edition was published, many more people have contributed to the subject and I could not have prepared this new edition without their published and private communications.

Readers in the United States will notice frequent references to *The Chemical Engineer* (*Chem. Eng. [U.K.]*), the biweekly magazine of the U.K. Institution of Chemical Engineers, for the publication has consistently reported new developments in the subject of this book. *Chem. Eng. (U.S.)* refers to the U.S. magazine *Chemical Engineering*. When no reference is given, the information is usually a private communication from the company or individual concerned.

Trevor Kletz

Preface to the Second Edition

The principles of inherently safer design are somewhat analogous to the laws of conservation of mass and energy. They have been around for a long time, and they have not changed since they first came into existence. But unlike mass and energy conservation laws, inherently safer design principles have been formalized in the process-safety community for only about the past 30 years. Individual practitioners undoubtedly applied the principles of intensification, substitution, and the like in their projects, but it was Trevor Kletz who first provided a road map for inherently safer design application in the late 1970s. Since that time, the concepts of inherent safety and inherently safer design have matured to the point where their widely acknowledged *common sense* nature has made them *common knowledge*. There remains a strong need, however, for books such as the current one to continue moving inherently safer design into the realm of *common application and practice*.

This second edition of *Process Plants: A Handbook for Inherently Safer Design* builds upon the first edition by a process of moderate revision of the original text and updating with developments over the past 10 years. New examples are given of application of the fundamental principles of inherently safer design—for this is first and foremost a "how-to" book. Emphasis has also been placed in this second edition on the role of inherent safety in process-safety management systems and in ensuring an appropriate process-safety culture. A comprehensive approach for developing case studies is provided to demonstrate the prevention and mitigation of process incidents by timely incorporation of inherently safer design. Dust explosion risk reduction by means of inherently safer approaches is also addressed. Finally, details on inherent-safety metrics—how to measure the degree or level of inherent safety—are also provided.

Three key features of the original text are also found in this second edition. First, the term *user friendly* has been retained in addition to the term *inherently safer* (although the latter appears more predominantly in the updated material included in the current edition). Second, the inherent-safety principles of *intensification* and of *attenuation* or *limitation of effects* have been retained as such rather than renaming them throughout the text as *minimization* and *moderation*, respectively. Although these latter labels are perhaps more commonly used in North America, there are advantages to identifying minimization with process intensification and to separating moderation into the concepts of attenuation and limitation of effects. Additionally, the essential character of the first edition is thus retained without diluting its scientific and engineering rigor. The observant reader will notice that in parts of the updated material, the user-friendly idea of *simplification* has been adopted more as a principle of inherent safety.

The third feature of the original text that has been retained is the use of the first person to refer to the work and thoughts of my coauthor in this second edition—Trevor Kletz—who, of course, is the sole author of the first edition. In general, then (with the exception of Section 16.1), "I" and "my" refer to Trevor and his pioneering work in the field of inherently safer design. Readers should, however, be aware of two facts. First, I made the decision to retain this feature because of the high regard in which I hold my coauthor; Trevor had suggested to me that "we" replace "I" throughout the text. Second, in spite of declining his gracious offer, I agree with each and every one of Trevor's personal observations. How could I not? These are the very points that have shaped my own understanding of what inherent safety means in practice.

The approach taken to revising the first edition has been to incorporate much of the work of my own research team over the past decade. I therefore want to gratefully acknowledge the many graduate students and research assistants who have so wonderfully enriched my knowledge of process safety. In particular, I greatly appreciate the friendship and partnership of Professor Faisal Khan of Memorial University, St. John's, NL, Canada. I also wish to acknowledge the huge debt of gratitude I owe to my colleagues in industry who have shared with me their valuable and practical insights; many of their examples appear in this second edition.

Ruth Domaratzki was a tremendous help in researching the literature for new examples of inherent-safety application that have appeared since publication of the first edition. She dutifully and cheerfully searched through journals and conference proceedings. Our archival efforts in this regard were directed primarily at the *Journal of Loss Prevention in the Process Industries, Journal of Hazardous Materials, Loss Prevention Bulletin, Process Safety and Environmental Protection, Process Safety Progress*, and *Safety Science*. Relevant conference papers were identified by a selective review of proceedings from, for example, the *American Institute of Chemical Engineers* suite of symposia (*Annual Loss Prevention Symposium, Center for Chemical Process Safety International Conference*, and *Process Plant Safety Symposium*), *Annual Conference of European Safety and Reliability Association, International Probabilistic Safety Assessment and Management Conference, International Symposium on Hazards, Prevention, and Mitigation of Industrial Explosions, International Symposium on Loss Prevention and Safety Promotion in the Process Industries*, and *Mary Kay O'Connor Process Safety Center Annual Symposium*.

There are two others without whom this edition would not have been possible. It should be well understood by the reader that this is Trevor's book. It is his knowledge and skillful writing that I have attempted to emulate with my words and my examples. If this edition helps to advance the field of process safety, and in particular inherent safety, full credit is extended to the lessons so ably taught by Trevor Kletz and to the man himself.

Finally, I would invoke the principle of simplification and say simply: My work on this book is dedicated to my wife, Peggy, without whom nothing in my life is possible.

Paul Amyotte
Dalhousie University, Halifax, NS, Canada

Authors

Trevor Kletz spent 38 years in the chemical industry, 7 in research, 16 in production, and the last 14 as a process-safety adviser. He then joined the Department of Chemical Engineering at Loughborough University, U.K., at first full-time and then as a visiting professor. He is also an adjunct professor at the Mary Kay O'Connor Process Safety Center at Texas A&M University. He has lectured and written widely and has written ten books, including *What Went Wrong? Case Histories of Process Plant Disasters and How They Could Be Avoided*, now in its fifth edition.

Paul Amyotte is a professor of chemical engineering at Dalhousie University in Canada. His research and practice interests are in the areas of inherent safety, process safety, and dust explosion prevention and mitigation. He has published or presented more than 100 papers on these topics. He is the current editor-in-chief of the *Journal of Loss Prevention in the Process Industries*.

Chapter 1

Introduction: What Are Inherently Safer and User-Friendly Plants?

We can pull and haul and push and drive,
We can print and plough and weave and heat and light,
We can run and race and swim and fly and drive,
We can see and hear and count and read and write....

But remember please, the law by which we live,
We are not built to comprehend a lie,
We can neither love nor pity nor forgive—
If you make a slip in handling us, you die.

—Rudyard Kipling, "The Secret of the Machines"

1.1 Introduction

The aim of this book is to show that the last line in the poem above need not always be true and that we can often design plants that can tolerate human error and equipment failure without serious effects on safety, output, and efficiency.

The chemical industry expanded rapidly during the 1960s, and two unforeseen results were a series of major fires and explosions, culminating in those at Flixborough, United Kingdom, and Seveso, Italy, and a consequent demand for

1

higher standards of safety. In response, the industry installed more and more safety equipment, and costs rose. People began to ask, "Where will it end?" "How much safety equipment should we add to our plants?" One answer was quantitative risk assessment: setting a target level of risk and not going beyond it (see Chapter 15). This approach has become widely accepted; it is an approach I use and advocate.[1] Nevertheless, it should be our second-best choice. Before we estimate the probability and consequences of a hazard and compare them with a target, we should ask whether the hazard can be eliminated.

In all industries, equipment failures and errors by operators and maintenance workers are recognized as major causes of accidents, and much thought has been given to ways of reducing them or minimizing their consequences.[2] However, it is difficult for operators and maintenance workers to keep up an error-free performance all day, every day. We may keep up a tip-top performance for an hour or so while playing a game or a piece of music, but we cannot keep it up continuously. Designers have a second chance, opportunities to go over their designs again, but operators and maintenance workers do not get second chances. Plants should therefore be designed, whenever possible, so that they are user-friendly, to borrow a computer term, and can tolerate departures from ideal performance by operators or maintenance workers without serious effects on safety, output, or efficiency.

Similarly, although much attention has been paid to the improvement of equipment reliability, 100% reliability is unattainable, and compromises have to be made between reliability and cost. Plants should therefore be designed, whenever possible, so that equipment failure does not seriously affect safety, output, and efficiency.

These arguments apply to all industry but particularly to the chemical and nuclear industries, where hazardous materials are handled and the consequences of failure—by people or equipment—are serious. The levels of reliability required are high and may be beyond the capabilities of people or materials. For example, a joint leaked after a shutdown in which 2000 joints were broken and remade. Only one was remade incorrectly, but it was the only one that anyone heard about. Afterward, the fiber gaskets in most of the 5000 joints in the unit—all those exposed to liquid—were replaced by friendlier spiral-wound gaskets.

A *hazard* is a situation that can lead to harm. A *risk* is the probability that the harm will occur. Traditional plant designs try to reduce the risk by adding protective equipment and following safe methods of working. Inherently safer and friendlier plants remove or reduce the hazards. Following traditional methods, we become more and more efficient at things we should not be doing, except as a last resort.

The ways by which friendliness in plant design can be achieved are summarized below and discussed in detail in later chapters. The characteristics are not sharply defined and merge into each other.

1. *Intensification or minimization.* Friendly plants contain low inventories of hazardous materials—so little that it does not matter if the entire inventory leaks out. What you don't have, can't leak. This may seem obvious, but until the explosion at Flixborough in 1974, little thought was given to ways of reducing the amount of hazardous material in a plant. Engineers simply designed a plant and accepted whatever inventory the design required, confident that they could keep it under control. At Bhopal, India, in 1984, the material that leaked, killing over 2000 people, was an intermediate that was convenient, but not essential, to store. Inventories can often be reduced in almost all unit operations as well as during storage (see Chapter 3).

2. *Substitution.* If intensification is not possible, then an alternative is substitution: using a safer material in place of a hazardous one. Thus, it may be possible to replace flammable refrigerants and heat-transfer media by nonflammable ones, hazardous products by safer ones, and processes that use hazardous raw materials or intermediates by processes that do not (see Chapter 4).

Intensification, when it is practicable, is better than substitution because it brings about greater reductions in cost. If less material is present, we need smaller pipes and vessels as well as smaller structures and foundations. Much of the pressure for intensification has come from those who are primarily concerned with cost reduction. In fact, friendliness in plant design is not only an isolated concept but a desirable one in a total package of measures, including cost reduction, lower energy usage, and simplification that the chemical industry needs to continue to adopt in the years ahead (Section 2.4).

3. *Attenuation or moderation.* Another alternative to intensification is attenuation by using a hazardous material under the least hazardous conditions. Thus, liquefied chlorine and ammonia can be stored as refrigerated liquids at atmospheric pressure instead of storage under pressure at ambient temperature. Dyestuffs that form explosive dusts can be handled as slurries (see Chapter 5).

Attenuation (moderation) is sometimes the reverse of intensification (minimization), for if we make reaction conditions less extreme, we may need a longer residence time and a larger inventory. In designing friendly plants, we may have to compromise by considering different possibilities (see the quotation by David Pye at the front of the book).

4. *Limitation of effects (a form of moderation) by changing designs or reaction conditions rather than by adding protective equipment that may fail or be neglected.* If friendly equipment does leak, it does so at a low rate, which is easy to stop or control. Spiral-wound gaskets, as already mentioned, are friendlier than fiber gaskets because, if the bolts work loose or are not tightened correctly, the leak rate is lower. A tubular reactor is friendlier than a pot (batch) reactor because the leak rate is limited by the cross section of the pipe and can be stopped by closing a valve in the pipe. Vapor-phase reactors are friendlier than liquid-phase reactors, for the mass flow rate through a hole of a given size is less.

By changing reaction conditions (e.g., the temperature or the order of operations), it is often possible to prevent runaways or make them less likely. By carrying out different stages of a batch process in different vessels, it may be possible to tailor the equipment more closely to the needs of each step. By using steam or oil as a heating medium, and limiting its temperature, it may be possible to prevent overheating (see Chapter 6).

Intensification, substitution, attenuation, and limitation of effects produce *inherently safer design* because they *avoid* hazards instead of controlling them by *adding* protective equipment. The term *inherently safer* implies that the process is safer because of its very nature and not because equipment has been added to make it safer. Note that we talk of inherently safer plants, not inherently safe ones, for we cannot remove all hazards. Note also that some writers use the term *inherently safer design* in a wider sense to include all the methods of making plants friendlier that I discuss in this book. Some go even further and expand the concept of inherently safer designs to include methods of locating or confining equipment so that the effects of fires and explosions are minimized. I regard such protective measures as add-ons. The inherently safer solutions use nonflammable materials or so little flammable material that a leak would hardly matter or, if that is impossible, to use the hazardous material in the least hazardous form. If we make the definition of any concept too broad, designers will say they are using it when they are barely doing so.

To use an analogy, How deep does the water have to be before we can say we are bathing? Walking in water up to our ankles may be better than nothing, but hardly allows us to claim we are bathing.

5. *Simplicity.* Simpler plants are friendlier than complex plants because they provide fewer opportunities for error and less equipment that can fail. They are usually also cheaper.

The main reason for complexity in plant design is the need to add equipment to control hazards. Inherently safer plants are therefore also simpler plants. Other reasons for complexity are as follows:

    a. Design procedures that result in a failure to identify hazards or operating problems until late in design. By this time, it is impossible to avoid the hazards, and all we can do is add complex equipment to control them (see Chapter 7).

    b. A desire for flexibility. Multistream plants with numerous crossovers and valves, so that any item can be used on any stream, have numerous leakage points, and errors in valve settings are likely (Section 8.2).

    c. Lavish provision of installed spares with the accompanying isolation and changeover valves (Section 8.1.4).

    d. Persistence in following rules or practices that are no longer necessary (Section 8.1).

    e. Our intolerance of risk. Do we go too far? (See Chapter 15.)

Equipment can, of course, combine more than one of the features of friendly plants, and they are interlinked. Thus, intensification and substitution often result in a simpler plant because there is less need for added safety equipment. At other times we may have to choose between, say, using a hazardous raw material in a reaction that cannot run away or using a less hazardous raw material in a reaction that may run away (Section 4.2.3).

6. *Avoiding knock-on effects.* Friendly plants are designed so that those incidents that do occur do not produce knock-on or domino effects. For example, friendly plants are provided with firebreaks between sections, like those in a forest, to restrict the spread of fire, or, if flammable materials are handled, the plants are built outdoors so that leaks can be dispersed by natural ventilation (Section 9.1).

7. *Making incorrect assembly impossible.* Friendly plants are designed so that incorrect assembly is difficult or impossible. For example, compressor valves should be designed so that inlet and exit valves cannot be interchanged (Section 9.2).

8. *Making status clear.* With friendly equipment, it is possible to see at a glance if it has been assembled or installed incorrectly or whether it is in the open or shut position. For example, check (nonreturn) valves should be marked so that installation the wrong way round is obvious (it should not be necessary to look for a faint arrow hardly visible beneath the dirt), and gate valves with rising spindles are friendlier than valves with nonrising spindles because it is easy to see whether they are open or shut. Ball valves are friendly if the handles cannot be replaced in the wrong position (Section 9.3).

9. *Tolerance of misuse.* Friendly equipment will tolerate poor installation or operation without failure. Thus, spiral-wound gaskets are friendlier than fiber gaskets, for if the bolts work loose or are not tightened correctly, the leak rate is much less. Expansion loops in pipework are more tolerant of poor installation than bellows. Fixed pipes, or articulated arms, if flexibility is necessary, are

friendlier than hoses. For most applications, metal is friendlier than glass or plastic (Section 9.4).

10. *Ease of control.* When possible, we should control by the use of physical principles rather than adding control equipment. Thus, one flow can be made proportional to another by using flow ratio controllers, which may fail or be neglected or, a better way, by letting one fluid flow through an orifice and suck in the other through a sidearm (Section 3.2.3).

Friendly processes have a slow and flat response to change rather than a fast or steep one; very accurate measurements are not necessary, and the control limits are not close to the safe operating ones. The control and safety systems are resilient; i.e., they do not interfere with operations or maintenance to the extent that there is a temptation to bypass them.

Processes in which a rise of temperature decreases the rate of reaction are friendlier than those with a positive temperature coefficient, but this is a difficult ideal to achieve in the chemical industry. However, there are a few examples of processes in which a rise in temperature reduces the rate of reaction (Section 9.5).

11. *Computer control.* Friendly software is scrutable; i.e., it is easy to see whether or not it will do what we want it to do. Most software is not. In addition, software should be designed by people who understand the process. The software should allow for foreseeable hardware failures and should not overload operators with too much information, such as numerous alarms sounding at the same time. It also should have been tested thoroughly (though testing of every possible combination of conditions is impracticable). Old software should be reused with caution (Section 9.6).

12. *Instructions and other procedures.* Instructions should not try to cover every conceivable condition that might arise; otherwise, they will be so long and complex that no one will read them. Another reason for complexity is a desire to protect the writer rather than help the reader.

If many types of gaskets or nuts and bolts are stocked, sooner or later the wrong type will be installed. It is better, and cheaper in the long run, to keep the number of types stocked to a minimum, even though more expensive types than are strictly necessary are used for some applications (Section 9.7).

13. *Life-cycle friendliness.* We should consider the problems of construction and demolition as well as operation and maintenance (Section 9.8).

14. *Passive safety.* If we cannot avoid hazards and have to add protective measures, then whenever possible, they should be passive rather than active or procedural. For example, to prevent or reduce damage by fire, an active method is water spray turned on automatically by a flame or heat detector. A procedural

method is water spray turned on by an operator. The equipment may fail or be switched off. The operator may fail. Fire insulation is passive; it is immediately available as a barrier to heat input and does not have to be commissioned. In nuclear reactors, pumped cooling is active; convective cooling is passive (Section 9.11).

It is the theme of this book that—instead of designing plants, identifying hazards, adding equipment to control the hazards, or expecting operators to control them—we should make more effort to choose basic designs, and design details, that are user-friendly. The chapters that follow give examples of what has been or might be done and discuss the action required. They also address the reasons that progress has not been more rapid than it has been and suggest that friendliness in plant design should be included in the training of chemical engineers (see Chapter 10).

Some of the equipment and processes I describe are well established; others are still under development, and some have still to be developed. Some may fall by the wayside. Higee distillation (Section 3.3.2) has not been as successful as I thought it would be when it became available, though I remain hopeful. Published accounts of new inventions are usually based on information from inventors and manufacturers and do not always list the disadvantages. The primary aim of this book, however, is not to provide a catalog of ready-made solutions, but to stimulate thought, and even unsuccessful ideas may do that. If the relevant inherent-safety questions are not asked when attempting to facilitate user-friendliness of process plants, then potential process or product alternatives cannot be explored.

Chapter length gives a very rough indication of the extent to which the various techniques can be used.

A few examples from industries besides the chemical industry are included—particularly nuclear power (Section 9.11 and Chapter 12). The particular issue of dust explosions is addressed in Chapter 13.

Table 1.1 summarizes the ways in which plants can be made user-friendly, and the appendix at the end of the book illustrates the principal ways in a more striking form. Note that there are always exceptions to Table 1.1.

Though this book is primarily concerned with safety, most of what is said applies also to the prevention of waste and pollution and the avoidance of those small continuous leaks into the atmosphere of the workplace that are the subject of industrial hygiene. It is better to avoid the production of pollutants and waste than to remove them by adding end-of-pipe equipment. Although simpler plants contain fewer joints and valve glands through which leaks can occur, whenever possible we should substitute safer solvents for toxic ones, such as benzene.

There has been an increasing recognition over the past decade that the principles of inherent safety have a key role to play in the field of environmental process engineering. More broadly, the integration of environmental (E) with health (H) and safety (S) concepts is clearly a topical issue in engineering research and practice.

Table 1.1 Ways of Making Plants Friendlier

	Examples of	
Feature	Friendliness	Hostility
Intensification (Chap. 3)		
Reactors	Continuous	Batch
	Well mixed	Poorly mixed
	High conversion	Low conversion
	Internally cooled	Externally cooled
	Vapor phase	Liquid phase
	Tubular	Pot
Nitroglycerin manufacture	NAB process	Batch process
Distillation	Higee	Conventional
Heat transfer	Miniaturized	Conventional
Intermediate storage	Small or nil	Large
Substitution (Chap. 4)		
Heat-transfer media	Nonflammable	Flammable
Solvents	Nonflammable and nontoxic	Flammable or toxic
Chlorine manufacture	Membrane cells	Mercury and asbestos cells
Carbaryl production	Alternative process	Bhopal process
Attenuation (Chap. 5)		
Liquefied gases	Refrigerated	Under pressure
Explosive powders	Slurried	Dry
Runaway reactants	Diluted	Neat
Any material	Vapor	Liquid

(continued)

Table 1.1 Ways of Making Plants Friendlier (Continued)

Feature	Examples of	
	Friendliness	*Hostility*
Limitation of effects (Chap. 6)		
Gasket	Spiral wound	Fiber
Rupture disc	Normal	Reverse buckling
Tank dikes	Small and deep	Large and shallow
Batch reactions	Several vessels	One vessel
Available energy	Energy level limited	Energy level high
Simplification (Chaps. 7 and 8)		
Fewer leakage points or opportunities for error	Hazards avoided	Hazards controlled by added equipment
	Single stream	Multistream with many crossovers
	Dedicated plant	Multipurpose plant
	One big plant	Many small plants
	Distributed to point of use	One big plant
	Operations combined	One piece of equipment for each operation
Spares	Uninstalled	Installed
Equipment	Able to withstand pressure and temperature	Protected by relief valves, etc.
	One vessel, one job	One vessel, two jobs
	Fluidic	Contains moving parts
Flow	Convection or gravity	Pumped

(continued)

Table 1.1 Ways of Making Plants Friendlier (Continued)

Feature	Examples of Friendliness	Examples of Hostility
Avoiding knock-on effects (Sec. 9.1)		
Buildings	Open-sided	Enclosed
Large plants	Firebreaks	No firebreaks
Tank roof	Weak seam	Strong seam
Horizontal cylinder	Pointing away from other equipment	Pointing at other equipment
Incorrect assembly impossible (Sec. 9.2)		
Compressor valves	Noninterchangeable	Interchangeable
Device for adding water to oil	Cannot point upstream	Can point upstream
Hose connections	All the same	Different for different duties
Status clear (Sec. 9.3)		
Valve	Rising spindle	Nonrising spindle
	Ball valve with handle that cannot be replaced wrongly	Ball valve with handle that can be replaced wrongly
Blinding device	Figure-eight plate	Spade
Tolerance (of maloperation or poor maintenance) (Sec. 9.4)		
	Continuous plant	Batch plant
	Spiral-wound gasket	Fiber gasket
	Expansion loop	Bellows
	Fixed pipe	Hose
	Articulated arm	Hose
	Bolted joint	Quick-release coupling

(continued)

Table 1.1   Ways of Making Plants Friendlier (Continued)

Feature	Examples of	
	Friendliness	*Hostility*
	Metal	Glass, plastic
	Ductile materials	Brittle materials
Type of control (Sec. 9.5)		
Controlled by	Physical principles	Added equipment
Ease of control (Sec. 9.5)		
Effect of errors in model	Small	Large
Response to change	Flat	Steep
	Slow	Fast
Safe operating limits	Wide	Narrow
Feedback	Negative (e.g., processes where rise in temperature produces reaction stopper)	Positive (most chemical reactions)
Resilience	Control system does not interfere with operations	Temptation to bypass
Computer control (Sec. 9.6)		
Errors easy to detect and correct	Some PES[a]	Some PES[a]
Scrutability of software	Scrutable	Inscrutable (most)
Age of software	Designed for job	Old, reused
Software designer	Understands process	Is outsider
Hardware failure	Foreseen and allowed for	Not foreseen or allowed for
Instructions and other procedures (Sec. 9.7)		
Instructions	Simple	Complex
	Flexible	Always followed

(continued)

Table 1.1   Ways of Making Plants Friendlier (Continued)

Feature	Examples of	
	Friendliness	*Hostility*
	Helps reader	Protects writer
Gaskets, nuts, bolts, etc.	Few types stocked	Many types stocked
Nuclear reactors (Chap. 12)		
Negative power coefficient	Most	Chernobyl
Slow response	AGR[a]	PWR[a]
Less dependent on added safety systems	AGR, FBR, HTGR, PIUS[a]	PWR[a]
Other industries (Sec. 9.9)		
Continuous movement[b]	Rotating engine	Reciprocating engine
Helicopters with two rotors	Cannot touch	Can touch
Chloroform inhaler	Reverse connection impossible	Reverse connection possible
Sliding windows	Horizontal	Vertical
Escape from buildings	Lath and plaster walls	Modern building
Passive safety (Sec. 9.11)		
Fire protection	Insulation	Water spray
Cooling	Convective	Pumped
Protection against high or low pressure	Vessel designed to withstand	Relief device
Analogies (Sec. 9.10)		
	Lamb	Lion
	One-story house	Staircase
	Staircase with landings	Continuous staircase
	Tricycle	Bicycle

(continued)

Table 1.1 Ways of Making Plants Friendlier (Continued)

| Feature | Examples of | |
	Friendliness	Hostility
Marble on saucer	Concave up	Convex up
Boiled egg	Pointed end up	Blunt end up
	Hard-boiled	Soft-boiled
	Medieval egg cup (egg horizontal)	Standard egg cup
Convenience foods	Boil-in-bag	Grill

Note: See chapter or section referenced for more information. Note that there may be exceptions to some of these general rules.

a   AGR = advanced gas-cooled reactor, HTGR = high-temperature gas reactor, PIUS = process-inherent ultimate safety reactor, FBR = fast breeder reactor, PWR = pressurized-water reactor, PES = programmable electronic system.

b   In practice, reciprocating internal combustion engines are not less friendly than rotating engines, though one might expect equipment that continually starts and stops to be less reliable.

As noted in the previously mentioned item 13, it is important to adopt a life-cycle approach when dealing with EHS integration throughout all phases of plant design, construction, operation, and shutdown. A publication of the American Institute of Chemical Engineers[3] highlights the benefits to be gained by integrating the concepts afforded by various paradigms within the EHS field (Table 1.2).

Recent work at ETH Zurich[4] has focused on comparing various methods for assessing EHS hazards in early-stage chemical process design. Environmental hazard categories considered in this work include persistence in air, water, and soil; solid waste; bioaccumulation; and other issues such as global warming potential

Table 1.2 Paradigms Available for Integration within an EHS Framework[3]

Paradigm	Safety	Health	Environment
Inherent safety	✓✓✓	✓✓	✓
Pollution prevention		✓	✓✓✓
Green chemistry	✓✓	✓	✓✓✓
Green technology	✓		✓✓✓
Design for the environment		✓	✓✓✓

Note: ✓✓✓ = primary focus; ✓✓ = secondary focus; ✓ = directly linked benefit.

and ozone depletion potential. Health hazard categories include dispersion, acute toxicity, irritation, and chronic toxicity; safety hazard categories consist of mobility (e.g., vapor pressure), fire/explosion, reaction/decomposition, and acute toxicity. Interestingly, the various assessment methods yielded results in general agreement, with no one method having a unique advantage in terms of EHS coverage. Linking of safety and environmental concerns with the additional consideration of optimized process scheduling has similarly been achieved by a team at Texas A&M University.[5]

1.2 Inherently Safer and Friendlier Design as Part of an Overall Approach

The steps to be followed are often summarized as *identify, prevent, control, and mitigate hazards*. Inherently safer and friendlier design might be described as part of prevention, but it is better to use the additional guide words *avoid* or *reduce* because designers seem to respond to the word *prevent* by adding protective equipment (Table 10.3).

This book describes various ways of making plants friendlier. On the whole, the earlier chapters, summarized in items 1–5 in the preceding section (i.e., the key principles of inherent safety), will help us avoid major hazards, whereas the later ones, items 6–14, will help us avoid the less serious ones.

We cannot, of course, avoid every hazard, and then we have to prevent, control, and mitigate. The methods we can use are, in the usual order of preference, *passive, active,* and *procedural*, as discussed previously and in Table 10.3. An early introduction to the ideas expressed in Table 10.3 is given at the end of this chapter in Section 1.3.

Inherently safer and friendly features should be introduced during design. It is often difficult and expensive to incorporate them into an existing plant. However, stocks can always be reduced (Section 3.6), and Section 3.2.9 describes the way an old plant was made inherently safer.

As stated at the beginning of this chapter, inherently safer and user-friendly designs are not merely desirable but isolated features but rather parts of a total package of improvements that the process industry needs: a move toward plants that are simpler, cheaper, and safer and that use less energy, need less maintenance, and produce less waste and pollution (Section 2.4).

As already stated, friendly plants are often cheaper than hostile ones. To quote a misprint in an English newspaper, we can have "Wealth and Safety at Work."[6] Table 1.3 summarizes the principal ways in which friendliness can be achieved and the effects on costs.

Table 1.3   Why and When Friendly Plants Are Cheaper

Feature	Effect on Cost Saving	Reason
Intensification	Large	Smaller equipment and less need for added safety equipment
Substitution	Moderate	Less need for added safety equipment
Attenuation	Moderate	Less need for added safety equipment
Limitation of effects	Moderate	Less need for added safety equipment
Simplification	Large	Less equipment
Avoiding knock-on effects		
• Layout	Negative	More land needed; some increase in cost
• Open construction	Moderate	Buildings not needed
• Weak roof tank	Nil	Safer design no more expensive
Incorrect assembly impossible	Nil	Good design usually no more expensive than bad
Status clear	Nil	Good design usually no more expensive than bad
Tolerance	Modest	Fixed pipe cheaper than hoses or bellows
Ease of control	Moderate	Less control equipment needed; less maintenance
Software	Nil	Good design usually no more expensive than bad

1.3 Hierarchy of Controls

To conclude this introductory chapter, the order of preference for safety control measures (referenced in Section 1.2) is presented as a partial answer to the question posed by the chapter title: "What are inherently safer and user-friendly plants?" Simply put, such plants are designed and operated according to the systematic approach to loss prevention illustrated in Figure 1.1. With this approach,

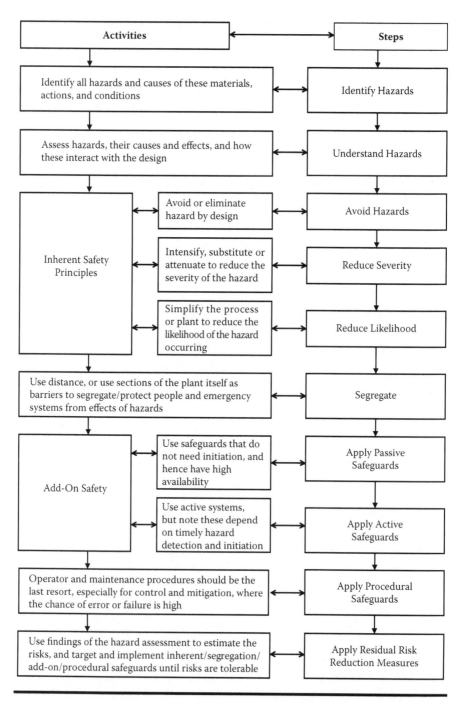

Figure 1.1 A systematic approach to loss prevention (hierarchy of controls).

the preferred order of consideration for risk-reduction measures is—from most to least effective—inherent, passive engineered, active engineered, and procedural safety. This is akin to the layer of protection analysis (LOPA) concept in which inherently safer process design sits at the central core of the layers.[7]

Andrew Hopkins, in his book *Safety, Culture and Risk: The Organizational Causes of Disasters*,[8] uses the phrase *hierarchy of controls* to describe essentially the same idea; i.e., there is a hierarchical ordering of controls to deal with hazards and the ensuing risk. This hierarchy covers the spectrum from elimination (at the top of the hierarchy) through engineering and administrative (procedural) controls, to PPE (personal protective equipment) at the bottom of the hierarchy. Manuele[9] calls this sequence the *safety decision hierarchy.*

It is important to keep in mind two critical features of the hierarchy of controls. First, this concept clearly indicates that inherent safety is not a stand-alone concept. Inherent safety works through a hierarchical arrangement in concert with engineered (passive and active) and procedural safety to reduce risk. Second, the hierarchy of controls does not invalidate the usefulness of engineered and procedural safety measures. Quite the opposite: The hierarchy of controls recognizes the importance of engineered and procedural safety by highlighting the need for careful examination of the reliability of both mechanical devices and human actions. These considerations must therefore be incorporated into the overall process of risk assessment.

The systematic, hierarchical approach displayed in Figure 1.1 affords practitioners the opportunity to bring their own interpretations to the quest for user-friendly process plants. One person's *intensify* may be another's *minimize*. Someone's *attenuate* may be someone else's *moderate*. You may wish to consider *segregate* (i.e., unit segregation) as a measure separate from inherent safety; your colleague may consider it as a form of *limitation of effects*. As previously mentioned, the characteristics of a user-friendly plant are sometimes not sharply divided and may merge into one another. Process design, like life, is seldom linear.

References

1. T. A. Kletz, *Hazop and Hazan: Identifying and Assessing Process Industry Hazards*, 4th ed. (Rugby, U.K.: Institution of Chemical Engineers, 1999).
2. T. A. Kletz, *An Engineer's View of Human Error*, 3rd ed. (U.K.: Institution of Chemical Engineers, 2001).
3. Center for Waste Reduction Technologies/Center for Chemical Process Safety, *Making EHS an Integral Part of Process Design* (New York: American Institute of Chemical Engineers, 2001).
4. I. K. Adu, H. Sugiyama, U. Fischer, and K. Hungerbuhler, "Comparison of Methods for Assessing Environmental, Health and Safety (EHS) Hazards in Early Phases of Chemical Process Design," *Process Safety and Environmental Protection* 86, no. 2 (2008): 77–93.

5. E. M. Al-Mutairi, J. A. Suardin, S. M. Mannan, and M. M. El-Halwagi, "An Optimization Approach to the Integration of Inherently Safer Design and Process Scheduling," *Journal of Loss Prevention in the Process Industries* 21, no. 5 (2008): 543–549.

6. *The Daily Telegraph,* 1983.

7. A. M. Dowell, "Layer of Protection Analysis and Inherently Safer Processes," *Process Safety Progress* 18, no. 4 (1999): 214–220.

8. A. Hopkins, *Safety, Culture and Risk: The Organizational Causes of Disasters* (Sydney: CCH Australia Limited, 2005).

9. F. A. Manuele, "Risk Assessment & Hierarchies of Control," *Professional Safety* 50, no. 5 (2005): 33–39.

Chapter 2

Inherently Safer Design: The Concept and Its Scope and Benefits

The main obstacle to innovative problem solving in petrochemical companies is over-conservatism in management.... Imaginative solutions to problems always run counter to the conventional wisdom of the firm.

—**C. H. Kline**[1]

2.1 The Concept

The essence of the inherently safer approach to plant design is the avoidance of hazards rather than their control by added protective equipment. Although we generally think of safety in a comparative sense, one experienced practitioner[2] has made the distinction between *safe* design and *safer* design. He explains that with *safe* design, there are active safeguards to prevent the occurrence of hazardous events and to protect people and plant from the effects. With *safer* design, there are fewer hazards, fewer causes, and fewer people to be exposed to the effects.

All the ideas described in this book might be considered as examples of inherently safer design. However, as explained in Chapter 1, I use the phrase in a narrower sense to describe ways of eliminating large inventories of hazardous materials and hazardous equipment and operations, and this aspect is discussed in this and the next four chapters.

Since the explosion at Flixborough in 1974 in which 28 people were killed, there has been a profusion of papers on ways of preventing similar incidents from happening again. At one extreme, because Flixborough was a plant for the manufacture of nylon, the "back-to-nature" enthusiasts have suggested we abandon human-made fibers and use natural materials such as wool and cotton instead. They overlook the fact that the "accident content" of natural fibers is higher than that of human-made fibers, although in agricultural accidents people are killed one at a time and receive less publicity than those who are killed in the chemical industry. Flixborough may have been the price of nylon (to quote a television reporter), but the price of wool and cotton is higher. (The cost of any article is the cost of the labor used to make it, and the capital costs are other people's labor. Agriculture is a low-wage industry, and thus there will be more labor in a woolen or cotton garment than in a synthetic one of the same cost. Because agriculture is also a high-accident industry, there will be more fatal accidents per woolen or cotton garment than per synthetic garment.)

Similarly, Bhopal, where over 2000 people were killed by a toxic gas release in 1984, was a plant for the manufacture of insecticides, and people understandably ask whether the price paid for insecticides is too high. But insecticides, by increasing food production, have saved far more lives than were lost at Bhopal.[3]

However, Flixborough and Bhopal were not inevitable prices for nylon and insecticides, and there are many ways of making these plants safer and many lessons to be drawn from these tragedies.[4] Most of the papers and reports on the accidents have suggested the need for installation of more and better protective equipment such as gas detectors, emergency isolation valves, trips and alarms, scrubbers and flare stacks, fire protection and firefighting equipment, stronger buildings, and so on. The safety adviser has acquired a reputation as someone who adds to the cost and complexity of the plant. The added equipment is necessary, we do not doubt, but it is also expensive and complex.

Other papers have suggested that more attention be paid to plant layout and location and that there should be more regulations, a system of licensing, more and better safety advisers, and so on. All but a few of the papers and committees have overlooked the possibility of a better and cheaper solution to the problem.

Many years ago Henry Ford said, "What you don't fit costs nothing and needs no maintenance." Similarly, what you don't have can't leak. If we could design our plants so that they use so little hazardous material that it does not matter if it all leaks out, or use safer materials instead, or use the hazardous ones at lower temperatures and pressures or diluted by a safe solvent, we would avoid rather than solve many of our problems. These changes—described as intensification, substitution, and attenuation—produce plants that are inherently or intrinsically safer, whereas a conventional plant in which the hazards are kept under control is extrinsically safer.

The earlier papers on the subject used the term *intrinsically safer*, but *inherently safer* is now preferred because it avoids confusion with *intrinsically safe*, as used by electrical engineers, to describe a circuit with insufficient power to ignite a mixture of flammable gas, vapor, or dust with air.

I do not, of course, suggest that there should be any ban on plants that contain large inventories of hazardous materials—only discouragement. The development of alternative processes and equipment will take time and may sometimes prove impossible. When large inventories are unavoidable, we can, by good design and operation, keep them under control, but I do suggest that, in designing plants, we make a low inventory one of our aims. Usually, in the past, we have given little or no thought to the amount of material in the plant. We have accepted whatever inventory was called for by the design. If we set out to reduce the inventory, we will find that in many cases we can do so, as the remaining chapters show.

To use an analogy, if the meat of lions were good to eat or their skins made very good clothes, our farmers would be asked to farm lions, and they could do so. They would need cages around their fields instead of fences, but by good design and operations, they could make the chance of an escape very small. Only occasionally, as at Flixborough and Bhopal, will the lions break loose. But why keep lions when lambs will do instead?

Or, to use another analogy, the most dangerous items in our homes are stairs. More people are killed and injured by falling down stairs than in any other way. Traditional or extrinsically safer ways of controlling the hazard are to add handrails, train people to use them, make sure the carpet is secure, and keep the stairs free from junk. The inherently safer solution is to buy a one-story house. If this is too expensive, then a halfway course of action is possible: we can install a staircase with frequent landings so that no one can fall more than a few steps.

This simple analogy illustrates a point that will be emphasized later (Section 10.1.1): Many decisions about inherent safety have to be made early in design. It is too late to tell the builder when our house is complete that we do not want any stairs; it is too late to tell the architect when the drawings are complete. The decision to avoid stairs must be made right at the beginning of the design process, when we are first instructing our architect. The decision may affect the amount of land we need and thus the location of our new home.

Similarly, the greatest opportunities for development of an inherently safer chemical manufacturing process exist at the earliest stages of research and development, in the selection of the product synthesis route.[5] Inherent-safety solutions must be checked for implementation at the basic design stage, but they must first be evaluated during conceptual design.[6] Notwithstanding the preference for early consideration, it is never too late to consider inherently safer alternatives.[7] There have been reports of inherent-safety enhancements to existing plants, such as the modification of a bromine raw-material-handling facility by replacing an existing bulk storage tank with cylinders.[8]

2.2 Defense in Depth

The prevention of accidents is based on defense in depth. If one line of defense fails, there are others in reserve. To put inherently safer design in context, the main lines of defense against leaks of flammable materials are described below. The first four steps also help prevent leaks of toxic or corrosive materials or minimize their effects.

1. *Avoidance.* Avoid the hazard by one of the methods already described; i.e., use so little hazardous material that leaks do not matter (intensification), use safer materials instead (substitution), or use hazardous materials in a safer form (attenuation).
2. *Containment.* We should design, construct, operate, and maintain equipment so that the chance of a leak is minimal.

Most of the materials that we handle in the oil and chemical industries are not flammable or explosive in themselves, but only when mixed with air (or oxygen) in certain proportions. To prevent fires and explosions, we have therefore to keep the air out of the plant and keep the fuel in the plant. The former is easy, for most equipment operates at pressure; nitrogen blanketing is widely used to keep air out of low-pressure equipment such as tanks, stacks, and centrifuges. Unfortunately, containment normally has to be broken during maintenance, which is the source of many accidents.

3. *Detection and warning.* Promptly detect any leaks that occur and warn people who are not needed to deal with the leak so that they can leave the area.
4. *Isolation.* Use emergency isolation valves to isolate the leaks. We cannot install valves on the lines leading to all equipment that might leak, but we can and should install them when experience shows that the chance of a leak is high or the amount of material that could leak out is large. An example of the former is lines leading to or from very hot or cold pumps; an example of the latter is lines leading from the bases of large distillation columns.[9]
5. *Dispersion.* Disperse the leaking material by open construction supplemented, if necessary, by steam or water curtains. Secondary containment (for example, by a building) is sometimes used to confine leaking material. This is the right policy for radioactive materials, for pollutants, or for toxic materials that may not disperse before they reach people who do not know how to react to them. This is not a suitable policy for flammable materials because the concentration in the building may soon rise above the lower flammable limit, and an explosion may then occur.
6. *Removal, as far as possible, of known sources of ignition.* Removal of known sources of ignition is one of the weaker lines of defense, and we should never rely on it. So little energy is needed to ignite a mixture of flammable vapor

and air that explosions may still occur even though we have tried to remove all known sources of ignition.[10]

7. *Protection of people and equipment against the effects of fire and explosion.* As discussed in Chapter 1, passive methods of protection are better than active ones. For example, it is better to protect equipment from fire by insulation than by water spray. The insulation does not have to be commissioned and offers an immediate barrier against fire, whereas someone or something has to activate a water spray system and may not do so. If emergency cooling is needed, convection or gravity flow is better than pumped flow.

8. *Provision of firefighting facilities.* A castle falls to the enemy. The moat had silted up and the outer wall had crumbled, but no one worried because the inner wall was impregnable. When it failed or someone left a gate open, there was nothing to fall back on.

Similarly, in industry, a common failing is to ignore the outer lines of defense because the inner ones are considered impregnable. When they fail, there is nothing to fall back on. Some companies have not worried about leaks of flammable gas because they had, they believed, removed all sources of ignition so that leaks could not ignite. When an unsuspected source of ignition turned up, an explosion occurred.[10] More often, companies have not worried about large inventories of hazardous materials because they knew, they thought, how to prevent leaks and deal with any that occurred. When they were proved wrong, a fire or explosion occurred or people were killed by toxic gas or vapor.

Effective loss prevention is based on strong outer defenses, i.e., on getting rid of our hazards when we can rather than controlling them by added equipment that may fail or may be neglected. "It is hubris to imagine we can infallibly prevent a thermodynamically favoured event."[11]

2.3 The Scope

Applying the concept of inherently safer design at the very beginning of a project, we may be able to choose a safe product instead of a hazardous one. When a process is being chosen, we may be able to choose a route that avoids the use of hazardous raw materials or intermediates (Section 4.2).

Once the chemistry has been decided and we are developing the flowsheet, we may be able to choose or develop intensified equipment such as reactors, distillation columns, and heat exchangers that do not require large quantities of materials in progress (Sections 3.1 through 3.5). It may be possible to manage without intermediate storage, possibly by siting production and consuming plants near each other (Section 3.6), and to avoid flammable heat-transfer fluids or refrigerants (Section 4.1).

With detailed design, we may be able to reduce inventories by the application of well-known methods (Section 3.7). The most important inventories to reduce or avoid are those of flashing flammable or toxic liquids, i.e., liquids under pressure above their atmospheric pressure boiling points. Liquids below their boiling points produce very little little vapor; gases leak at a lower mass rate than liquids through a hole of a given size and are often dispersed by jet mixing. Flashing liquids, however, leak at about the same rate as cooler liquids and then turn into a mixture of vapor and spray (Figure 2.1). The spray, if fine, is just as flammable, explosive, or toxic as the vapor and can be spread as easily as the wind. Most vapor-cloud explosions and most major toxic incidents have been the results of leaks of flashing flammable liquids.[12] Flixborough and Bhopal were both due to leaks of flashing liquids. At Bhopal, the liquid was not normally above its boiling point, but addition of water to a storage tank caused a runaway reaction to occur.

The scope of inherently safer design is thus seen to extend throughout the process-design life cycle. This is a recurrent theme in these pages, and also in the numerous inherently safer design resource publications that have appeared in the literature over the past 10 years or so.[13–22] These various review articles and texts contain a wealth of examples of inherent-safety application, some of which appear in later chapters of this book.

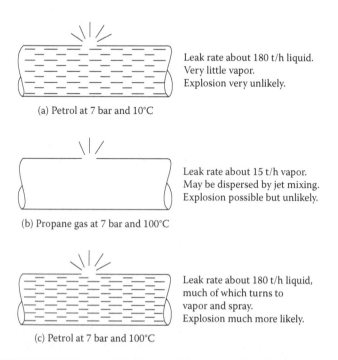

(a) Petrol at 7 bar and 10°C

Leak rate about 180 t/h liquid.
Very little vapor.
Explosion very unlikely.

(b) Propane gas at 7 bar and 100°C

Leak rate about 15 t/h vapor.
May be dispersed by jet mixing.
Explosion possible but unlikely.

(c) Petrol at 7 bar and 100°C

Leak rate about 180 t/h liquid,
much of which turns to
vapor and spray.
Explosion much more likely.

Figure 2.1   Leak rates through a 2-in. hole. (*Petrol* is the U.K. term for *gasoline*.)

2.4 The Benefits of Inherently Safer Design

Inherently safer plants are usually cheaper than conventional ones because they do not need so much additional protective equipment. It is difficult to say how much cheaper, for designers do not keep records of safety costs. Safety should not be something added afterward to a plant, like a coat of paint, but an integral part of design. If a design engineer is asked to list the cost of the safety features on a plant, the designer usually lists the newer features that were the subject of debate but not the cost of well-established safety features such as relief valves. These are considered basic engineering and taken for granted.

At a guess, we would save at least 5%, often 10%, of the capital cost of most new plants if we could reduce our inventories of hazardous materials and thus reduce expenditure on added protective equipment such as trips and alarms, fire and leak detectors, emergency isolation valves, fire insulation, water spray, and firefighting equipment. On some plants the savings would be much greater.

Equally important would be reductions in the cost of testing and maintaining the equipment. For instruments, this is roughly equal to the installed capital cost. As a result, instruments cost twice what you think. In addition, much management time is, or should be, taken up by monitoring to make sure that the protective equipment is operated, tested, and maintained correctly. If plants can be made inherently safer, this effort can be reduced. Also, it will be easier to persuade the authorities and the public that the plant will not blow up or poison the neighborhood. It will be easier to find a site for the plant, and the equipment may not have to be spread so far apart (Section 10.6).

These and several other cost parameters were considered in a recent analysis of the "real cost of process safety."[23] This hypothetical study was conducted for a simple process of making salt solution in open vats by dissolving sodium chloride in a highly hazardous chemical, *dihydrogen monoxide* (i.e., water). Although obviously somewhat tongue in cheek, the cost benefits of inherent-safety applications are well made by the authors.[24]

If inventories can be reduced, the biggest saving will probably come from a reduction in the size of the plant items (reactors, distillation columns, heat exchangers, storage vessels, etc.) and a corresponding reduction in the size of the pipework, structures, and foundations. Much of the capital cost for a chemical plant is related to its civil engineering structural design and components.[25] Compare the reductions in cost achieved in other industries by reductions in size. Compare modern computers and radios with earlier models or a modern electric pump with a steam-driven beam-pumping engine. Much of the pressure for intensification has come from those primarily interested in reducing costs, and because intensification produces such big reductions in cost, it should be our first choice before substitution or attenuation. The finance director and the process-safety engineer can be allies! Consider the title of Reference 25: *Process*

Intensification Magnifies Profits; surely such a headline would catch the eye of any business manager.

The plants we are now designing, with their large vessels and inventories, may seem to us to be in the forefront of technology. Perhaps they are really the chemical engineering equivalent of a beam engine, and our grandchildren will be starting a society to preserve the few remaining large distillation columns so that, like the surviving beam engines, they can be shown to the public on weekends and circulated on public holidays.

Finally, another possible advantage of substantial reductions in size and cost, by analogy with computers, is that if large reductions can be achieved, then overcapacity may be economic and will simplify control problems (Section 10.1.4).

Inherently safer design is not simply an isolated improvement, but a desirable part of the total packet of changes that the chemical industry needs to make for the twenty-first century. Our plants are too hazardous, too expensive, and too complicated and use too much energy, need too much maintenance, and produce too much pollution and waste. Inherently safer design, especially intensification, can help us to reduce these limitations (Figure 2.2). Simplification, and therefore reduction in cost and maintenance, comes about because less additional safety equipment is needed. If we can intensify, the smaller size of the equipment reduces cost directly. If we can reduce the recycles of unconverted raw material ("free rides"), we will save energy and further reduce cost (Section 3.2.4). The principles of inherently safer design can be applied to prevention of pollution and waste reduction. As described in Chapter 1, many companies talk about inherent SHE (safety, health, and environment).

Sir Maurice Hodgson, former chairman of Imperial Chemical Industries (ICI), has written, "Certainly in ICI we are clear that our future plants will have to contain less ironmongery for a given capacity, be less energy-consuming and be more

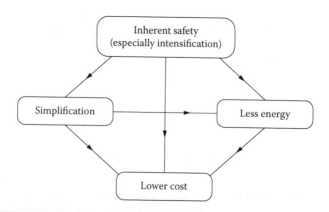

Figure 2.2   Inherent safety is not an isolated (although desirable) change but, rather, part of a package of improvements needed by the process industries.

productive per unit of everything. Process intensification, energy-saving technology, better design and control systems are the means."[26]

When I worked in industry, my research colleagues sometimes complained about the high cost of extrinsic safety features. They suggested that by asking for so much added safety I was making their designs uneconomic. I replied that the added safety was necessary because they invented such poor processes—poor in other ways besides safety. It is not economic to push large inventories round and round, giving them a "free ride." If engineers could invent better processes, we would not need to add so much safety equipment. See the comments from an experienced process designer at the end of Chapter 4.

In the fourth century a theologian wrote that "all great controversies depend on both sides sharing one false premise."[27] A false premise is shared by the operating staff who want to add more protective equipment to their plants and the designers who say, "We've spent enough; the plant is now as safe as it needs to be." We do not have to spend more money to make the plant safer; if we consider safety early in design, we can avoid many hazards. Expressed from a probabilistic risk perspective, inherently safer design is a way to decrease the uncertainty about whether harmful events will take place.[28]

Appendix: "The Lady or the Tiger?"—A New Version of an Old Tale

A king offered a challenge to three young men. Each would be put in a room with two doors and could open both. If he opened one, a hungry tiger would come out—the fiercest and most cruel that could be obtained—that would tear him to pieces. If he opened the other, a young lady would come out, the most suitable to his years and station that His Majesty could select from amongst his fair subjects.

> The first young man refused the challenge. He lived safe and died chaste.
> The second young man hired risk-assessment consultants. They collected all the available data on lady and tiger populations. They brought in sophisticated equipment to listen for growling and detect the faintest whiff of perfume. They completed checklists. They developed a utility function and assessed the young man's risk aversion. Naturally this took time (and money). The young man, now no longer quite so young, began to worry that he would soon be no longer able to enjoy the lady. Finally, he asked the consultants to recommend a course of action. He opened the optimal door and was eaten by a low-probability tiger.
> The third young man took a course in tiger handling.[29]

The moral of the story (for those who like to have parables explained) is as follows:

The young men represent us all, the tiger the hazards of industry, and the lady the benefits industry brings to humanity. Like the first young man, society can leave the game. We can manage without chemical plants, the benefits they bring, and the risks they carry.

Like the second young man, we can (and do) try to assess the risks and open the safest doors, but we can never be completely sure that our assessments are correct and that an accident will not occur.

When possible, we should try, like the third young man, to change the work situation and to choose designs or methods of working that minimize the hazard.

References

1. C. H. Kline, "Surviving the Petrochemical Collapse," *Hydrocarbon Process* 73, no. 2 (1983): 84A–84H.

   This is not an isolated opinion. I might equally well have quoted the following:

   R. B. Olney: "We have failed to provide a fertile climate in academia and industry for the growth of innovators, developers and inventors." (Letter: "Profession Stagnating," *Chem. Eng. Prog.* 80, no. 4 [1984]: 4.)

   C. Ramshaw: "The ideas abound but the difficulty of getting acceptance of new engineering developments in our conservative environment represents a nearly insuperable obstacle to innovation." (Paper presented at Institution of Chemical Engineers Research Meeting, Manchester, U.K., 18–19 April 1983.)

   R. E. Smith: "As technologists we have failed to get across the case for continuing improvement and innovation in a language that the decision makers can understand." (Paper presented at Institution of Chemical Engineers Research Meeting, Manchester, U.K., 18–19 April 1983.)

   D. N. Frey: "The strangeness of a seminal innovation to the organization can be so profound that the company may have to form separate development teams backed by powerful sponsors to prevent the internalized, change-resistant bureaucracy from killing the innovation early on." (In D. W. T. Rippin, J. C. Hale, and J. F. Davis, eds., *Foundations of Computer-Aided Process Operations* [Austin, TX: CACHE, 1993].)

   Japanese proverb: "The stake that sticks up will be hammered down."
2. G. A. Dalzell, "Inherently Safer Design: Changing Attitudes and Relationships," in *Seventh International Conference on Health, Safety, and Environment in Oil and Gas Exploration and Production* (Calgary, Canada: Society of Petroleum Engineers, 2004).
3. T. A. Kletz, *Dispelling Chemical Engineering Myths*, 3rd ed. (Washington, DC: Taylor & Francis, 1996), 101.
4. T. A. Kletz, "Protective System Failure," in *Learning from Accidents*, 3rd ed. (Oxford, U.K.: Butterworth–Heinemann, 2001), chap. 2.

5. D. C. Hendershot, "The Impact of Chemistry on Process Safety," in *Mary Kay O'Connor Process Safety Center Annual Symposium* (College Station: Mary Kay O'Connor Process Safety Center, Texas A&M University, 1998).

6. A. Hale, B. Kirwan, and U. Kjellen, "Safety by Design: Where Are We Now?" *Safety Science* 45, no. 1-2 (2007): 305–327.

7. D. C. Hendershot, "Designing Safety into a Chemical Process," in *Fifth Asia Pacific Responsible Care Conference and Chemical Safety Workshops* (Shanghai, PRC, 1999).

8. D. C. Hendershot, J. A. Sussman, G. E. Winkler, and G. L. Dill, "Implementing Inherently Safer Design in an Existing Plant," *Process Safety Progress* 25, no. 1 (2006): 52–57.

9. T. A. Kletz, "Emergency Isolation of Chemical Plants," *Chem. Eng. Prog.* 71, no. 9 (1975): 63.

10. T. A. Kletz, "A Gas Leak and Explosion: The Hazards of Insularity," in *Learning from Accidents*, 3rd ed. (Oxford, U.K.: Butterworth-Heinemann, 2001), chap. 4.

11. P. G. Urben, "Learning from Accidents in Industry," *Journal of Loss Prevention in the Process Industries* 2, no. 1 (1989): 55.

12. T. A. Kletz, "Will Cold Petrol Explode in the Open Air?," *Chem. Eng. (U.K.)* 426 (1986): 62. Reprinted in *Loss Prevention Bulletin* 188, April 2006: 9.

13. D. C. Hendershot, "Inherently Safer Chemical Process Design," *Journal of Loss Prevention in the Process Industries* 10, no. 3 (1997): 151–157.

14. J. P. Gupta and D. W. Edwards, "Inherently Safer Design: Present and Future," *Process Safety and Environmental Protection* 80, no. 3 (2002): 115–125.

15. F. I. Khan and P. R. Amyotte, "How to Make Inherent Safety Practice a Reality," *Canadian Journal of Chemical Engineering* 81, no. 1 (2003): 2–16.

16. T. A. Kletz, "Inherently Safer Design: Its Scope and Future," *Process Safety and Environmental Protection* 81, no. 6 (2003): 401–405.

17. D. W. Edwards, "Are We Too Risk-Averse for Inherent Safety? An Examination of Current Status and Barriers to Adoption," *Process Safety and Environmental Protection* 83, no. 2 (2005): 90–100.

18. D. C. Hendershot, "An Overview of Inherently Safer Design," *Process Safety Progress* 25, no. 2 (2006): 98–107.

19. D. Moore, A. M. Dowell, D. Hendershot, J. F. Murphy, B. Moore, and B. Rosen, "Solicited Letters to the Editor Regarding Inherent Safety," *Process Safety Progress* 25, no. 4 (2006): 263–267.

20. D. W. Green and R. H. Perry, *Perry's Chemical Engineers' Handbook*, 8th ed. (New York: McGraw-Hill, 2008), chap. 23, pp. 38–39.

21. D. A. Moore, M. Hazzan, M. Rose, D. Heller, D. C. Hendershot, and A. M. Dowell, "Advances in Inherent Safety Guidance," *Process Safety Progress* 27, no. 2 (2008): 115–120.

22. Center for Chemical Process Safety, *Inherently Safer Chemical Processes: A Life Cycle Approach*, 2nd ed. (Hoboken, NJ: John Wiley & Sons, 2009).

23. J. P. Gupta, D. C. Hendershot, and M. S. Mannan, "The Real Cost of Process Safety: A Clear Case for Inherent Safety," *Process Safety and Environmental Protection* 81, no. 6 (2003): 406–413.

24. D. W. Edwards, "Editorial: Special Topic Issue—Inherent Safety: Are We Too Safe for Inherent Safety?" *Process Safety and Environmental Protection* 81, no. 6 (2003): 399–400.

25. A. Green, B. Johnson, and A. John, "Process Intensification Magnifies Profits," *Chemical Engineering* 103, no. 13 (1999): 66–73.
26. M. A. E. Hodgson, "Making More Out of Less," *Chem. Eng. (U.K.)* 380 (1982): 163.
27. N. MacGregor, "Scholarship and the Public," *J. Royal Soc. Arts* 139, no. 5415 (1991): 191–194.
28. N. Moller and S. O. Hansson, "Principles of Engineering Safety: Risk and Uncertainty Reduction," *Reliability Engineering and System Safety* 93, no. 6 (2008): 776–783.
29. W. C. Clark, "Witches, Floods and Wonder Drugs: Historical Perspectives on Risk Management," in *Societal Risk Assessment*, ed. R. C. Schwing and W. A. Albers (London and New York: Plenum Press, 1980), 302–313.

    For the original version of the story, see F. Stockton, *Century Magazine* Nov. 1882. Reprinted in 1968 as *The Lady or the Tiger* (New York: Airmont Publishing, 1968).

Chapter 3

Intensification

> In a healthy society, engineering gets smarter and smarter; in gangster
> states it gets bigger and bigger.
>
> **—Peter Beckman, *A History of π***

> From big acorns small trees grow.
>
> **—Anonymous**

This chapter describes ways in which the amount of hazardous material in plants
and storage has been, or could be, reduced with consequent increase in safety and
reduction in cost, as discussed in Chapter 2. Different unit operations are discussed
in turn. However, intensification (minimization) can also come about by combining operations (Sections 3.2.8 and 7.8).

3.1 Process Intensification

The term *process intensification (PI)* has become firmly established in the vocabulary
of chemical process engineering researchers and practitioners. Process intensification results in substantial improvements in the scale, cleanliness, energy efficiency,
cost effectiveness, and safety of units designed to effect physical changes or molecular transformations in materials. Size reduction of unit operations and chemical
plants is a key goal of process intensification.

Quoting from the publisher's description of the latest book in this area, "PI ... is now
reaching a maturity that is seeing [these] concepts applied to a wide range of processes
and technologies."[1] Evidence of the maturity of process intensification can be found in

the existence of a journal devoted to advances in the field (*Chemical Engineering and Processing: Process Intensification*), as well as a Process Intensification Network,[2] organized by the authors of Reference 1 and hosted by Newcastle University, U.K.

As mentioned in Chapter 2, the motivation for process intensification often comes from the economic side of the business equation. In today's world, the economic need for PI is increasingly framed within the context of global competitiveness and sustainable development.[3,4] Sometimes, however, the driving force behind PI is the attendant safety benefits, as seems the case with the recent IMPULSE project (Integrated Process Units with Locally Structured Elements).[5] In this work, smaller and more efficient devices such as microreactors and compact heat exchangers are expected to present a lower risk of fire and explosion, as well as fewer harmful emissions, than traditional plant equipment.

Reference 5 concludes with the statement that, while major releases and hazards are less likely for IMPULSE plants, the potential for operator exposure and minor problems (e.g., pump leakage) may be greater. This remark is a helpful reminder that inherent safety involves both change, which must be managed, and tradeoffs, which must be assessed. No single approach can be expected to cover all issues that arise during process design and operation. In this regard, Etchells[6] has provided a useful listing of the potential problem areas, as well as the benefits, of process intensification.

Potential safety benefits of process intensification include the following:[6]

1. Reduction of inventories of hazardous materials and the consequences of process failures.
2. Reduction in the number of process operations, leading to fewer transfer operations and less pipework (thus reducing leakage).
3. Containment of explosion overpressures in smaller vessels, so that passive and active devices (e.g., rupture disks and automatic suppression systems, respectively) may not be needed.
4. Reduction in the number of process incidents during transient conditions because of fewer startups and shutdowns with continuous, intensified plant.
5. Less variable (and easier to control) heat evolution than in batch reactors when dealing with exothermic reactions.
6. Easier-to-achieve heat transfer because of the enhanced specific surface area of continuous, intensified plant, thus reducing the potential for runaway reactions.

Potential safety issues related to process intensification include the following:[6]

1. Requirement for high temperatures and pressures, and high-energy inputs. For example, the use of microwaves has been reported in intensified desorption processes for the regeneration of adsorbents used in the control of organic vapor emissions.[7]
2. Increase in process complexity and an accompanying increase in the complexity of control systems.

3. Heightened concerns with respect to control and monitoring due to shorter residence times for many intensified processes.
4. Higher potential for equipment failure or operator error if the process pipework becomes more complex.
5. Increased rate of energy release due to enhanced reaction rates as a result of improved mixing.
6. Introduction of a new ignition source with the combination of rotating equipment and friction-sensitive materials.
7. Overheating of thermally unstable materials on complex heated surfaces subject to fouling.
8. High throughput leading to the possibility of rapid accumulation downstream of off-spec products.
9. Closer proximity of people to smaller plant.

Not all of the stated benefits or concerns will occur in a given application of process intensification. The purpose in raising these points here is twofold. First, they remind us that assessment and management of process risks are comprehensive undertakings that are crucial to any commercial endeavor. Second, they offer a challenge to the reader while reviewing the examples in the remainder of this chapter to bear in mind the tradeoffs often presented by inherent-safety application.

3.2 Reaction

No unit operation, except storage, offers more scope for reduction of inventory than reaction. Many continuous reactors, such as liquid-phase oxidation reactors, contain large inventories of highly flammable liquids, and leaks from these reactors have caused many fires and explosions,[8,9] including that at Flixborough.[10,11]

Reactors can be arranged in the following hierarchy of (usually) increasing inventory and decreasing inherent safety:

Vapor-phase reactors	
Liquid-phase reactors	Inventory
Thin-film reactors (Section 3.2.3)	decreases
Tubular reactors (once-through and loop)	
Continuous pot (stirred tank) reactors	Safety
Semibatch reactors	increases
Batch reactors	

In a batch reactor, all the reactants are added before reaction starts. In a semi-batch reactor, one (or more) of the reactants is added and the final reactant(s) added gradually as reaction proceeds; reaction takes place at once, and an unreacted mixture cannot accumulate unless mixing fails or the catalyst is consumed (Sections 3.2.7 and 6.2.2).

Each process should be considered on its merits, for there are exceptions to the general rules. For example, according to Englund,[12] for the copolymerization of styrene and butadiene, a semibatch reactor is safer than a batch one, but a continuous reactor may not be the safest because some designs contain more unreacted raw material than semibatch reactors.

When possible, vapor-phase reactors should be developed in place of liquid-phase ones because the density of the vapor is less and the leak rate through a hole of a given size is lower (Figure 3.1 and Section 3.2.6). Of course, a gas at very high pressure with a density similar to that of a liquid is as hazardous as a liquid.

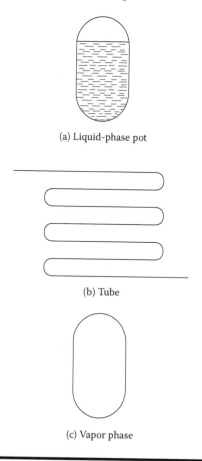

(a) Liquid-phase pot

(b) Tube

(c) Vapor phase

Figure 3.1   Types of reactors.

As a rule, reactors, of all types, are large not because a large output is desired, but because conversion is low, reaction is slow, or both.

3.2.1 Conversion Is Low

When conversion is low, most of the throughput has to be recovered and recycled, which increases the plant inventory.

In theory, there should be no need for large reactors. If 20,000 tonnes/yr of product are needed, it can all pass through a 2-in. (5 cm) diameter pipe if one assumes a specific gravity of 0.8 and a linear velocity of 0.5 m/s, which is not particularly fast[13] (Figure 3.2). Because conversion is low, most plants for the manufacture of 20,000 tonnes/year contain far bigger pipelines.

If 1000 tonnes/year of product are needed, then in theory it can all pass through a 0.4-in. (1 cm) diameter line. Instead of a batch reactor we should consider a continuous reactor made from narrow-bore pipe. Heat transfer and mixing are effective in small pipes.

The size of process pipes is a measure of the improvement theoretically possible. The ratio

$$\frac{5000 \ (\text{line diameter in inches})^2}{(\text{output in tonnes/yr})}$$

has an ideal value of 1. The actual value (the "free-ride" or free-load index) is often 10 or even 20.

We have become so used to giving our raw materials "free rides" that we may no longer see that it is inefficient and unsafe or that it reflects on our abilities as chemists or chemical engineers. The "free ride" indices for a company's plants should be displayed at the entrance to its research department as a reminder of the scope for possible improvement, and the indices ought to be known to managers and

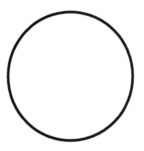

Figure 3.2 A pipe of this internal diameter can handle the passage of 20,000 tonnes/year.

directors. Regulators should ask for the value when visiting a plant or commenting on application for permission to build a new plant.

When one of my sons was 4 years old, he played miniature golf for the first time. He was halfway round before he realized that the winner is the one with the lowest score, not the highest, which is so different from all the other games he had played. Similarly, the company that will win in safety and economy is the one with the lowest "free ride" index.

3.2.2 Reaction Is Slow

Reactions are slow either because the mixing is poor or because the reaction is inherently slow. The former is the more common situation, and thus better mixing will reduce reaction volume, as discussed in this section. If the reaction is inherently slow, it may be possible to speed it up by increasing pressure, temperature, or both, or by developing a better catalyst. In theory, if this cannot be done, a tubular reactor can be made very long, though there may be practical difficulties. Tubular reactors should always be considered as an alternative to pot reactors, continuous or batch. Their integrity is high, and leaks can be stopped by closing a remotely operated isolation valve in the line. If desired, several such valves along the length of a tubular reactor can limit a spillage to any desired amount, say 5 or 10 tonnes, and the chance of a serious fire, explosion, or toxic incident is reduced (Section 3.2.7 and Figure 3.3). Continuous pot reactors rarely leak, but they can empty quickly if large inlet and exit lines fail.

Another advantage of tubular reactors is that, compared with stirred reactors, there is less variation in mixing rate and thus in product quality. In stirred vessels, mixing rates vary widely. The energy dissipation, a measure of the efficiency of mixing, can be a hundred times greater near the agitator blade than at the edge of the vessel.[14] Liquids as thick as porridge with entrained solids can be passed through tubular reactors if the flow rate is adjusted to maintain plug flow.[15]

An alternative to a very long tubular reactor is a loop reactor, a tubular reactor with continuous addition of reactants and removal of products. Loop reactors have been used for manufacturing polymers from aqueous suspensions of monomers[16] and for chlorinating an organic compound. In the latter case, compared with a batch process, 43% more output was obtained from 30% of the reactor volume, with increased efficiency.[17] There is another example in Section 5.1.2.

The remaining sections include examples of processes that have been intensified by a change to once-through tubular reactors.

The heat loss from tubular reactors could be increased by using flattened tubes. For example, a flat tube with 4-in. × 0.8-in. internal dimensions has the same cross section as a tube of 2-in. bore but with 50% more surface area. Some animals, such as tapeworms, have increased their surface area per unit mass in this way.[18] (Others have done so by developing lungs; see Table 3.1.)

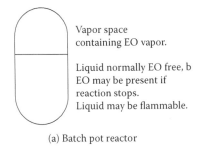

Vapor space containing EO vapor.

Liquid normally EO free, b
EO may be present if
reaction stops.
Liquid may be flammable.

(a) Batch pot reactor

Emergency valves close automatically if a leak occurs.

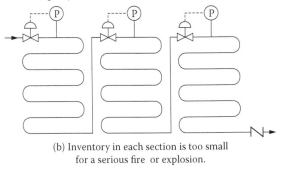

(b) Inventory in each section is too small
for a serious fire or explosion.

Figure 3.3   Processes for the manufacture of ethylene oxide derivatives.

In trying to reduce the size of liquid-filled continuous or batch pot reactors, we should beware of compromises that are worse than either extreme. A very small reactor operating at high temperature and pressure may be inherently safe because it contains so little material; even if all of it leaks out, a serious incident is impossible or unlikely. A large reactor operating at or near atmospheric temperature and pressure, or at a lower temperature, may be safe for different reasons. Because the pressure is low, leaks will be small and infrequent, and because the temperature is low, the leaking liquid will produce little vapor. This is an example of attenuation, not intensification. A compromise solution (moderate temperature, pressure, and volume) may combine the worst features of the extremes. If there is a leak, the pressure may be high enough to produce a large leak rate, the temperature high enough to cause significant evaporation, and the inventory high enough to create a serious fire, explosion, or toxic incident (Figure 3.4). Compromises are not always better than extremes. (When I lectured to students, I used to add that this applies to engineering, not politics.)

The high temperature and pressure in the first reactor may increase the chance of a small leak, but this may be a price worth paying to make large leaks impossible.

Table 3.1   **Surface Compactness of Heat Exchangers**

Type of Exchanger	Surface Compactness (m²/m³)	Reference
Shell and tube	70–500	89
Plate	120–225	89
	Up to 1,000	90
Spiral plate	Up to 185	89
Shell and finned tube	65–270	91
	Up to 3,300	89
Plate fin	150–450	91
	Up to 5,900	89
Printed circuit	1,000–5,000	88
Regenerative		
Rotary	Up to 6,600	89
Fixed	Up to 15,000ᵃ	89
Twinned screw extruder	High	92
Human lung	20,000	89

ᵃ Some types have a compactness as low as 25 m²/m³.

3.2.3 Nitroglycerin Production

Nitroglycerin production provides one of the best examples of the reductions in reactor inventory that can be achieved by redesign. Nitroglycerin (NG) is made from glycerin and a mixture of concentrated nitric and sulfuric acids according to the following equation:

$$C_3H_5(OH)_3 + 3HNO_3 = C_3H_5(NO_3)_3 + 3H_2O$$

The sulfuric acid does not take part in the reaction. The reaction is very exothermic, and, if the heat is not removed by cooling and stirring, an uncontrollable reaction is followed by explosive decomposition of the NG.

The reaction was originally carried out batchwise in large stirred pots containing about a tonne of material. The operators had to watch the temperature closely, and to make sure they did not fall asleep they sat on one-legged stools (Figure 3.5). If they fell asleep, they fell off. (The stools might usefully be revived in lecture theaters and

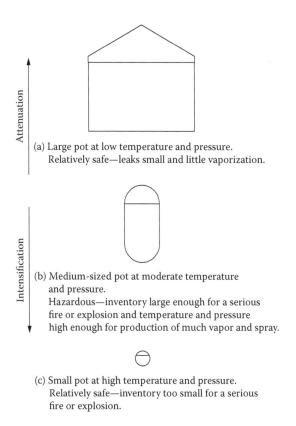

(a) Large pot at low temperature and pressure.
Relatively safe—leaks small and little vaporization.

(b) Medium-sized pot at moderate temperature
and pressure.
Hazardous—inventory large enough for a serious
fire or explosion and temperature and pressure
high enough for production of much vapor and spray.

(c) Small pot at high temperature and pressure.
Relatively safe—inventory too small for a serious
fire or explosion.

Figure 3.4   The effect of varying the conditions in a liquid-phase pot reactor.

conference halls.) The photograph was taken about 1890, but the process continued in use until the 1950s in the United Kingdom and even longer in other countries.

If we were asked to make this process safer, what would most of us do? We would add instruments to the reactor for measuring temperature, pressure, flows, rate of temperature rise, and so on, and then use these measurements to operate valves that stop flows, increase cooling, open vents and drains, and so on. By the time we had finished, the reactor would hardly be visible beneath the added protective equipment (Figure 7.1). However, when the NG engineers were asked to improve the process, they did not proceed in this way. They asked why the reactor had to contain so much material. The obvious answer was because the reaction is slow. But the chemical reaction is not slow. Once the molecules come together they react quickly. It is the chemical engineering—the mixing—that is slow. The engineers therefore designed a small, well-mixed reactor holding only about a kilogram of material that achieves about the same output as the batch reactor. The new reactor resembles a laboratory water pump. The rapid flow of acid through it creates a partial vacuum that sucks in the glycerin through a sidearm. Very rapid mixing occurs, and by the

Figure 3.5   The old batch process for the manufacture of nitroglycerin.

time the mixture leaves the reactor, reaction is complete (Figure 3.6). The residence time in the reactor has been reduced from 2 h to 2 min, thus enabling the operator to be protected by a blast wall of reasonable size. Similar changes were made to the later stages of the plant where the NG is washed and separated. The reactor, cooler, and centrifugal separator together contain 5 kg of NG.[19]

The control system on the reactor is also inherently safer than conventional control. If the flow of acid falls, the flow of glycerin must fall in proportion without the intervention of a flow ratio controller or other instruments that might fail or be neglected.

Figure 3.6 shows the simple but effective way in which the reactor can be shut down. If a fault is detected, the solenoid is deenergized; the lead weight opens the plug valve; air enters the reactor, which destroys the partial vacuum created by the flowing acid; and the flow of glycerin stops and thus stops the reaction.

Another advantage of the continuous process is that heat evolution is uniform and thus easier to control. In the batch process, heat changes during the batch and is affected by the quality of the mixing.

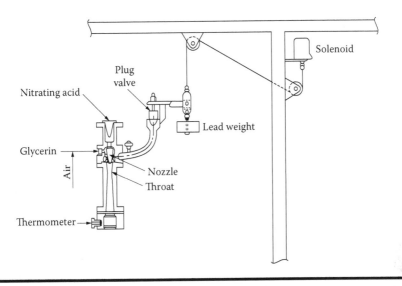

Figure 3.6 Nitration injector in the Nobel AB process for manufacture of nitro-glycerin. (From R. Bell, *Loss Prevention in the Manufacture of Nitroglycerin, Process Optimization Symposium Series No. 100* [Rugby, U.K.: Institution of Chemical Engineers, 1987].)

The change made in nitroglycerin manufacture illustrates the statement made earlier that continuous processes are usually safer than batch ones. Not only does a batch reactor contain more material for a given output than a continuous one, but many more operations have to be carried out; thus, there are more opportunities for error.

In modern nitroglycerin plants, the product is usually transferred as an emulsion rather than as a pure liquid, which is an example of attenuation.

Intensification does not, of course, guarantee immunity from all accidents. Though we remove some hazards, others may remain. Explosions have occurred in continuous NG plants, though not so often as in batch plants. In one case, nitroglycerin and acid were separated in a centrifuge. The plastic NG exit line from the centrifuge swelled and choked, and the NG went down the acid line and settled on top of the acid. Two explosions occurred: one in the acid tank and the other in the recycle line leading out of the acid tank. The first explosion was probably triggered by vibration, and the second by the heat of the sun. Several men were killed. If we intensify equipment, pipelines may contain more inventory than the equipment; in any plant, materials may become hazardous if we let them get into the wrong place. Because NG detonates so easily, most explosives manufacturers have now discontinued its production.

In Hong Kong, the storage of explosives has been discouraged. If needed for quarrying, the raw materials are transported to the site and then mixed immediately before use. This technique is now widely used elsewhere.[20]

For other nitration reactions, see Section 5.1.2.

3.2.4 Liquid-Phase Oxidation

The explosion at Flixborough in 1974 occurred in a continuous plant for the oxidation of cyclohexane with air at about 150°C and a gauge pressure of about 10 bar (150 psi) to a mixture of cyclohexanone and cyclohexanol, usually known as KA (ketone–aldehyde) mixture (Figure 4.6[a]), which is a stage in the manufacture of nylon. The inventory in the plant was large (200–500 tonnes has been quoted[21]) because reaction was slow and conversion low (about 6% per pass[22]). Much of the inventory was held in six large continuous pot reactors operated in series, and the rest was held in the equipment for recovering the product and recycling the unconverted raw material.

The first stage of the reaction (hydroperoxide formation) was slow because mixing was poor. Conversion was low for the same reason. If the oxygen concentration in the liquid is high, it can easily get too high (where it leaves the gas sparger), and then unwanted side reactions, including further oxidation of the cyclohexane, occur.[22] This cannot be tolerated because, in addition to purification costs, hydrocarbon costs are typically 80% of production costs.[23]

Litz has described a method of improving the mixing in gas–liquid reactors.[24] In most such reactors, the gas is added through a sparger, and the liquid is stirred by a conventional stirrer (Figure 3.7). In the Litz design (Figure 3.8), the gas is added to the vapor space and a down-pumping impeller sucks the gas down and mixes it

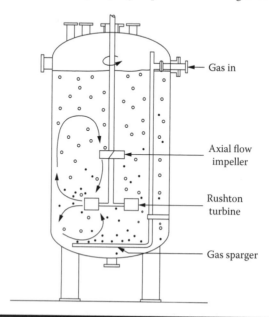

Gas in

Axial flow impeller

Rushton turbine

Gas sparger

Figure 3.7 Conventional gas–liquid stirred tank reactor. (From L. M. Litz, "A Novel Gas–Liquid Stirred Reactor," *Chem. Eng. Prog.* 81, no. 11 [1985]: 36–39. Reproduced by permission of the American Institute of Chemical Engineers.)

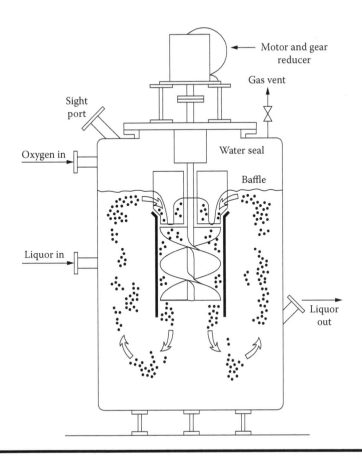

Figure 3.8   Advanced gas reactor. (From L. M. Litz, "A Novel Gas–Liquid Stirred Reactor," *Chem. Eng. Prog.* **81, no. 11 [1985]: 36–39. Reproduced by permission of the American Institute of Chemical Engineers.)**

intimately with the liquid. Unreacted gas that escapes back into the vapor space is recirculated. The reactor was designed to increase output and efficiency, but, compared with a conventional reactor, less inventory is needed for a given output. Roby and Kingsley have described a development of this reactor.[25]

Another method of improving the mixing is to add air, oxygen, or other gas to the liquid at several different points. This has been done by making the wall of the reactor from a porous ceramic membrane and allowing oxygen to diffuse through it.[26]

Other methods of improving mixing are suggested in Section 3.2.8.

A problem that must be overcome in a new design of a reactor with higher conversion would be heat removal. The heat of reaction is considerable and is removed by evaporation and condensation of unconverted cyclohexane and its return to the reactor. About 7 tonnes have to be evaporated for every tonne that reacts. A higher conversion process would require a high heat-exchange area, which is a property of long, thin tubes. Looking further ahead, could some other method of heat removal be devised, such as electricity generation in fuel cells?[27]

The second stage of the cyclohexane oxidation reaction, decomposition of the hydroperoxide, is inherently slow, and a long residence time is therefore required. Could this be achieved in a tubular reactor as discussed above? Higher temperatures would increase the rate of decomposition.

The leak at Flixborough was due to the failure of a bellows, which had been installed incorrectly in a way specifically forbidden in the manufacturer's literature. A bellows is unfriendly equipment because it cannot tolerate incorrect installation (Section 9.4). Nevertheless, if a large inventory had not been present, the explosion could not have occurred.

Similar processes to those used at Flixborough are operated by many companies worldwide. Conversion rarely, if ever, exceeds 10%. We have become so accustomed to low conversions on this and other processes that we do not realize that it is fundamentally poor engineering to give 90%–95% of our material a "free ride." What would we think of an airline on which only 5%–10% of the passengers got off at the end of each trip, the rest staying on board to enjoy the movies?

Other liquid-phase oxidation processes such as the oxidation of cumene to phenol are similar. Reaction is slow, conversion is low, and recycle is large (Section 4.2.4).

Beyond the hazard presented by large inventories, there is an additional hazard of oxidation plants: The vapor space above the liquid or the off gas can enter the explosive range. In the oxidation of *o*-xylene to phthalic anhydride, a new catalyst makes it possible to operate farther from the explosive region.[28] Similarly, in cumene oxidation plants, the chance of a runaway reaction can be reduced by lowering the reaction temperature, though this increases the reaction volume. Intensification and attenuation are often alternatives.

If oxygen is used instead of air in an oxidation reactor, off-gassing is reduced, and the separation unit is smaller; on the other hand, explosive limits are wider, and steel can burn in an oxygen atmosphere.[23]

3.2.5 Adipic Acid Production

The KA mixture produced in the process just described is oxidized to adipic acid with nitric acid as a further stage in the manufacture of nylon. For many years, the reaction took place in a reactor fitted with a stirrer, an external cooler, and a pump (Figure 3.9[a]). (This figure is much simplified. In practice, various combinations of series and parallel operation were tried for reactors and coolers.[29]) Ultimately, an internally cooled plug-flow reactor was designed and constructed (Figure 3.9[b]). It contains less material for a given output, and the pump, external cooler, connecting lines, and stirrer gland—all of them possible sources of leaks—were eliminated. Mixing is achieved by the gas given off the reactor.[29]

All the acid is added to the first compartment of the reactor, and the KA is added to each compartment via sparge pipes. The reduction in inventory results

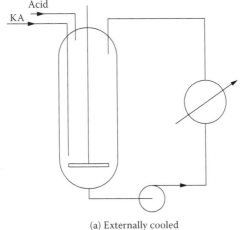

(a) Externally cooled

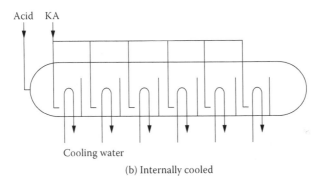

(b) Internally cooled

Figure 3.9   Adipic acid reactors.

from the elimination of the external equipment and the prevention of back-mixing. Distillation can also be intensified by eliminating back-mixing (Section 3.3.2).

Over a period of 25 years, many improvements were made to the detailed design of the externally cooled reactors, but no change was made to the basic design. Nylon production was very profitable, and this discouraged innovation; it was more important to increase output than experiment with new designs that might have had unforeseen snags or taken longer to bring on line.

This does not explain why the original design used an external cooler. Perhaps it was used on the pilot plant and simply copied. Perhaps the reactor was designed by a reaction specialist and the cooler by a heat-transfer specialist, and the two engineers never got together.

A reactor similar to that shown in Figure 3.9(b) has been used for cyclohexane oxidation.[30]

3.2.6 Polyolefin Production

At one time, polypropylene was manufactured using a flammable hydrocarbon as solvent, but all new plants now use pure propylene. In some of them, the propylene is in the vapor phase and in others in the liquid phase, but there is little to choose so far as safety is concerned because the vapor-phase propylene is near its critical point and has a density not very different from that of the liquid.[31]

The solvents used were, of course, no more flammable than the propylene, but the newer processes do not give a solvent a "free ride" and avoid the need for separation equipment.

Ethylene can now be polymerized to low-density polyethylene (LDPE) in the vapor phase at low pressure (Section 5.1.3). The older processes use gas at high pressures so that the ethylene has a density similar to that of the liquid.[32,33] Examples of high-pressure ethylene polymerization to LDPE still exist; the tubular reactor is enclosed in a concrete bunker. This is an example of a passive barrier, not inherent safety.

3.2.7 Ethylene Oxide Derivatives

Ethylene oxide derivatives are usually manufactured in semibatch reactors. The other reactant is usually put into the reactor first, and the ethylene oxide is added gradually. The mixture is circulated by stirring or by circulation through a cooler. The ethylene oxide reacts quickly, and its concentration remains low. However, ethylene oxide is present in the vapor space and, if a source of ignition turns up, ethylene oxide can explode without the presence of air. In addition, if reaction is slow or fails to occur, for example, because the stirrer or circulation pump stops, the temperature is too low, or the catalyst is inactive, ethylene oxide can accumulate in the liquid. If reaction then starts, it can accelerate out of control.[34] The other reactant is often flammable and can burn or explode if there is a leak.

In an alternative process, reaction takes place in a tubular reactor. There is no vapor space, and leaks can be stopped by closing a valve. If desired, several valves, automatically operated, can be used to limit the size of any leak (Figure 3.3).

Section 8.3.1 describes another hazard of reactions involving ethylene oxide.

3.2.8 Devices for Improving Mixing

As already emphasized, many reactions are slow and need a large reaction volume because mixing is poor. Some methods of improving mixing have already been described: mixing in an injector (Section 3.2.3) and using a down-pumping impeller (Section 3.2.4). Another method is mixing in a pump (Section 5.1.2). Figure 7.13 shows a fluidic mixer, i.e., one without any moving parts in contact with the liquid. Figures 3.10 and 3.11 show some methods suggested

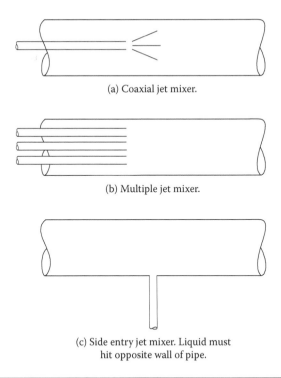

(a) Coaxial jet mixer.

(b) Multiple jet mixer.

(c) Side entry jet mixer. Liquid must
hit opposite wall of pipe.

Figure 3.10   Some methods of intensifying liquid–liquid reactions.

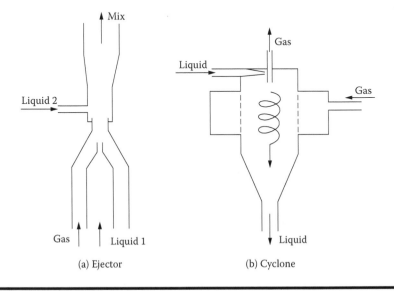

(a) Ejector

(b) Cyclone

Figure 3.11   Some methods of intensifying gas–liquid reactions.

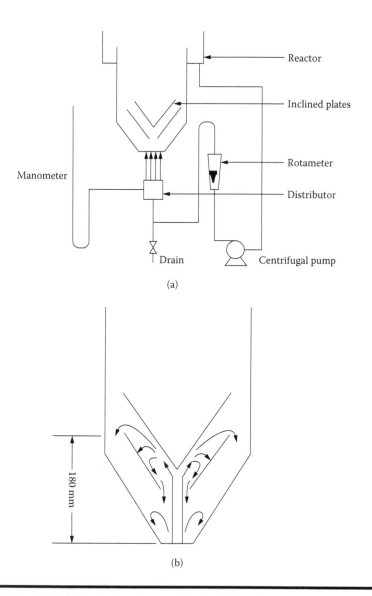

Figure 3.12 (a) Schematic flowsheet of inclined-plate jet reactor system; (b) the flow pattern for the preferred plate arrangement.

by Middleton and Revill[13] for improved mixing of two liquids or a liquid and a gas.

Leigh and Preece[35] have described a system that combines reaction (or extraction) and separation in one unit (Figure 3.12). Turbulent jets of liquid are discharged into a second immiscible liquid, thus generating high velocities between the two phases. Inclined plates provide a large surface area for the heavier phase to coalesce and flow back to the reentrainment zone (the base of the vessel), while the lighter phase rises and is drawn off by the pump.

In addition to the inherent advantage of combining the extraction and reaction processes, the increased efficiency of the mass transfer process per unit volume results in a smaller inventory for a higher throughput. The device has a simple construction and no internal moving parts or sealing problems other than at the pump gland, and the pump that transfers the process liquid between the units also provides the agitation power. The device can easily be modified for continuous operation and the internal plates used, if necessary, for heat transfer.

Several authors[36,37] have shown how reaction and distillation can be combined.

Butcher[38] describes a research program for improving mixing in reactors. The important parameter is power input per unit mass, and the following figures are quoted:

Straight pipe	1 W/kg
Stirred vessel	1–5 W/kg
Static mixer	Several hundred Watts per kilogram
Pump	Several hundred Watts per kilogram at highest point (average is lower); a pump that is a good mixer is not efficient as a pump
Ejectors	About 1000 W/kg (Section 3.2.3)
Rotor–stator mixers for dispersing powders	Much higher

Another method of improving mixing in a reactor is to mix the reactants thoroughly under conditions in which they cannot react and then start reaction by raising the temperature or pressure or adding a solvent or catalyst. When reaction occurs as soon as reactants are mixed, high local concentrations may produce unwanted side reactions (Section 3.2.4). If mixing is complete before reaction starts, higher conversions may be possible.

Other technologies can sometimes suggest solutions to our problems. In older beverage vending machines, the various constituents of the drinks (tea or coffee powder or concentrate, milk, sugar, etc.) are added through the same pipe. The equipment is complex, and contamination can result and cause the drink to taste of the previous drink. This problem is overcome through designs in which the constituents are already mixed in the cup, and all the machine has to do is select the right cup and add water.

After the explosion at Flixborough, another company that operated a similar process started a research program on alternative designs. A process using a tubular reactor employing the principle described earlier showed promise. The program was abandoned because there seemed little chance that a new plant would be needed in the foreseeable future. The researchers were disappointed, but commercially the decision was right. So far as I am aware (as of the late 1990s), no new plant has built in the West since Flixborough; all the new capacity has been focused in Eastern Europe or the Far East.

Another possible idea is to replace some of the raw material that will be recycled by an inert safe material that is easily separated. This is done in fluidized beds for the combustion of coal where only 1 particle in 200 needs to be a coal particle and the rest can be sand.[39] The use of chemically inert powders as a form of attenuation for dust explosion prevention is discussed in detail in Section 13.5.1.

Centrifugal mixers are explained in Section 3.3.3.

When a reaction is complete, the product has to be separated from unconverted raw material and unwanted by-products. Intensification of distillation is discussed in the next section, but the most effective way of intensifying separation is to increase conversion and specificity in the reaction section of the plant. What you don't have to separate won't leak while it is separated.

3.2.9 Other Processes

See Section 3.4 for a method of combining heat exchange and reaction.

Caro's acid is a mixture of sulfuric acid, water, and peroxymonosulfuric acid (H_2SO_5) made by reacting concentrated sulfuric acid and hydrogen peroxide. Its use as a powerful oxidizing agent has been impeded by the cost and difficulty of its manufacture, transport, and storage. It decomposes unless cooled within a second of manufacture. These problems have been overcome by the design of a well-mixed tubular reactor and immediate cooling of the product in the solution to be treated.[40]

As commented earlier, it is far easier to introduce inherently safer design in new plants than in existing ones, though storage can often be reduced on the latter (Section 3.6). However, Gowland[41] has described the way in which Dow improved the inherent safety of an acquired plant that contained a unit for the manufacture of phosgene. Improving the reliability of the control equipment made it possible to reduce in-plant storage; buying purer carbon monoxide, one of the raw materials, made it possible to eliminate parts of the purification section (thus simplifying as well as intensifying); and changing the chlorine supply from liquid to gas reduced the chlorine inventory by 90%.

Carrithers et al.[42] have also described the ways in which an old plant was made more user friendly. One is described in Section 5.2.

An interesting example of intensification, though one that improves the environment rather than safety, is the deep-shaft process developed by ICI for the treatment of liquid industrial and domestic waste. A mixture of air and waste is passed through a buried U-tube about 100 m long. The high pressure at the bottom of the tube causes the air to dissolve, and the fermentation rate is high. The process requires only a fraction of the land needed by conventional treatment plants.[39,43]

3.2.10 Recent Developments in Reactor Technology

Chapter 5 in Reference 1 begins with the statement that, in the context of intensification, the reactor has received more attention than other unit operations. This is similar to the earlier remark in the current book that no unit operation, except storage, offers more scope for reduction of inventory than reaction. One need only examine the number of references at the end of this chapter to appreciate the veracity of both comments.

It is not the intention here to review the field of reactor intensification; this has been comprehensively done by Reay et al.[1] Rather, several examples from the recent process-development and process-safety literature are given to illustrate the breadth of this important field.

Benaissa et al.[44] describe a type of continuous heat exchanger-reactor (see also Section 3.4) known as an open plate reactor (OPR) and having the flow pattern shown in Figure 3.13.[44] The reaction system studied was the esterification of propionic anhydride by 2-butanol to yield butyl propionate and propionic acid. They used a combination of kinetic modeling and experimentation to determine temperature and concentration profiles during the synthesis reaction and to decide on optimal operating conditions that could be adequately controlled.

By changing from a batch to a continuous processing mode, this heat exchanger-reactor design[44] leads to smaller inventories and therefore minimization of the consequences of a hazardous release. A hazard and operability (Hazop) study was also used to investigate accident scenarios leading to runaway reaction. The study revealed that in the case of simultaneous stoppage of both process and utility flows, the stainless steel sandwich and reaction plates helped to dissipate some of the energy released by the reaction (thereby offering additional thermal mass as an inherent characteristic of the apparatus).

A similar improvement in inherent safety can be accomplished by a switch from batch to continuous processing in a spinning disc reactor (SDR).[45] In this particular type of thin-film reactor, styrene was polymerized using centrifugal acceleration to create highly sheared films of thickness 100–200 µm on a rotating surface (see also Section 3.3.3). From the perspective of the simplification principle, there is additional complexity because of the introduction of moving parts. While this may on the one hand be viewed as a disadvantage,[1] Boodhoo and Jachuck[45] comment that the rotational speed can also be seen as an additional parameter for control of reaction rate and product quality.

The safety aspects of membrane reactors have recently been reviewed by Chiappetta et al.,[46] who describe these devices as being capable of taking advantage of the synergistic effects of separation and reaction in a chemical process. In some applications, the membrane serves to provide a controlled feed of a given reactant, thus improving reagent distribution, maintaining a low concentration level, and avoiding side reactions.

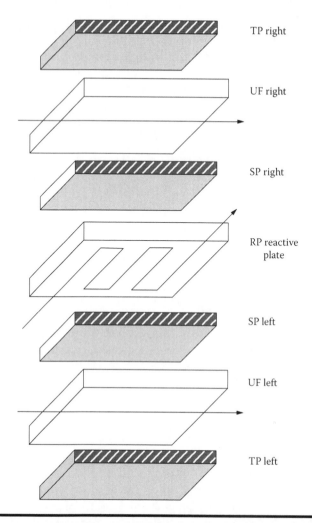

Figure 3.13 Flow pattern in an open-plate reactor. TP (transition plate), SP (sandwich plate), RP (reaction plate), UF (utility fluid). From W. Benaissa, N. Gabas, M. Cabassud, D. Carson, S. Elgue, and M. Demissy, "Evaluation of an Intensified Continuous Heat-Exchanger Reactor for Inherently Safer Characteristics," *Journal of Loss Prevention in the Process Industries* 21, no. 5 (2008): 528–536. With permission.)

This is the case for gas-phase oxidation of hydrocarbons, in which the gradual feeding of oxygen through a membrane avoids the formation of hot spots and limits the likelihood of reaction runaway.[47] In the porous-wall ceramic membrane reactor (PWCMR) described by Coronas et al.,[47] oxygen is supplied as it is consumed, and thus the reactor can be operated outside the flammability range of the oxygen/hydrocarbon mixture while maintaining a sufficiently high conversion. Attenuation of reaction conditions is therefore achieved in addition to intensification of hazardous inventories.

Another reactor classification discussed by Reay et al.[1] is the microreactor. Microreactor technology dates from the 1970s,[48] but only over the last decade have these devices made the transition from laboratory tools to production reactors in the chemical process industries.[49] Burns and Ramshaw[50] describe a benzene nitration reactor in which rapid mass transfer can be accomplished by stable parallel flow in capillaries of bore size between 127 and 254 µm. The integration of heat exchange with chemical reaction has also been achieved with a microstructure combination used for fuel processing.[51]

Inherently safer operation of microreactors is, of course, brought about by reduction of the inventory of the reacting mixture. This advantage can manifest itself in different ways. For example, microreactors are generally characterized by a high surface-area-to-volume ratio, which if large enough can significantly increase the time to reaction runaway.[52] Additionally, the narrow diameter of flow channels facilitates use of the quenching-distance concept for determination of maximum safe capillary diameters in gas-phase oxidation reactors, as evidenced by the work of Fischer et al.[49] with stoichiometric ethane/oxygen mixtures.

3.3 Distillation

Much of the following applies also to other liquid–vapor contacting processes such as absorption and scrubbing.

3.3.1 Conventional Distillation

There is a large inventory of boiling liquid, often under pressure, in the base of a distillation column, and a larger quantity, often several times more, is held up in the column. In an atmospheric-pressure column, the liquid is just above its normal boiling point, but in a pressure column much of the liquid will flash if it is released (Section 2.3).

In choosing a packing or tray design, many factors have to be taken into account. Low inventory should be one of our objectives. There is a big variation between different trays and packings in the holdup per theoretical plate.[53] Most tray designs have a holdup between 40 and 100 mm per theoretical plate; for most packings it is 30–60 mm; for film trays it is less than 20 mm. Unfortunately, comparative information is not readily available to the designer.

Figure 3.14 shows some other ways of reducing inventory. The amount of material in the base can be reduced by narrowing the base so that the column appears to balance on the point of a needle (Figure 3.14[a]). This is done when the bottoms product degrades when kept hot, but it could be done more often. However, although column bases can be designed for low inventory, it may be hazardous to

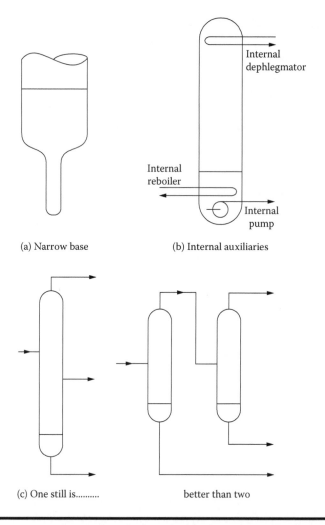

Figure 3.14 Some ways of reducing inventory in a distillation column.

reduce the level in an existing column. If the top of the reboiler is uncovered, the vapor may be overheated. Indeed, this led to an explosion in an ethylene oxide still in Texas in 1992 (Section 6.2.4).[54]

Internal calandrias and dephlegmators require a lower inventory than external reboilers and condensers; there is less equipment to leak; and there is no large overhead vapor line. Perhaps even the bottoms pump could go inside the column (Figure 3.14[b]). An alternative to a dephlegmator is an air cooler mounted on the top of the column. When possible, two distillation columns should be combined, as shown in Figure 3.14(c).

Besides the hazard due to the large inventory, a large column base can give rise to a more subtle hazard. In a still heated by live steam, calculations have shown that when reflux is lost, the rise in pressure should be sufficient to prevent steam from entering the column. This was taken into account in the design of the relief system. However, the designer overlooked the fact that the light ends from the top of the still will be dumped into a large quantity of hot bottoms, which will vaporize them and increase the pressure. The quantity of hot bottoms could be reduced by narrowing the base.

Low-inventory distillation equipment such as Luwa and other agitated thin-film evaporators[55,56] have been used for exceptionally hazardous materials such as cumene hydroperoxide, an intermediate in the manufacture of phenol and acetone from cumene, for it can decompose explosively. Perhaps such distillation equipment could be used more often.

Other methods of separation could be considered in place of distillation. Most chemical engineers accept distillation as the default choice and consider alternatives only when distillation is impracticable. Membrane separation or liquid–liquid extraction might be considered more often. The latter needs large inventories, but the liquids are usually at or near ambient temperatures.

3.3.2 Higee Distillation

The reductions in inventory obtainable by the methods just described are small compared with those attainable by the use of the Higee rotating equipment[57–60] (Figure 3.15) invented by Ramshaw (at the time with ICI) and marketed by Glitsch of Dallas, Texas. This equipment can reduce residence times in the distillation equipment to about a second and can reduce inventories by a factor of 1000.

In the Higee process distillation takes place in a rotating packed bed at an acceleration of 10^4 m/s^2. The packing has a voidage of 90%–95% and a specific surface area of 2000–5000 m^2/m^3. It has the shape of a flat cylinder with a hole in the middle; the radius is typically about 1 m, and the height a little less. Vapor or gas is fed to the case of the machine, enters the packing through the cylindrical outer surface, and travels inward. Liquid is introduced through a stationary distributor in the center of the rotor, enters the inner surface of the packed bed, and moves outward.

The radius of the packed bed corresponds to the height of a normal column and determines the number of theoretical plates, and the height of the packed bed corresponds to the diameter of a normal column and determines the capacity. Because it would be difficult to add liquid partway through the packing, two units are needed: one for the stripping section and one for the fractionation section.

The Higee unit is very compact, and photographs show it arriving on-site on the back of a truck (Figure 3.16). However, the condenser and reboiler are not part of the unit and must be supplied separately.

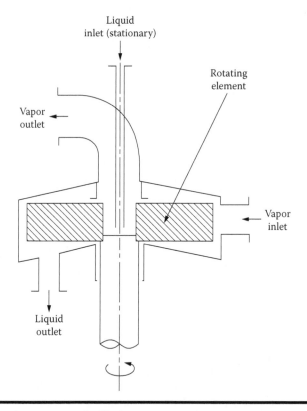

Figure 3.15 The ICI Higee distillation unit. (From C. Ramshaw, "Higee Distillation: An Example of Process Intensification," *Chem. Eng. [U.K.]* 389 [1983]: 13–14.)

The Sherwood flooding correlation (Figure 3.17) for a packed bed shows that, for a given packing, an increase in *g* allows the gas and liquid flows to be increased. Alternatively, the packing size can be reduced and the area increased.

However, this description does not give us a picture of what is happening. To get such a picture, think of the increased "gravity" as reducing back-mixing, which, in a normal packed column, prevents the achievement of the degree of separation that is theoretically possible. (Section 3.2.5 showed how a decrease in back-mixing in a reactor produced a decrease in reactor volume.)

Although the chemical engineering of the Higee unit is novel, the mechanical engineering is conventional. The speed of rotation is similar to that of a centrifuge.

Though the Higee is a brilliant invention that has been shown to work satisfactorily on the full scale, very few were in use in the late 1980s,[58] even in the company where it was invented, and most of the applications have been for stripping liquids with gases or for treating gases with liquids,[58,59,61,62] not for distillation. A recent industry report[63] states that there are only about 30 units in operation worldwide,

Figure 3.16   A Higee installation. (From Bulletin 394, Glitsch, Inc. Reproduced by permission of Glitsch, Inc.)

with many of these being experimental columns at various universities (e.g., Indian Institute of Technology Kanpur, India,[64] and University of Campinas, Brazil[65]).

Why has the uptake by industry of Higee technology been slow?

Higee technology can reduce the capital cost of distillation equipment, but the saving, compared with the total capital cost of a project, is small. Project engineers and business managers, including those at ICI, where Higee was invented, are reluctant to invest in new technology for a small percentage saving in cost in case there are unforeseen difficulties that prevent or delay the achievement of design output. According to Ramshaw,[66] "The main problem to be overcome is the dubious perception, by most plant operators, of the reliability of rotating machines.... We must convince skeptics that hazardous process fluids can be retained in the equipment." These same thoughts have been echoed by Green et al.[67] In addition, since Higee first became available in late 1982, there have been periods of recession when few new plants were built.

Nevertheless, the major oil and chemical companies spend large sums every year on distillation equipment. The possible advantages to them, in cost savings and increased safety, are large. It is surprising that companies have not invested in one or two Higee units each, just to gain experience with them, with any unexpected costs or delays underwritten by the company as a whole rather than by specific projects or businesses. Unfortunately, one of the results of the movement toward business-centered rather than functional organization is that there may be no central organization that can fund (or even recommend) such an investment. This is a good example of the effects of organization on technology (Section 10.1.5).

A recent review of equipment improvements in distillation technology supports these comments on the motivation for change within industry. Olujic et al.[68] give

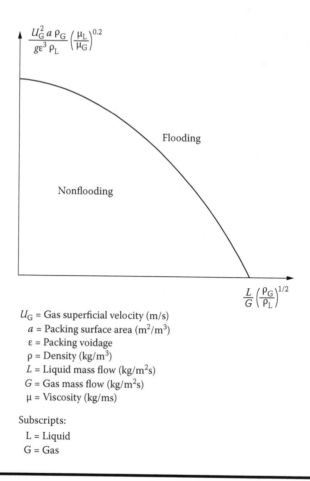

U_G = Gas superficial velocity (m/s)
 a = Packing surface area (m²/m³)
 ε = Packing voidage
 ρ = Density (kg/m³)
 L = Liquid mass flow (kg/m²s)
 G = Gas mass flow (kg/m²s)
 μ = Viscosity (kg/ms)

Subscripts:
 L = Liquid
 G = Gas

Figure 3.17   The Sherwood flooding correlation for a packed bed.

the main reason for implementation of innovations in distillation technology as the demand to stay competitive by minimizing capital expenditures and operating costs while increasing production capacity. Activities often therefore focus on improving the performance characteristics of vapor–liquid contactors such as trays and packings. Higee and other process-intensification alternatives to large-column distillation do not figure prominently in their review of the distillation field.[68]

The advantages of Higee are, however, so great that ultimately it will surely succeed. (However, worthwhile inventions are not always exploited. In pre-Columbian Central America, wheels were known but used only as toys.) The large distillation columns that we think of as advanced technology are, perhaps more than any other equipment, the chemical engineering equivalents of the beam engines referred to in Section 2.4. In the short term, Higee seems most likely to be employed when expensive materials of construction have to be used (so that the savings are greater than usual), when space is scarce (as on offshore oil platforms, but see Section 10.7), when there are height limitations, and for other gas–liquid-contacting applications.

Higee technology is being developed at the "Higrav" research center at the University of Chemical Technology, Beijing, China. Processes developed or being developed include deaeration of water, absorption of sulfur dioxide from flue gas, HOCl production, and preparation of nanoparticles. Zhao et al.[69] indicate that long-time running of rotating packed beds using Higee to gain commercial data and experience will be key to market acceptance of the technology.

The use of Higee in a nanotechnology application[69] is interesting, in that inherently safer designs have been cited as a key factor in eliminating or reducing hazards associated with emerging technologies such as nanotechnology.[70] Potential concerns with nanoparticles are fires and explosions should large quantities of combustible powder be generated during reactions or production;[71,72] intensification through inventory reduction is again seen to be advantageous. Binding nanoparticles to a substrate (a form of attenuation or moderation) can help reduce the toxicity of some materials.[72]

The British company AEA Technology developed a fluidic gas–liquid contactor somewhat similar to Higee, but the gas moves rather than the packing. The gas enters at the side, as in Figure 3.15, but tangentially through a series of vanes around the rim, and leaves from the center. The liquid is sprayed into the center. The intensification factor is much lower than for Higee, but there are no moving parts. There is also no packing to foul. Like Higee, the contactor has been used mainly for gas–liquid-contact processes other than distillation.[73]

The term *fluidics* is used to describe equipment, such as mixers, pumps, and valves, that is driven by liquid pressure and contains no moving parts. Widely used by the nuclear industry, this equipment is robust and reliable. Such mixers, pumps, and valves are often intensified, compared with conventional alternatives, and are also friendlier in another way: They contain no moving parts that can malfunction and start to leak (Section 7.9).

3.3.3 Other Applications of Centrifugation

This is a convenient place to mention other operations in which intensification can be achieved by rotation, often using lower speeds than Higee. Interphase mass transfer and phase separation can be improved by an increase in the relative velocity of the phases. The applications, most using rotating plates or packed beds, have been reviewed by Ramshaw[66,74] and include the following:

1. Scrubbing benzene and other impurities from coke-oven gas.
2. Using liquid–liquid extraction.[75]
3. Demisting gases, in particular by using a rotating mop irrigated with liquid.[76] Such a device has been used successfully to remove lime dust from air. It acts as a fan to move the gas as well as a scrubber and is thus a good example of intensification achieved by combining two operations in one item of equipment.

4. Deaerating seawater.[77]
5. Mixing.[78,79]
6. Convection drying of solid particles.[80,81]
7. Improving gas disengagement at electrodes.[74]
8. Using absorption heat pumps.[82]
9. Polymerizing on a spinning disc.[45,83,84]
10. Intensifying heat transfer (Section 3.4).

Slow rotation has been used to improve the wetting in thin-film evaporators.[85] Keyvani and Gardner[86] have reviewed gas–liquid contacting in rotating beds.

3.4 Heat Transfer

If Higee (Section 3.3.2) is used instead of conventional distillation, then the inventory in the reboiler and condenser is far greater than the inventory in the distillation unit. The team that developed Higee therefore turned their attention to ways of intensifying heat transfer. One suggestion made was to place the reboiler on the periphery of the Higee unit because centrifugal fields improve heat transfer when a phase change is involved.[87]

Another idea from the same team for heat transfer when there is no phase change has been developed to the stage that the equipment is now available commercially. It uses parallel plate exchangers with very narrow spaces (fractions of a millimeter) between the plates. To manufacture the exchangers, fluid-flow passages are etched into metal plates by techniques similar to those used to produce printed circuits. The plates are then diffusion-bonded together to form blocks. Fouling does not seem to be a problem if intermittent reverse flow can be used.[87,88]

Table 3.1 summarizes the surface compactness—the ratio of heat-transfer area to fluid volume—of various types of heat exchangers. Note that the inventory of hazardous material in a shell-and-tube heat exchanger can be reduced by putting the more hazardous material in the tubes. (Water-tube boilers contain much less water under pressure than old-fashioned boilers with the water in the shell, and their failure is correspondingly less hazardous.) In all exchangers, the inventory can be reduced by higher flow rates, extended surfaces, and higher temperature differences. A compromise of inventory, efficiency, and pressure drop may be necessary.

Most plate exchangers have a high surface compactness that fins can increase further. At one time, the difficulty of finding suitable gaskets restricted the use of plate exchangers, but these difficulties now seem to have been overcome.[93,94] Gasket-free all-welded designs are now available (though only for use at pressures up to 100 bar).[95] Weld-free (diffusion-bonded) titanium plate exchangers are now made using technology developed for jet engine blades.[96]

Jachuk and Ramshaw[97] have suggested the use of corrugated high-performance plastic films in heat exchangers. Polyether ether ketones and polyimides could be used up to 250°C.

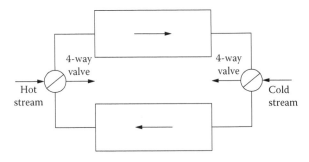

Figure 3.18   Regenerative heat exchanger.

In both plate and shell and tube exchangers, heat transfer can be enhanced by increasing turbulence and thus reducing fouling.[98] The U.K. Energy Efficiency Office has described several applications of intensified heat exchangers.[99]

Figure 3.18 illustrates a fixed-bed regenerative heat exchanger. The hot and cold streams flow alternately through two beds of matrix, which store the heat.[100] Heat transfer in shell-and-tube exchangers can be improved, and the inventory reduced, by inserting a matrix of wire into the tubes. The wire promotes turbulence, particularly near the walls.[101,102] If the increased pressure drop can be tolerated, this is one of the few ways in which existing equipment can be intensified. Turbulence is also higher in twisted-tube heat exchangers.[103] Oscillating flow increases heat transfer and thus reduces inventory.[104]

Unfortunately, information on the compactness of commercially available heat exchangers is not readily available to the designer because manufacturers do not, as a rule, include this information in their catalogs.

Aqueous liquids can be heated by direct injection of steam, thus simplifying the plant and saving the cost of an exchanger. For examples, see Sections 4.1.3 and 6.2.4. Collapse of steam bubbles can cause excessive vibration, but this can be overcome by mixing steam and water before injection into the bulk liquid.[105]

Ramshaw has suggested combining heat exchangers and reactors by coating the surfaces of a compact heat exchanger with catalyst.[106] An example is the production of hydrogen by the combustion and subsequent steam re-forming of methane. The combustion is exothermic and is carried out on one side of the heat exchanger while the endothermic re-forming reaction takes place on the other side. A thousandfold reduction in size is possible.[107]

3.5 Other Unit Operations

Liquid–vapor-contacting processes such as gas scrubbing and liquid stripping have been discussed under distillation (Section 3.3), and many of the devices described under reaction (Section 3.2) can be used in other liquid–liquid- and liquid–gas-contacting processes. The volume of liquid in liquid–liquid separators can be reduced by devices designed to promote coalescence of the dispersed phase, such as extended surfaces or packings[108] or electrostatic coalescers.[109] For example, a study of coalescers for the removal of water from aviation jet fuel brought about a "significant" reduction in equipment size.[108] Section 3.2.6 describes a device that achieves reaction and settling in a single unit.

Ramshaw has discussed intensification of crystallization[110] and drying,[111] and Gillett[112] has discussed the drying of pharmaceuticals. According to Williams,[113] vortex mixing can intensify membrane separation.

It thus seems that every unit operation can be intensified if we look for ways of achieving that objective; the recent text by Reay et al.[1] provides more examples. Common features, according to Ramshaw, are increased acceleration and laminar flow and shorter mixing paths. In 1980, Merck Sharp & Dohme's biochemical engineering consultants were asked if it was worth trying to improve fermentation performance through improved mixing. The answer was no. "You will perhaps not be surprised that over the last couple of years Merck has reported significantly improved performance through enhanced mixing. Merck was able to do this because the company had a pilot plant with the quality of instrumentation and control that allowed a realistic assessment of what the changes in agitation were achieving."[114]

In a fascinating paper published in 1993, Benson and Ponton[115] point out that, although electronic equipment had become smaller and cheaper during the past 50 years, chemical plants looked and cost much the same as they did 50 years before. They speculate on some of the ways in which the chemical industry could follow the trend toward miniaturization and at the same time reduce the need for storage and transportation by manufacture at the point of use. They give three examples that remain relevant today:

1. Manufacturers who buy small amounts of chlorine in cylinders could instead manufacture it on-site in small packaged units by the electrolysis of brine. The by-product hydrogen could be used to generate electricity in a fuel cell. In situ ozone production is suggested for water supplies and swimming pools.
2. Methanol for use in place of gasoline could be made at home in small packaged units from natural gas by re-forming it to carbon monoxide and hydrogen and then combining the products. The excess hydrogen could be burned to provide heat for the re-forming step and domestic hot water.
3. Manufacturers who need hydrogen cyanide could make it on-site from ammonia and natural gas in a small turbine that could also produce electricity. Packaged units for manufacturing phosgene at the point of use are already available[116] (Section 3.6).

Rinard says that chemical plants today are like the mainframe computers of the 1950s and 1960s: large, expensive, awkward, centralized facilities runs by staffs of experts. Miniaturized plants, like personal computers, are intended for distributed use by nonexperts.[117]

If this seems impossible, remember that most of us, though unskilled in plant operation, regularly operate plants in which controlled explosions are carried out thousands of times every minute. Though accidents are frequent, they are not caused by chemistry getting out of control. The operators often find it difficult to control the mechanics, not the process.

In an entirely different field, pesticides can be spread more safely and economically by using more powerful and effective agents and more effective methods of dispersion.[118]

Some of the methods of simplifying designs described in Sections 7.8 and 7.10 also produce a reduction in inventory.

3.6 Storage

The worst disaster in the history of the chemical industry occurred in Bhopal, India, in 1984 when a leak of methyl isocyanate (MIC) from a storage tank killed over 2000 people and injured many more. There are many lessons to be learned from Bhopal[119] (Section 4.2.1), but the most important, ignored by many commentators, is that the material that leaked was not a raw material or product but an intermediate that it was convenient, but not essential, to store. After Bhopal, Union Carbide, the company concerned, and many other companies announced that MIC and other hazardous intermediates would not be stored but would be used as soon as they were produced. Instead of 50 or 100 tonnes in a tank, there would be only a few kilograms in a pipeline.[120,121]

In 1994, 10 years after Bhopal, one U.S. company still stored up to 120 tons of methyl isocyanate, probably one of the last companies in the world to store such large quantities of the material. A report sponsored by concerned citizens suggested ways of reducing the amount stored. Producing it more uniformly throughout the year and storing more of the final products could reduce the maximum inventory, they said, to about 70 tons. Converting all four processes that use methyl isocyanate to direct feed could reduce it, in stages, to 5–50 kg in the pipelines feeding the four units. In the long term, a change to other process routes could eliminate it entirely. The report admitted that much money had been spent since 1984 on upgrading the plant and installing safety features, but "a lesson of Bhopal is that safety features are prone to failure."[122]

The chemical industry has reduced or eliminated stocks of other intermediates such as phosgene[116,123] (Section 4.2.3), hydrogen cyanide,[116] ethylene and propylene oxides,[124] sulfur trioxide, and chlorine.

Ciba–Geigy developed skin-mounted units for the production of 15 tonnes/day of phosgene; they could produce as little as 1.5 tonnes/day if required. Imperial Chemical Industries stopped supplying phosgene in 1989, preferring its customers to produce this intermediate on-site as required.[116]

DuPont developed a process for the distributed production of small amounts of hydrogen cyanide. Two reactions were studied: the reaction of ammonia with solid carbon and the reaction of ammonia with methane over a catalyst. Both reactions are endothermic, and heat is supplied by microwave heating. The reactions can be stopped quickly by turning off the heating and cooling with inert gas.[125]

Small amounts of storage may be necessary in some cases so that intermediates can be analyzed, and in other cases diversion tanks for off-specification intermediates may be necessary. The need for such storage should be questioned. Do not ask whether it is essential to have such storage. It is too easy to say yes. Ask, "If I had to manage without it, how would I do so?" (See Section 10.2.3.)

One company, located in a built-up area, moved its ethylene oxide storage and consuming plant to another site.[121] Another company reduced its chlorine storage from several thousand tonnes to several hundred tonnes. Another plant eliminated chlorine storage entirely and, with it, the associated liquefaction and vaporization equipment. The reduction was made possible by changes in the chlorine manufacturing process that make it possible to increase and decrease rate much more rapidly than in the past when demand changes.

One plant had seven spheres for the storage of liquefied petroleum gas (LPG). These were necessary, the managers said, to cover interruptions in production. When new legislation required the spheres to be mounded or insulated, the staff found that it could manage with three. This is not unlike the conclusion reached by Leal and Santiago,[126] who determined that, for their study conditions, it was better to keep a smaller number of fully-filled spheres than many spheres only partially filled with LPG.

Note that intermediates are usually reactive chemicals and therefore likely to be hazardous, or they would not be used as intermediates.

Computer simulations based on variations in demand and the known pattern of shutdowns, planned and unplanned, in the producing units are often used to show that so much raw material, intermediate, or product storage is necessary. However, these simulations take no account of the fact that, if stocks are low, people improvise, change plans, carry out maintenance more urgently, and so on in order to maintain production in a way not foreseen at the design stage. "Grey hairs cost the company nothing." Holusha,[127] reviewing a book on Japanese industry, writes that

> the American practice of having buffer stocks of partly finished components all along the production line concealed problems. If a machine broke down, parts from a nearby buffer stock could keep the assembly line working until somebody got round to fixing the machine. But … if the security of the buffers was removed, if the whole plant were in jeopardy of being shut down by one malfunctioning machine, the workers would be forced to tend their machines more carefully, so that all worked correctly all the time. The discipline this system imposed brought surprising improvements in productivity and quality.

In short, the need for large stocks is a self-fulfilling prophecy. In the process industries, reduced stocks will not make operators treat their plants more carefully, but the reductions will encourage the maintenance organization to get the plant back on-line quickly after a shutdown.

Another example: After visiting a Porsche plant, a Toyota engineer said that workers were swamped by the piles of parts stacked in racks above the shop floor. "Where's the factory?" he asked. "This is a warehouse." A week later, the parts inventory in the engine room was reduced from a 28- to a 7-day supply.[128]

Note also that it may be possible to reduce storage requirements by making the plant 5% or 10% larger than required, the extra capacity being used to cover delays in the arrival of raw material, upsets in one part of the plant, delays in dispatch of product, and so on. It may be cheaper as well as safer to do this rather than supply the storage.

The need for intermediate storage (and transport) can often be reduced or eliminated by constructing storage and using plants on the same site. At one time, cyanogen chloride was manufactured on one site and transported several hundred miles in cylinders. A hundred journeys were made per year. Now it is manufactured near the point of use, and the inventory has been reduced from 20 tonnes under pressure to a few kilograms at atmospheric pressure.

The need to transport chlorine (or hypochlorite) to swimming pools and store it there can be avoided by generating it at the point of use by electrolysis of brine.

Production at the point of use also avoids the hazards of transport, usually by road. The probability that someone will be killed by an ordinary road accident involving a tank truck is much greater than the probability that he or she will be killed by the contents.[129] The risks associated with transportation of hazardous materials are the subject of intensive research worldwide (e.g., Bubbico et al.[130,131]).

Material in storage is less likely to be involved in a serious leak or fire than material in a plant because it is not being heated, cooled, pumped, or processed in other ways. On the other hand, when storage is involved in a major incident, the financial losses are higher. The largest losses in the period 1955–1984 occurred wholly or mainly in 24 of the 100 storage areas.[132] For the period 1965–1994, the figure was 19.[133] The final investigation report[134] of the Buncefield incident that occurred on December 11, 2005, in Hertfordshire, U.K., gives a figure of 1 billion £ for the economic impact (comprising compensation for loss, costs of emergency response and investigations, and costs to the aviation sector). The report comments that, although this may seem a rather large sum compared to other major incidents, it is not unique.[134]

In the European Union, the reduction in stocks of hazardous chemicals has been stimulated by the so-called Seveso Directive[135] (enacted following the accident at Seveso in Italy in 1976;[136] see Section 6.2.4), which requires all companies that handle more than defined quantities of hazardous chemicals to demonstrate that they are capable of handling them safely. In the United Kingdom, the directive was implemented by the CIMAH (Control of Industrial Major Accident Hazard) Regulations,[137,138] which owe their origin to Flixborough and would still have been enacted, though in a slightly different form if Seveso had never occurred. CIMAH has now been replaced by COMAH (Control of Major Accident Hazards) in the United Kingdom, commensurate with the Seveso II Directive.

In the United States, the Process Safety Management Regulations have had a similar effect (Section 10.3). New regulations in Canada under the Canadian Environmental Protection Act, Section 200, have legislated the requirement for an emergency response plan at facilities storing or using any of a number of listed substances at or above specified threshold quantities.[139] The concept of inventory reduction has therefore been introduced, albeit through an environmental regulatory route.

If time-consuming procedures and consultations with the authorities have to be followed when the stock of, say, chlorine, in storage exceeds a threshold quantity (in the U.K. case, 75 tonnes), companies find that they can manage with far less. The Seveso Directive and CIMAH/COMAII Regulations apply, of course, to hazardous materials in process as well as storage, but reduction of the inventory in process is difficult without a major rebuild, whereas reduction of stocks in storage is comparatively easy. (For chlorine in process, the threshold is 25 tonnes.)

Within companies, the finance director can be an ally of the process-safety adviser, for reduction in stocks saves capital. Design engineers know the cost of a storage tank but often do not know how much money they are tying up in the contents; it can exceed the cost of the tank.

3.7 Intensification by Detailed Design

Most of the methods of intensification discussed so far have involved major changes in equipment design or reductions in storage capacity. The following example shows that substantial reductions in capacity can be made by the application of well-known principles without the need for any new technology.

Figure 3.19(a) shows part of the first design of a distillation unit for separating liquefied petroleum gases (LPG). There were actually two more similar distillation columns in the plant. Figure 3.19(b) shows a revised design in which the inventory was reduced, to the extent shown in Table 3.2, by making the following changes:

1. The reflux drum was left out, with the reflux pump taking suction from the liquid level in the condenser. The design of the condenser was reversed so that the LPG was on the shell side and the refrigerant in the tubes.
2. Buffer storage for the raw materials and products was left out, with flows going directly to the main off-plot storage areas from small surge drums.
3. A low hold-up packing was used in the column, and the holdup in the base was reduced to a 2-min residence time by narrowing the base.

See also Section 7.8.

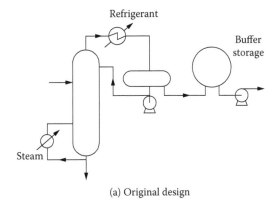

(a) Original design

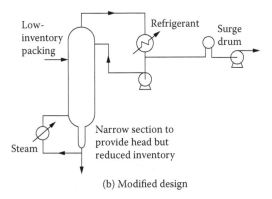

(b) Modified design

Figure 3.19   Two designs for liquefied petroleum gas separation plant.

Table 3.2   Reductions in Inventory Achieved by Attention to Detail (tonnes)

	Original		Revised	
	Working	*Maximum*	*Working*	*Maximum*
Storage	425	850	nil	nil
Plant	85	150	50	80

Note: See Figure 3.19.

3.8 Many Small Plants or One Big One?

Suppose we can choose between one large single-stream plant or two plants of the same or similar design, each half the size. The single large plant will usually be inherently safer because

1. It will contain a lower inventory than the two small plants combined (though there will be more of it in one place).
2. The two plants will contain more valves, flanges, pumps, sample points, pipes, and other sources of leak, and thus leaks will in total be twice as numerous.
3. The leaks will not be much smaller. Whereas the single plant contains a 4-in. pipeline, the smaller plants will contain 3-in. lines. A flange leak from a 3-in. line will not be much smaller than one from a 4-in. line.
4. The single-stream plant will usually be much cheaper than the two half-size plants. Some of the money saved will be available for extrinsic safety measures that might be considered too expensive if they had to be duplicated. If we have to put all our eggs in one basket, one basket is cheaper than two, and we can afford a good, strong one.

However, we may be able to break these scaling laws by adopting different design approaches for the smaller plants, as already described in Section 3.5. Smaller plants, for example, have a larger surface area and may require less additional cooling capacity. It may be possible to construct them, in whole or in part, off-site. They are usually more flexible. Sir John Harvey-Jones, former chairman of ICI, writes the following:[140]

> Among the entrenched beliefs with which I grew up as a businessman, one of the most sacred cows was that the advantages would always go to the largest organization, because the larger the scale and size of the plant, the lower would be the cost of production. The difficulty with this belief was that we built larger and larger plants, which were in fact less and less flexible and required us to hold larger and larger stocks.... Now most of us are looking for flexibility: production lines which can produce many different models at the same time; machining centres which can machine any item or product, parallel production systems which gain in speed and time though apparently sub-optimising in cost, are all developments which lie totally against the beliefs of yesterday.... The effects of these changes in the needs of manufacturing are very far-reaching. In the first place, the cost of variety is sharply reduced, enabling competition to develop in unexpected ways.

In the process industries, there is an increasing demand for small amounts of short-lived products, frequent changes in formulation, and faster response times. Large-scale continuous plants are unsuited to these requirements.[141] But the answer is not batch plants; small-scale continuous plants are possible (Section 3.2).

The ICI-intensified "Leading Concept" ammonia process is said to be as economic to build and run at the 500-tonnes/year scale as on the 1000- or 1500-tonnes/year scale.[142]

If a single large plant is multistream, it will be no safer than several smaller plants. Indeed, it may be less safe if large numbers of cross-connections are installed between the streams (Section 8.2).

When intensification results in less work—less material to be moved or handled—then there will be fewer accidents. A Bushman requires 525 acres (1300 hectares) to produce his food requirements. A British farmworker produces 50 times his requirements in 4 acres (10 hectares), giving an intensification factor of 6500 to 1. With less work, there will be fewer accidents.[143] The advocates of low-intensity organic farming do not seem to have considered the effects on agricultural accidents.

3.9 Some Thoughts on Intensification as Minimization

We most often think of intensification acting to reduce inventories of hazardous materials; this has been the primary theme of Chapter 3. It can also be helpful to consider aspects of intensification brought about by use of the concept of *minimization*.

As described by Goetsch,[144] there are various risk-reduction strategies for dealing with violence in the workplace. Access control, reporting policies, training, and monitoring devices are all commonly employed. An effective measure often overlooked is natural surveillance, which involves designing, arranging, and operating the workplace in a way that minimizes secluded areas.[144] Here, again, is a helpful reminder of the need for early consideration of inherent-safety principles in the design life cycle.

A newspaper article[145] reported on problems associated with biomedical wastes ending up in a local landfill. To protect workers at the site, three safety measures were introduced. Least effective in terms of the hierarchy of controls (Figure 1.1) was training the workers to pick up and sort only what they could actually see. Moving up the hierarchy to an add-on feature, personnel were immunized against hepatitis B and were also given tetanus shots. Most effective was the approach of identifying the source of the waste and taking steps to ensure that proper disposal was undertaken. This is clearly an attempt to minimize the hazard rather than accepting it and adding on control measures. All levels of the hierarchy are needed in this case, but there should be no doubt as to the order of precedence.

Appendix: Conference Report—New Technology

Chapter 3 has discussed the difficulties inventors experience in getting their ideas accepted. It also mentioned that in pre-Columbian Central America, wheels were known but were used only as toys. In the following we assume that wheels have just been invented and were described at a conference. What follows is an extract from the discussion. All the arguments have been used before.

There was much interest at a recent conference on new technology in the description by International Chemicals, Inc. (ICI), of the WHEEL, a new device WHich spEEds traveL. Because there would be no advantage in putting chemical plants on WHEELs, ICI intends to fit them to fire engines so that the engines can get to the scenes of fires more quickly than current technology allows.

Although there was praise for the company's ingenuity, most speakers expressed reservations. Joe Brown, speaking for the construction industry, thought caution was needed. The value of the WHEEL could not be fully assessed, he said, until several years' experience had been obtained. There might be unforeseen snags that would not become apparent until the device had been in use for some time. He drew attention to the unforeseen effects of other changes, such as the temporary bellows at Flixborough.

Thomas Dowting, of the Chemical Industries Federation, regretted that the device had been made public before the views of other companies had been obtained. The government might expect other factories to adopt the WHEEL. Although it might be useful on large sites (although this was not yet proven), it was not appropriate to the needs of smaller factories in which fire engines did not have to travel so far.

Dr. Werner Hackenschmidt (Gesellschaft für Unsinnfabrikat) asked how WHEELs would be fabricated. The production of continuous rotating load-bearing devices presented difficult metallurgical problems. What materials would be used? Little was known about the behavior of metals when subjected to such unusual forces.

Professor Patrick Murphy, a member of the faculty of the University of Ballybunion, asked if maintenance had been considered. How could a WHEEL be removed for repair without the vehicle tipping over?

Fred Bloggs, speaking for the fire departments, felt that fire appliances should not be used as subjects for experimentation. Had firefighters been consulted? He was sure their view would be that safety equipment should stick to well-proven designs. If smoother travel was needed, why not dig canals between the fire station and the plants?

Dr. Angus McGregor, from Crianlarich Polytechnic, said that, because WHEELs could operate only on smooth surfaces, he could not see how they would be economic when the cost of road improvements was taken into account.

Bill Muddle (consultant) said that it was a mistake to assume that speedier travel to the scene of a fire is always desirable. Using present methods of travel, firefighters had time during the journey to formulate their plan of attack. There would be no gain if firefighters rushed in unprepared.

Myfanwy Price, of University College, Blanau Ffestiniog, said that the idea was not new. A similar device was in use at the Annisgrifiudwy-Cymysglyd factory when she worked there over 30 years ago, but it had fallen into disuse because no one had been able to devise a satisfactory way of stopping the vehicles. Hexagonal

WHEELS had been found to assist braking but were disadvantageous in other respects; the ride was no longer smooth.

In summing up, the chairman said that the trials will be watched with interest; in the meantime, other organizations seem to prefer waiting.

References

1. D. Reay, C. Ramshaw, and A. Harvey, *Process Intensification: Engineering for Efficiency, Sustainability and Flexibility* (Oxford, U.K.: Butterworth-Heinemann, 2008).
2. Process Intensification Network, http://www.pinetwork.org/.
3. J. C. Charpentier, "In the Frame of Globalization and Sustainability, Process Intensification, a Path to the Future of Chemical and Process Engineering (Molecules into Money)," *Chemical Engineering Journal* 134, no. 1-3 (2007): 84–92.
4. J. C. Charpentier, "Among the Trends for Modern Chemical Engineering, the Third Paradigm: The Time and Length Multiscale Approach as an Efficient Tool for Process Intensification and Product Design and Engineering," *Chemical Engineering Research and Design* (2009): in press. It is now a corrected proof with the additional notation: doi:10.1016/j.cherd.2009.03.008
5. P. Khoshabi and P. N. Sharratt, "Inherent Safety through Intensive Structured Processing: The IMPULSE Project," in *12th International Symposium on Loss Prevention and Safety Promotion in the Process Industries*, IChemE Symposium Series No. 153 (Edinburgh, U.K., 2007).
6. J. Etchells, "Process Intensification: Safety Pros and Cons," *Process Safety and Environmental Protection* 83(B2), no. 2 (2005): 85–89.
7. R. Cherbanski and E. Molga, "Intensification of Desorption Processes by Use of Microwaves: An Overview of Possible Applications and Industrial Perspectives," *Chemical Engineering and Processing: Process Intensification* 48, no. 1 (2009): 48–58.
8. T. A. Kletz, "Causes of Hydrocarbon Oxidation Unit Fires," *Loss Prevention* 12 (1979): 96–102.
9. T. A. Kletz, "Fires and Explosions of Hydrocarbon Oxidation Plants," *Plant/Operations Progress* 7, no. 4 (1988): 226–230.
10. R. J. Parker, *The Flixborough Disaster: Report of the Court of Inquiry* (London: Her Majesty's Stationery Office, 1975).
11. T. A. Kletz, "Flixborough," in *Learning from Accidents*, 3rd ed. (Oxford, U.K.: Butterworth-Heinemann, 2001), chap. 8.
12. S. M. Englund, "Design and Operate Plants for Inherent Safety, Part 1," *Chem. Eng. Prog.* 87, no. 3 (1991): 85–91.
13. J. C. Middleton and B. K. Revill, "The Intensification of Chemical Reactors for Fluids" (paper delivered at the Institution of Chemical Engineers Research Meeting, Manchester, U.K., 18–19 April 1983).
14. M. Mulholland, "Engineering a Reaction," *Chem. Eng. (U.K.)* 618 (1996): 14–15.
15. D. Shore, "From Beer to Operating Table," *Chem. Eng. (U.K.)* 468 (1990): 38–40.
16. M. Wilkinson and K. Geddes, "An Award Winning Process," *Chem. Brit.* 29 (1993): 1050–1052.

17. D. C. Hendershot, "Inherently Safer Plants," in *Guidelines for Engineering Design for Process Safety* (New York: American Institute of Chemical Engineers, 1993), 5–52.

18. S. J. Gould, *The Flamingo's Smile* (New York: Norton and London: Penguin Books, 1985), 235.

19. N. A. R. Bell, "Loss Prevention in the Manufacture of Nitroglycerin," in *Loss Prevention in the Process Industries*, Symposium Series No. 34 (Rugby, U.K.: Institution of Chemical Engineers, 1971), 50–53.

20. "Bulk Explosives in Hong Kong," *Downline—A Periodical for Explosives Users* 12 (1990): 2–4.

21. D. J. Lewis, "Loss Estimation," *Chem. Eng. (U.K.)* 462 (1989): 4.

22. R. Scott, "Optimal Cost Design of a Cyclohexanol Plant," in *Process Optimization*, Symposium Series No. 100 (Rugby, U.K.: Institution of Chemical Engineers, 1987).

23. G. H. Shahani, H. H. Gunardson, and N. C. Easterbrook, "Consider Oxygen for Hydrocarbon Oxidations," *Chem. Eng. Prog.* 92 (1996): 66–71.

24. L. M. Litz, "A Novel Gas–Liquid Stirred Tank Reactor," *Chem. Eng. Prog.* 81, no. 11 (1985): 36–39.

25. A. K. Roby and J. P. Kingsley, "Oxidise Safely with Pure Oxygen," *Chemtech* 26 (1996): 39–45.

26. J. Coronas, M. Menéndez, and J. Santamaria, "The Porous-Wall Ceramic Membrane Reactor," *Journal of Loss Prevention in the Process Industries* 8, no. 2 (1995): 97–101.

27. C. Ramshaw, private communication, 1989.

28. T. Sato, Y. Nakashini, and Y. Haruna, "Recycling Vent Gas Improves Phthalic Anhydride Process," *Hydrocarbon Process.* 62, no. 10 (1983): 107–110.

29. F. Hearfield, "Adipic Acid Reactor Development—With Benefits in Energy and Safety," *Chem. Eng. (U.K.)* 316 (1980): 625–633.

30. A. Krzysztoforki, P. Gasiorowski, S. Maciejczyk, and Z. Wojcik, "Cyclohexane Oxidation Using Oxygen Enriched Air" (paper no. 64 delivered at the 7th International Symposium on Loss Prevention and Process Safety, Taormina, Italy, 4–8 May 1992).

31. C. Butcher, "Polypropylene—Still Forging Ahead," *Chem. Eng. (U.K.)* 459 (1989): 32–40.

32. S. M. Crimmin, "How Competitive Is Linear Low Density Polyethylene?" *Hydrocarbon Process.* 61, no. 12 (1982): 75–78.

33. C. R. Raufast, "Process Details of BP Chimie's LLDPE," *Hydrocarbon Process.* 63, no. 5 (1984): 105–109.

34. T. A. Kletz, *What Went Wrong!—Case Histories of Process Plant Disasters*, 5th ed. (Houston, TX: Gulf, 2009), sec. 3.2.8.

35. A. N. Leigh and P. E. Preece, "Development of an Inclined Plate Jet-Reactor System," *Plant/Operations Prog.* 5, no. 1 (1986): 40–43.

36. M. Doherty and G. Buzad, "Reactive Distillation by Design," *Chem. Eng. (U.K.)* 525 (1992): s17–s19.

37. D. A. Crowl, ed., *Inherently Safer Chemical Processes: A Life Cycle Approach* (New York: American Institute of Chemical Engineers, 1996), sec. 3.1.

38. C. Butcher, "High-Intensity Mixing," *Chem. Eng. (U.K.)* 468 (1990): 17.

39. J. F. Davidson, "Planned and Unplanned Research," *Chem. Eng. Res. Des.* 74A, no. 2 (1996): 281–300.

40. M. J. L. Whiting, "The Benefits of Process Integration for Caro's Acid Production," *Chem. Eng. Res. Design* 70A (1992): 195–196.

41. R. Gowland, "Applying Inherently Safer Concepts to a Phosgene Plant Acquisition," *Process Safety Progress* 15, no. 1 (1996): 52–57.

42. C. G. Carrithers, A. M. Dowell, and D. C. Hendershot, "It's Never Too Late for Inherent Safety," in *International Conference and Workshop on Process Safety Management and Inherently Safer Processes* (New York: American Institute of Chemical Engineers, 1996), 227–241.

43. S. Hadlington, "Deep Thinking," *The Roundel* 71, no. 553 (1993): 24–27.

44. W. Benaissa, N. Gabas, M. Cabassud, D. Carson, S. Elgue, and M. Demissy, "Evaluation of an Intensified Continuous Heat-Exchanger Reactor for Inherently Safer Characteristics," *Journal of Loss Prevention in the Process Industries* 21, no. 5 (2008): 528–536.

45. K. V. K. Boodhoo and R. J. Jachuck, "Process Intensification: Spinning Disk Reactor for Styrene Polymerization," *Applied Thermal Engineering* 20, no. 12 (2000): 1127–1146.

46. G. Chiappetta, G. Clarizia, and E. Drioli, "Analysis of Safety Aspects in a Membrane Reactor," *Desalination* 193, no. 1-3 (2006): 267–279.

47. J. Coronas, M. Menedez, and J. Santamaria, "The Porous-Wall Ceramic Membrane Reactor: An Inherently Safer Contacting Device for Gas-Phase Oxidation of Hydrocarbons," *Journal of Loss Prevention in the Process Industries* 8, no. 2 (1995): 97–101.

48. S. H. DeWitt, "Microreactors for Chemical Synthesis," *Current Opinion in Chemical Biology* 3 (1999): 350–356.

49. J. Fischer, C. Liebner, H. Hieronymous, and E. Klemm, "Maximum Safe Diameters of Microcapillaries for a Stoichiometric Ethane/Oxygen Mixture," *Chemical Engineering Science* 64, no. 12 (2009): 2951–2956.

50. J. R. Burns and C. Ramshaw, "Development of a Microreactor for Chemical Production," *Transactions of Institution of Chemical Engineers, Part A* 77 (1999): 206–211.

51. G. Kolb, J. Schurer, D. Tiemann, M. Wichert, R. Zapf, V. Hessel, and H. Lowe, "Fuel Processing in Integrated Micro-Structured Heat-Exchanger Reactors," *Journal of Power Sources* 171, no. 1 (2007): 198–204.

52. R. A. de Graaf and R. Tikku, "Inherently Safe Reactor Design Using Micro Reactors," in *12th International Symposium on Loss Prevention and Safety Promotion in the Process Industries*, IChemE Symposium Series No. 153 (Edinburgh, U.K., 2007).

53. D. Bradley and U. Buehlmann, "Column Internals: Selection and Performance," *Chem. Eng. (U.K.)* 440 (supplement, 1987): 6–7.

54. G. A. Viera and P. H. Wadia, "Ethylene Oxide Explosion at Seadrift, Texas: Part 1—Background and Technical Findings" (paper presented at the 27th Annual Loss Prevention Symposium, Houston, TX, 29 March–1 April 1993).

55. J. Tyzack, "Applications for ATFEs: Drying and Concentration," *Chem. Eng. (U.K.)* 485 (1990): 33–38.

56. J. Tyzack, "Applications for ATFEs: Distillation," *Chem. Eng. (U.K.)* 490 (1991): 20–23.

57. C. Ramshaw, "Higee Distillation: An Example of Process Intensification," *Chem. Eng. (U.K.)* 389 (1983): 13–14.

58. R. Fowler, "Higee: A Status Report," *Chem. Eng. (U.K.)* 456 (1989): 35–37.

59. Leaflets available from Glitsch Inc., Dallas.

60. J. R. Burns and C. R. Ramshaw, "Process Intensification: Visual Study of Liquid Maldistribution in Packed Beds," *Chem. Eng. Sci.* 51 (1996): 1347–1352.

61. R. J. Mohr and A. S. Khan, "Higee: A New Approach in Groundwater Clean-Up" (paper presented at the AQTE Conference, Montreal, Canada, 1987).

62. F. Fowler and A. S. Khan, "VOC Removal with a Rotary Air Stripper" (paper presented at the American Institute of Chemical Engineers Annual Meeting, New York, 15–17 November 1987).

63. R. C. Costello, "Process Intensification Can Enhance Distillation, Heat Transfer and Other Operations," http://www.chemicalprocessing.com/articles/2008/166.html.

64. K. J. Reddy, A. Gupta, and D. P. Rao, "Process Intensification in a HIGEE with Split Packing," *Ind. Eng. Chem. Res.* 45 (2006): 4270–4277.

65. J. V. S. Nascimento, T. M. K. Ravagnani, and J. A. F. R. Pereira, "Experimental Study of a Rotating Packed Bed Distillation Column," *Brazilian Journal of Chemical Engineering* 26, no.1 (2009): 219–226.

66. C. Ramshaw, "Opportunities for Exploiting Centrifugal Fields," *Chem. Eng. (U.K.)* 437 (1987): 17–21.

67. A. Green, B. Johnson, and A. John, "Process Intensification Magnifies Profits," *Chemical Engineering* 103, no. 13 (1999): 66–73.

68. Z. Olujic, M. Jodecke, A. Shilkin, G. Schuch, and B. Kaibel, "Equipment Improvement Trends in Distillation," *Chemical Engineering and Processing: Process Intensification* 48, no. 6 (2009): 1089–1104.

69. H. Zhao, L. Shao, and J.-F. Chen, "High-Gravity Process Intensification Technology and Application," *Chemical Engineering Journal* (2009): 156, no. 3 (2010), pp 588–593.

70. M. L. Myers, "Emerging Technologies: Inherently Safer Designs," *Professional Safety* 51, no. 10 (2006): 20–26.

71. J. J. Springston, "Nanotechnology: Understanding the Occupational Safety and Health Challenges," *Professional Safety* 53, no. 10 (2008): 51–57.

72. M. F. Hallock, P. Greenley, L. DiBerardinis, and D. Kallin, "Potential Risks of Nanomaterials and How to Safely Handle Materials of Uncertain Toxicity," *Journal of Chemical Health and Safety* 16, no. 1 (2009): 16–23.

73. N. Hanigan, "Solvent Recovery: Try Power Fluidics," *Chem. Eng. (U.K.)* 555/556 (1993): 19–22.

74. C. Ramshaw, "The Opportunities for Exploiting Centrifugal Fields," *Heat Recovery Systems & CHP* 13, no. 6 (1996): 493–513.

75. Literature available from Perkins Inc., Saginaw, MI, and Robatel SLPI, Genas, France.

76. C. Ramshaw, "Process Intensification: A Game for *n* Players," *Chem. Eng. (U.K.)* 416 (1987): 30–33.

77. V. Balasundaram, J. E. Porter, and C. Ramshaw, "Process Intensification: A Rotary Seawater Deaerator" (paper presented at the 5th BOC Priestley Conference on Separation of Gases, Birmingham, U.K., 19–21 September 1989).

78. J. Redman, "Getting Out of a Mess in Mixing," *Chem. Eng. (U.K.)* 489 (1991): 13.

79. T. Sparks, "Rotor/Stator: The Art," *Chem. Eng. (U.K.)* 606 (1996): 17–19.

80. "New Fluidized Bed Drier Shows Higee Style Intensification," *Chem. Eng. (U.K.)* 416 (1985): 25.

81. Literature available from Krauss Maffei AG, Munich, Germany.

82. C. Ramshaw, "An Intensified Absorption Heat Pump," *Proc. Institute Refrig.* 89, no. 2 (1988–9): 1–6.

83. K. V. K. Boodhoo, R. J., Jachuk, and C. Ramshaw, "Process Intensification: Spinning Disc Polymeriser for the Manufacture of Polystyrene," in *Proceedings of the International Conference on Process Intensification in the Chemical Industry, Antwerp, Belgium, December 1995*, ed. C. Ramshaw (London: Institution of Mechanical Engineers, 1995).

84. R. J. Jachuck and C. Ramshaw, "Process Intensification: Potentials of Spinning Disc Reactor Technology," in *Process Safety: The Future*, Symposium Series No. 141 (Rugby, U.K.: Institution of Chemical Engineers, 1997).

85. R. W. King, "Inclined Rotary Fin Distillation Column," *Ind. and Eng. Chem.* 61, no. 9 (1969): 66–78.

86. M. Keyvani and N. C. Gardner, "Operating Characteristics of Rotating Beds," *Chem. Eng. Prog.* 85, no. 9 (1989): 48–52.

87. W. T. Cross and C. Ramshaw, "Process Intensification: Laminar Flow Transfer," *Chem. Eng. Res. Des.* 64, no. 4 (1986): 293–301.

88. T. Johnston, "Miniaturized Heat Exchangers for Chemical Processing," *Chem. Eng. (U.K.)* 431 (1986): 36–38.

89. R. K. Shah, "Classification of Heat Exchangers," in *Heat Exchangers: Thermo-Hydraulic Fundamentals*, ed. S. Kakac, A. E. Bergles, and F. Mayinger (Washington, DC: Hemisphere, 1981).

90. E. Gregory, "Plate and Fin Heat Exchangers," *Chem. Eng. (U.K.)* 440 (1987): 33–39.

91. A. P. Fraas, *Heat Exchanger Design*, 2nd ed. (New York: John Wiley and Sons, 1989), 271, 487.

92. "Heat Exchangers Looking for Applications," *Chem. Eng. (U.K.)* 427 (1986): 27.

93. S. Sjogren and W. Grueiro, "Applying Plate Heat Exchangers in Hydrocarbon Service," *Hydrocarbon Process.* 62, no. 9 (1983): 133–136.

94. L. Sennik, "Selecting Elastomers for Plate Heat Exchanger Gaskets," *Chem. Eng. (U.K.)* 406 (1984): 41–45.

95. K. M. Lunsdorf, "Understand the Use of Brazed Heat Exchangers," *Chem. Eng. Prog.* 92, no. 11 (1996): 44–53.

96. "No-Welds Titanium Heat Exchangers Enter Production," *Chem. Technol. Eur.* 3, no. 2 (1996): 11–12.

97. R. J. J. Jachuk and C. Ramshaw, "Process Intensification: Polymer Film Compact Heat Exchanger (PFCHE)," *Chem. Eng. Res. and Des.* 72, no. 2A (1994): 255–262.

98. G. Gardner, "No Fouling and King of the Castle," *Chem. Eng. (U.K.)* 571 (1994): s9–s15.

99. Energy Efficiency Office, *Good Practice Case Studies 12, 22, 29, and 109* (London: Department of Energy, 1990–1992).

100. P. J. Heggs, "Regenerative Heat Exchangers" (paper presented at the Institution of Chemical Engineers Research Meeting, Manchester, U.K., 18–19 April 1983).

101. T. R. Bott, M. J. Gough, and J. V. Rogers, "Heat Transfer Enhancement" (paper presented at the Institution of Chemical Engineers Research Meeting, Manchester, U.K., 18–19 April 1983).

102. J. Redman, "Compact Future for Heat Exchangers," *Chem. Eng. (U.K.)* 452 (1988): 12–16.

103. D. Butterworth, "A Twist in the Tale," *Chem. Eng. (U.K.)* 626 (1997): 21–24.

104. M. Mackley, "Using Oscillatory Flow to Improve Performance," *Chem. Eng. (U.K.)* 433 (1987): 18–20.

105. S. Dutt, "Safe Design of Bulk Direct Steam Injection Heaters" (paper presented at the 5th World Congress of Engineering, San Diego, CA, 14–18 July 1996).

106. C. Butcher, "New Wrinkles in Heat Exchange," *Chem. Eng. (U.K.)* 500 (1991): 11.

107. R. Charlesworth, A. Gough, and C. Ramshaw, "Combustion and Steam Reforming of Methane on Thin Layer Catalysts for Use in Catalytic Plate Reactors," in *Proceedings of the 1st International Conference on Intensive Processing*, ed. G. Akay and B. J. Azzopardi, Paper No. S510/099/95 (London: Institution of Mechanical Engineers, 1995).

108. G. A. Davies, "Development and Design of Intensified Coalescence Equipment" (paper presented at Institution of Chemical Engineers Research Meeting, Manchester, U.K., 18–19 April 1983).

109. P. J. Bailes, "Enhanced Liquid Disengagement by Electrostatic Coalescence" (paper presented at the Institution of Chemical Engineers Research Meeting, Manchester, U.K., 18–19 April 1983).

110. C. Ramshaw, "Industrial Crystallization Research: The Next Steps," *Chem. Eng. (U.K.)* 349 (1979): 691–694.

111. A. Murphy, R. Sibbet., and C. Ramshaw, "Process Optimization: Fluid Bed Heat Transfer," in *Process Optimization*, Symposium Series No. 100 (Rugby, U.K.: Institution of Chemical Engineers, 1987).

112. J. E. Gillett, "Intensification of Pharmaceutical Granule Drying Using Microwaves" (paper presented at the Institution of Chemical Engineers Research Meeting, Manchester, U.K., 18–19 April 1983).

113. D. Williams, *Research in Medical Engineering* (Swindon, U.K.: Science and Engineering Research Council, 1985).

114. A. Nienow, "Stirred Bioreactors: Fact or Fiction," *Chem. Eng. (U.K.)* 459 (1989): 73–75.

115. R. S. Benson and J. W. Ponton, "Process Miniaturisation: A Route to Total Environmental Acceptability," *Chem. Eng. Res. Des.* 71, no. 1A (1993): 1–9. For a summary, see P. Varey, "Small but Environmentally Formed," *Chem. Eng. (U.K.)* 538 (1993): s6–s7.

116. "Phosgene on demand," *Chem. Eng. (U.K.)* 485 (1990): 7.

117. I. Rinard, quoted by C. Caruana, "Still a Way to Go," *Chem. Eng. Prog.* 92, no. 4 (1996): 14.

118. D. McDonald, "The Promise of a Healthier World," *The Roundel* 5 (1989): 105–107.

119. T. A. Kletz, "Bhopal," in *Learning from Accidents*, 3rd ed. (Oxford, U.K.: Butterworth–Heinemann, 2001), chap. 10, 95–106.

120. "Major Reductions Made in Toxic Gas Storage," *Plant/Operations Prog.* 5, no. 2 (1986): A3.

121. D. E. Wade, "Reduction of Risk by Reduction of Toxic Material Inventories," in *Proceedings of the International Symposium on Preventing Major Chemical Accidents*, ed. J. L. Woodward (New York: American Institute of Chemical Engineers, 1987).

122. M. Lapkin, *Report on Inventory Reduction Opportunities for Methyl Isocyanate* (Waverly, MA: Good Neighbor Project for Sustainable Industries, 1994).

123. W. Reganass, U. Osterwalder, and F. Brogli, "Reactor Engineering for Inherent Safety," in *Eighth International Symposium on Chemical Reactor Engineering*, Symposium Series No. 87 (Rugby, U.K.: Institution of Chemical Engineers, 1984).

124. W. Orrell and J. Cryan, "Getting Rid of the Hazard," *Chem. Eng. (U.K.)* 439 (1987): 14–15.

125. T. A. Koch, K. R. Krause, and M. Mehdizadeh, "Improved Safety through Distributed Manufacturing of Hazardous Chemicals," *Process Safety Progress* 16, no. 1 (1997): 23–24.

126. C. A. Leal and G. F. Santiago, "Do Tree Belts Increase Risk of Explosion for LPG Spheres?" *Journal of Loss Prevention in the Process Industries* 17, no. 3 (2004): 217–224.

127. J. Holusha, review of *The Japanese Automobile Industry*, by M. Cusumano, *New York Times*, 2 April 1986.

128. S. Wagstyl, review of *Lean Thinking*, by J. P. Womack and D. T. Jones, *Financial Times (London)*, 30 September 1996, 12.

129. T. A. Kletz, *Hazop and Kazan: Identifying and Assessing Process Industry Hazards*, 4th ed. (Rugby, U.K.: Institution of Chemical Engineers, 1999), sec. 3.4.8.

130. R. Bubbico, S. Di Cave, and B. Mazzarotta, "Risk Analysis for Road and Rail Transport of Hazardous Materials: A Simplified Approach," *Journal of Loss Prevention in the Process Industries* 17, no. 6 (2004): 477–482.

131. R. Bubbico, S. Di Cave, and B. Mazzarotta, "Risk Analysis for Road and Rail Transport of Hazardous Materials: A GIS Approach," *Journal of Loss Prevention in the Process Industries* 17, no. 6 (2004): 483–488.

132. *One Hundred Largest Losses*, 8th ed. (Chicago: Marsh & McLennan, 1985).

133. *Large Property Damage Losses in the Hydrocarbon-Chemical Industries: A Thirty Year Review*, 16th ed. (New York: Marsh & McLennan, 1995).

134. Lord Newton of Braintree, chair of the Buncefield Major Incident Investigation Board, *The Buncefield Incident 11 December 2005: The Final Report of the Major Incident Investigation Board* (London: HSE Books, 2008).

135. European Community, "Council Directive of 24 June 1982 on the Major Accident Hazards of Certain Industrial Activities," *Official Journal of the European Communities* L230 (1982): 1–18.

136. T. A. Kletz, "Seveso," in *Learning from Accidents*, 3rd ed. (Oxford, U.K.: Butterworth–Heinemann, 2001), 88–94.

137. "Control of Industrial Major Accident Hazard Regulations," Statutory Instrument No. 1902 (London: Her Majesty's Stationery Office, 1984).

138. Health and Safety Executive, *A Guide to the Control of Industrial Major Accident Hazard Regulations 1984*, publication HS(R)21 (Sudbury, U.K.: HSE Books, 1990).

139. J. Shrives, "Environment Canada's New Environmental Emergency Regulations," *Canadian Chemical News* 57, no. 2 (2004): 17–21.

140. J. Harvey-Jones, *Managing to Survive* (London: Heinemann, 1993), 11.

141. A. Friedman and J. Norbury, "Contain Yourself," *Chem. Eng. (U.K.)* 617 (1996): 10–11.

142. H. Short, "NH_3 Breakthrough," *Chem. Eng. (U.S.)* 96, no. 7 (1989): 41–45.

143. M. B. Green, quoted by S. F. Bush, "Order and Precision in the Design of Polymerization Reactors" (paper presented at the Institution of Chemical Engineers Research Meeting, Manchester, U.K., 18–19 April 1983).

144. D. L. Goetsch, *The Safety and Health Handbook* (Englewood Cliffs, NJ: Prentice Hall, 2000) 213–215.

145. "Dead Pets, Wallets End Up at Landfill." *The Mail Star*, (Halifax, Canada), 10 February 2000.

Chapter 4

Substitution

Caffeine is removed naturally by gently soaking the beans in water,
then treating them with the same elements that put the effervescence
in sparkling water.

—Advertisement for decaffeinated coffee

Intensification is not always possible, and an alternative is substitution, which entails using a safer material in place of a hazardous one. Intensification and substitution decrease the need for additional protective equipment and thus decrease plant cost and complexity, but intensification, in addition, as mentioned in Chapter 2, brings about a reduction in plant size and further reduction in cost. Intensification is therefore preferred to substitution if both are possible.

Substitution is not, of course, a new idea and has been used in many industries for many years, as the following examples show:

1. The use of thatch for new buildings was forbidden in London as early as 1212.[1]
2. In the eighteenth century, attempts were made to illuminate a gassy coal mine with the phosphorescent glow from putrefying fish skins. In the nineteenth century, the safety lamp provided a more effective solution.
3. The early anesthetics had disadvantages. Ether was explosive, whereas chloroform formed phosgene when it came into contact with the gas flames used for lighting. "You could only choose between poisoning everybody with phosgene or blowing them up with an ether explosion."[2] These hazards disappeared with the introduction of new anesthetics.

4. In 1896 in the United Kingdom, the minimum flash point for oil for use in oil lamps was 73°F. If the oil was upset or the lamp broken, the oil ignited so furiously that it was said to explode. In London in 1893, oil lamps caused 456 fires and killed 48 people. Someone suggested raising the flash point to 100°F, at which temperature the lamp would go out without harm or with only a small flame on the wick. It will come as no surprise to learn that some people opposed the suggestion; users, they said, should take more care of their lamps.[3]

5. In recent years, many people have died because they could not escape from burning buildings or football stadiums. Yet in 1613, when the Globe Theatre in London caught fire, all of the 3000 people inside escaped unharmed, although there were only two small exits from a wooden building with a thatched roof. The walls were made from lath and plaster, and it was easy to knock escape holes through them. In one respect—flammability—the materials of construction were inherently less safe than those we use today, but in another respect they were inherently safer.[4]

6. After the R101 airship disaster in 1930, the use of helium instead of hydrogen was advocated.[5,6] There would, at the time, have been supply problems to overcome.

7. Bulls can be a hazard to farmers and the public; the inherently safer solution is to use artificial insemination.

8. In hospitals, the use of polyester sheets and blankets in place of cotton has reduced fires.

In this chapter, we discuss first the use of safer materials as nonreactive agents, such as heat-transfer agents and solvents, and then the substitution of different chemistry so that we avoid the use of hazardous raw materials or intermediates. Some of the applications described are now in regular use, whereas others are still in the development stage.

4.1 The Use of Safer Nonreactive Agents

4.1.1 General

Flammable hydrocarbons or ethers are often used as heat-transfer media for cooling or heating a process. Sometimes the medium is heated in a furnace and supplies piped heat to a reactor or distillation column reboiler; sometimes the medium removes heat from a reactor and gives it up to water in a cooler, possibly raising steam. In some plants, the inventory of flammable liquid in the heat-transfer system exceeds that in the process.

When possible, we should use a liquid of high boiling point or, better still, water. With water, the pressure will be higher than in a hydrocarbon or hydrocarbon–ether system, but the technology is well understood. If the heat-transfer medium is used to remove heat from a reactor, the heat is immediately available as steam, which can usually be used elsewhere on the site. It is not necessary to cool the heat-transfer oil, and the cost of a heat exchanger is saved. Examples of the successful replacement of oil by water (and of one unsuccessful replacement of water by oil) are described below in this section.

If the heat-transfer medium is used to supply heat, it may be possible to use direct heating instead, but this usually has disadvantages: Several furnaces may be needed instead of one, and temperature control is more difficult.

A wide variety of heat-transfer oils are available commercially with initial boiling points between 275°C (a mixture of diphenyl and diphenyl oxide) and 400°C and working ranges from –50°C to 500°C. These oils have been reviewed by several authors;[7–11] the last reference describes some explosions that have occurred. Vapor-phase media have been reviewed by Frikken et al.[12] Molten salts can be used for high-temperature applications (200°C–2000°C).[13] Refrigerants are discussed in Section 4.1.4.

Many transformer oils burn readily, but oils are now available that burn slowly and produce little smoke. A nonflammable fluid is also available.[14,15] Further improvement in the inherent safety of transformer oils has been made with the introduction of natural oil seed triglycerides (e.g., soybean oil with antioxidants) in place of oils of mineral origin. These new fire-resistant oils are not ignited by arcing because their dielectric strength exceeds 40 kV, and their use is permitted in some jurisdictions without the requirement of sprinklers in transformer buildings.[16]

Polymer quenchants, diluted with water, can now be used instead of oil in heat treatment plants. They are not flammable under the conditions of use.[17]

When chlorofluorocarbons (CFCs) first became available, they were hailed as wonder solvents, nonflammable and with low toxicity, much safer than older solvents such as trichloroethylene. Now that they are known to affect the ozone layer, their manufacture has been discontinued, and some users have started using flammable solvents again. However, other materials are available, such as hydrofluorocarbons, hydrochlorofluorocarbons, and hydrofluoroethers. Some of them are flammable, but much less so than hydrocarbons. If they are not suitable on their own, some of them can be used as carriers for flammable solvents; the mixtures may not be flammable. Hydrochlorofluorocarbons (HCFCs) also have an ozone-depleting potential (ODP), and their use is regarded as an interim measure. A recent international agreement calls for developed nations to begin reducing production and consumption of HCFCs by 2010 and phase them out by 2020.[18] Methyl siloxanes have no ODP and can be blended with other solvents.[19]

According to a press report, CFC-113 (trifluorotrichloroethane) is essential for detecting fingerprints on porous materials. As of the late 1990s, some countries still allowed its use for this purpose, but supplies were known to be limited. Three hundred alternative solvents were tested before a new hydrofluorocarbon was found to be suitable.[20]

According to Kagensberg,[19] "Manufacturers may be restricted from using even very low ODP materials and therefore effectively be driven to use solvents which may have more acute environmental and toxicological consequences." It can only be hoped that regulators will take the challenge to consider the impact of a given proposed regulation on other environmental and safety issues, so that manufacturing can occur while maximizing safety for employees, the surrounding community, and the world.

Although some manufacturers of computer chips are using isopropyl alcohol for cleaning, others are using soap and water (at lower cost); a mixture of water, steam, and ultrasound (much more efficient); or processes have been changed so that cleaning is no longer needed.[21] Supercritical carbon dioxide can be used for industrial degreasing; it has the advantage that oil residues are easily separated from the solvent.[22]

Bombardment by pellets of solid carbon dioxide propelled by compressed air has also been used for cleaning, but this introduces some new hazards. The pellets can cause cold burns if they come into contact with skin, and good ventilation is necessary to avoid asphyxiation.[23]

In a number of cases, toxic solvents have been replaced by less hazardous ones. Thus, the less toxic cyclohexane can often be used instead of benzene. Supercritical carbon dioxide is often used instead of the flammable solvents hexane, ethanol, and ethyl acetate in the food industry, for example, for decaffeinating tea and coffee or extracting hops.[24] It can also be used as a partial replacement for solvents in paint spraying.[25]

A 1991 survey showed that 90% of painters and decorators who used solvent-based paints had suffered ill health from their use. With the growing use of water-based paints, ICI, which at the time supplied about half the U.K. market, made the decision to discontinue the production of solvent-based paints by the year 2000.[26] Burning and absorption have been used to prevent pollution when cars are sprayed with solvent-based paints, but waterborne paints avoid the problem.[27]

Instead of using solvent to apply coatings to paper, board, optical glass, or other substrates, the coatings can be applied as dry powders and then fixed by exposure to ultraviolet or electron beam radiation. The ancient Egyptians are said to have embalmed mummies by coating them and then exposing them to the sun.[28]

In one plant, ultraviolet light was used instead of a hazardous catalyst.

We should, when possible, avoid solvents that can undergo unwanted side reactions. Fierz gives an example.[29] A runaway reaction (not named) could be made worse by the decomposition of the solvent dimethylsulfoxide (and by its high viscosity, which prevents rapid mixing and cooling). Replacement of the solvent was therefore investigated. Possible alternatives would be a better vacuum, so that the temperature is lower, or low-inventory distillation (Section 3.3.1).

Evon et al.[30] recommend thermal screening of reaction mixtures, including solvent variations, early in the development cycle of new processes. This will hopefully identify material/solvent combinations that provide inherently safer properties, as well as those combinations that create greater hazards due to thermal decomposition and runaway.

Carbon dioxide can be used instead of mineral acid for controlling the pH of drinking water. The water cannot be overdosed, for if too much carbon dioxide is injected, the excess will not dissolve. This is an example of inherently safer control.[31]

Membrane cells for chlorine production, as well as having the advantages described in Sections 3.5 and 3.6 and being more economical, do not use the mercury or asbestos employed in the older mercury and diaphragm cells.[32]

A process stream was purified by ion exchange in two vessels, one working, one spare. The ion exchange resin was bought in an acidic form and neutralized before use. A new batch of resin had been charged to the spare vessel but not yet neutralized when the handle of the inlet ball valve was knocked open. The process liquid entered the vessel, the acid catalyzed a runaway reaction, the vessel ruptured, and the escaping liquid ignited. Everyone knew that acids catalyze a runaway reaction, but no one had realized that the catalyst would act as an acid. Afterward, the resin was bought already neutralized. It was just as cheap and just as readily available.

Helium is often purified by passing it over a carbon bed cooled to –196°C by boiling liquid nitrogen. If oxygen is present as well as nitrogen, it is absorbed preferentially, for its boiling point is higher (–183°C). If more oxygen than nitrogen is present, the carbon may absorb so much oxygen that it explodes. Following such an explosion, one company decided to use silica gel instead of carbon. It is less efficient but safer.[33]

Similarly, aluminum and brass packings used for the low-temperature distillation of air are liable to ignite in the presence of oxygen. Copper packing, however, will not do so.[34]

Ductile materials are usually safer than brittle ones. When brittle materials are used, it is usually because they have other properties such as resistance to corrosion or chemical attack, which ductile alternatives do not possess.

The inventor of the Safe Spade used in the mining industry in South Africa replaced the conventional wooden handle with a plastic urethane-molded handle that was stronger.[35] The new design eliminated splintering, cracking, and rotting that would ordinarily result in a loose handle subject to rotation and reduced control of the spade. The handle was also designed to automatically detach in the event it became entangled in moving machinery.

In their article describing electrostatic charging issues in process plants, Astbury and Harper[36] comment that these hazards arise from the materials of construction; the unavoidable presence of flammable, insulating, immiscible liquids; and the size of vessels and pipework. Intensification of material inventories and equipment size, as well as attenuation of flow velocities, are key measures to avoid ignitions by electrostatic discharge. Additionally, the use of electroconductive, corrosion-resistant plastics has been described.[36]

Fires in ventilation ducts are an underrated hazard. Noncombustible materials should be used when possible, not only for the ducts, but also for components such as flexible sections and insulation coverings.[37]

Insulation itself is recognized as playing a potentially active role in fire propagation.[38] According to Bovard,[38] inorganic insulation materials do not generally contribute to fire, but organics, in particular plastic foams such as polyurethane and polystyrene, can introduce serious concerns:

1. Rapid flame spread
2. Attainment of extremely high temperatures
3. Dense smoke
4. Toxic or flammable gases

A newspaper article[39] in the early days of the investigation of Swissair Flight 111, which crashed off Nova Scotia, Canada, in 1998, reported that that the U.S. Federal Aviation Administration had ordered metalized Mylar blanket insulation replaced in the MD-11 aircraft, calling it "fairly flammable." The U.S. National Transportation Safety Board also advised that Mylar use be reduced or eliminated.[39]

Restaurants often prepare flambé dishes at the table using butane heaters. On several occasions, the small butane cylinders used have snapped off at the thread while they were being changed, and diners and staff have been burned. The design of the cylinders has been improved, and restaurateurs have been advised not to reuse old washers and to change cylinders outdoors (what a hope), but it would be better to use another method of heating or to cook in the kitchen.

Smoldering cigarettes have started many house fires, and, according to a U.S. report, about 1500 people per year are killed by fires started in this way. Fire-resistant foams have reduced the risk, and the International Association of Fire Chiefs advocated the development of self-extinguishing cigarettes.[40]

Another case where substitution is needed is in the use of cyanide for the extraction of gold. In 1995, 1.25 million m^3 of water containing cyanide waste entered a river in Guyana, South America, after a dam failed.[41]

Aerosol propellants are discussed in Section 4.1.4.

4.1.2 Ethylene Oxide Manufacture

In most ethylene oxide plants, the catalyst tubes are cooled by heat-transfer oils— often by boiling kerosene under pressure (up to 400 tonnes of it). (The Flixborough explosion was caused by the ignition of a leak of about 50 tonnes of boiling hydrocarbon under pressure.) The piping at the top of the reactors is congested, and bellows (Section 9.4) are often used. The kerosene is a bigger hazard than the ethylene–oxygen mixture inside the reactor because this mixture is in the vapor phase, and when it has entered the explosive range and an explosion has occurred, the effects have been localized. At least 18 incidents have occurred, but in only one of them was a fatality reported. At least four coolant fires have occurred, one of which involved a fatality. No vapor-cloud explosions have occurred, but the potential for them is present.[42]

When one ethylene oxide plant was being designed in the late 1960s, the client's project engineer asked the design contractor if water could be used instead of kerosene. The contractor was willing to use water but could not give the usual guarantees as to output; the client accepted kerosene. By the time another new plant was needed 10 years later, Flixborough had occurred, the dangers of vapor cloud explosions were better known, and water was, therefore, used on the new plant.

A few older plants had already used water, whereas others used heat-transfer oils that have a higher boiling point than kerosene and are used below their boiling points. According to one paper,[43] water cannot be used as a coolant because variations in temperature are greater than with oil, but this is not supported by practical experience. (Section 6.2.4 discusses the hazards of ethylene oxide purification, and Section 4.2.5 describes an attempt to avoid the use of ethylene oxide as an intermediate in the manufacture of glycol.)

4.1.3 Heating an Aqueous Slurry

A slurry of an organic salt with water had to be heated to 300°C at a gauge pressure of 70 bar (1000 psi) to dissolve the salt. The slurry was heated with a heat-transfer oil in several shell-and-tube heat exchangers with the slurry in the tubes. It was realized that, if a tube ruptured, some of the water would vaporize instantly, causing a large increase in pressure. To prevent the oil from being blown out of the system, an elaborate protective system (Figure 4.1) was installed. When a tube burst, the rise

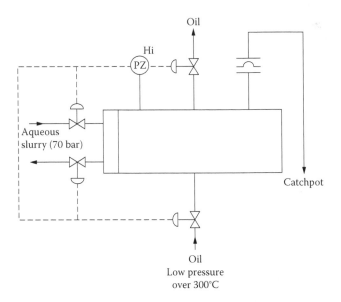

Figure 4.1 Protective system required when an aqueous slurry is heated by a heat-transfer oil. The valves must be quick acting.

in pressure in the shell operated the high-pressure switch PZHi, which closed four valves in the oil and slurry inlet and exit lines, thus isolating the exchanger. At the same time a rupture disc blew, the mixture of oil and steam was blown into a catch pot. When a tube actually burst, the system operated, but not quickly enough, and the oil was blown out of the plant and caught fire.

Of course, when the protective system was tested, its time of response should have been measured, but protective systems are easily neglected or can fail, and it is better, when we can, to avoid the use of flammable oils and our dependence on added protective equipment. In later designs of the plant, the slurry was heated by high-pressure steam, even though the pressure of the site steam supply was too low, and a special boiler had to be installed. In yet later designs, the slurry was partially heated by direct injection of steam, thus avoiding the need for some of the heat exchangers. Heating by steam injection alone would have made the slurry too dilute.

4.1.4 Refrigerants, Propellants, and Firefighting Agents

Some plants, such as olefin-separation plants, contain large inventories of flammable refrigerants like ethylene and propylene. Other plants use ammonia. These liquefied gases are usually readily available on-site and have the properties required for use as refrigerants, but nevertheless nonflammable and nontoxic materials should be considered. When chlorofluorocarbons (CFCs) were first introduced, they were widely welcomed as efficient nonflammable agents, but now that we know their effects on the ozone layer, their production has been prohibited by international agreement. There has been a move back to the use of hydrocarbons and ammonia, but other new agents are now available: hydrofluorocarbons that do not damage the ozone layer and hydrochlorofluorocarbons that cause less damage than CFCs (although, recall the discussion in Section 4.1.1 on phasing out of HCFCs). Some of these materials are flammable, but much less so than hydrocarbons[44,45] (Section 4.1.1).

Consistent with the above comment on a move back to the use of more traditional refrigerants, a recent paper[46] has advocated the expanded use of ammonia, given the mandated elimination of some of the newer environmentally harmful synthetic refrigerants. Does this represent a sort of "reverse substitution," or does it signal the need for careful consideration of all facets of safety, health, and environment hazards? In his article on chemical substitution, Glenn[47] gives an interesting example of the latter.

As reinforced in Section 4.1.5, chemical substitution should be based on the likelihood that a given material can cause harm in addition to its intrinsic hazardous properties (i.e., on a risk basis).[47] This point is illustrated in Table 4.1, which gives the example of monochlorobenzene, a process chemical used in the manufacture of dyestuffs and some pesticides. In terms of relative ranking with several other chlorinated products, it ranks in the middle for health concerns (toxicity), at the top for environmental issues (ecotoxicity), and at the bottom with respect to safety (volatility). Therefore, a chemical substitution would likely be in order to remove

Table 4.1 Chemical Substitution from an EHS Perspective

Ranking	Toxicity	Ecotoxicity	Volatility
Highest	Dichloromethane	Monochlorobenzene	Dichloromethane
	Trichloromethane	Trichloromethane	Acetone
	Monochlorobenzene	Toluene	Trichloromethane
	Toluene	Dichloromethane	Toluene
Lowest	Acetone	Acetone	Monochlorobenzene

Source: From W. M. Glenn, "Better the Devil You Know?" *OHS Canada* 22, no. 6 (2006): 52–57.

an environmental hazard. On the other hand, if occupational health and safety are primary concerns, a review of administrative controls and procedures may be sufficient to ensure an acceptable risk.[47]

In Europe, hydrocarbons have been used in domestic refrigerators because they contain no more than is found in a cigarette lighter. The United States does not allow the use of flammable materials—even those that can be ignited only by high-energy sources.

Liquid nitrogen and liquid carbon dioxide should be considered for some industrial applications. The latter has occasionally been used as a direct-contact refrigerant.[48]

In one ethylene liquefaction plant, a fluorinated hydrocarbon was used instead of propylene as a refrigerant even though propylene was available on-site. The result was a cheaper plant as well as a safer one. The use of a nonflammable refrigerant made a more compact layout possible, and the relief valves could discharge to atmosphere instead of a flare system. A flare system was still needed for the ethylene, but it was smaller than it would have been if propylene had been used as the refrigerant.

Refrigeration was probably the least harmful of the uses of CFCs because, if the plant were well maintained, the material would not be released to the atmosphere. In contrast, all the material used in aerosol propellants is intended for release. Many aerosol manufacturers have used butane (or butane–propane mixtures) instead of halogenated hydrocarbons, but unfortunately the change was followed by several fires in factories and warehouses,[49] for the hazards of the new propellants were not understood. They included the United States' largest warehouse fire, which occurred in the K-Mart warehouse in Morrisville, Pennsylvania, in 1982.[50]

A new design of aerosol cans avoids the discharge of either CFCs or butane to the atmosphere. The propellant, mainly carbon dioxide, is separated from the product by a flexible membrane or a piston and is not discharged during normal use but only if the can is punctured, and then it is, to quote the manufacturer's literature, "essentially nonflammable."[51]

Just as CFCs were considered wonder refrigerants, so bromochlorofluorocarbons (BCFs or halons) were considered wonder firefighting agents when first introduced. Now their production has ceased, for they also affect the ozone layer, though, as with CFCs, existing stocks could still be used as of the late 1990s. Newer halogenated hydrocarbons are available, but they are less effective, and larger quantities are required to extinguish fires.[52-54]

Let us hope there will not be a return to the use of carbon dioxide for the automatic protection of rooms containing electrical equipment. If the carbon dioxide is accidentally discharged while someone is in the room, he or she will be asphyxiated. Accidental discharge of halon and the newer alternatives will produce a concentration too low to cause serious harm. Of course, procedures require the carbon dioxide supply to be isolated before anyone enters the room, but these procedures have been known to break down.

Solid-propellant gas generators, used to inflate airbags in cars, are being developed without the bags for use as firefighting agents. They contain a fuel and an oxidizer and "burn" in the absence of air to produce a stream of inert gas that could be used directly to extinguish fires or as an agent to disperse liquids or dry powders.[55]

4.1.5 Changes in the Wrong Direction

As we have just seen, safer refrigerants, solvents, propellants, and firefighting agents cause effects on the environment. In other cases, changes made to improve the environment have had adverse effects on safety.[49] Few changes are 100% beneficial; there are usually some disadvantages, even though on balance the change is for the better. For this reason, we speak of inherently safer plants rather than inherently safe ones. For example, water-based cleaners are usually safer and more environmentally friendly than organic ones, but nevertheless the concentrated agent before it is dissolved may be toxic, corrosive, or dusty, and bacteria might grow in the solution. As we saw at the beginning of this chapter, although fires occurred more often in lath and plaster buildings, escape was easier when people could escape through the walls.

Here are two more examples of changes that were not for the better:

1. Several years ago the U.K. Mines Inspectorate discouraged the use of mineral oils as hydraulic fluids underground. Officials were concerned that leaks would ignite, and a change was made to "safer," less combustible fluids based on emulsions or phosphate esters. Unfortunately, these fluids are poor lubricants, and they caused excessive wear and overheating. This caused vaporization of the water in the emulsions and further damage; the new fluids also attacked the rubber seals, and it was therefore agreed that, on balance, the conventional oils are safer.[56] Schmidt has reviewed the less combustible oils.[57] They may be suitable for other applications.

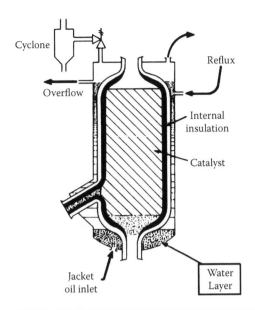

Figure 4.2 The water in the jacket was replaced by flammable oil. Some water contaminated the oil and vaporized, blowing some oil out of the system. (This figure is not drawn to scale.)

2. A reactor was insulated internally to prevent the steel shell from getting too hot. Cracks in the insulation and channeling in the catalyst bed could cause hot spots to develop on the steel, and thus a stainless steel liner was fitted inside the insulation. It was difficult to seal; therefore, when a new plant was built, the liner was left out, and instead a water jacket was installed on the outside of the steel shell. The water jacket was designed to operate at atmospheric temperature.

The reactor had to be operated at a higher temperature than intended, and thus the jacket temperature had to be raised to 120°C. The pressure in the jacket could not be raised; consequently, the water was replaced by an oil, a by-product of the process, of boiling point 170°C (Figure 4.2). Oil vapor from the top of the jacket was condensed and returned at 120°C.

To prevent degradation of the oil, there was a continuous purge and makeup. As a result of an upset at the plant, the makeup oil became contaminated with water, which settled in the base of the jacket and was prevented from boiling by the hydrostatic pressure. A minor disturbance caused some mixing of the oil and water, and some of the water turned to steam and blew some of the oil out of the jacket. A cyclone had been installed after the relief valve, but it was not designed for a high flow rate.

The oil was ignited by a neighboring furnace, and although the fire burned itself out in 5 minutes, damage to instruments and electric cables was extensive.

4.1.6 The Past Seems Incredible

Once we have changed to an inherently safer material, the original choice of material often seems incredible. Here are two examples:

1. Until about 1930, storage tanks for gasoline were often made with wooden roofs.
2. In the nineteenth century, at least one railway tunnel was lined with wood.[58] This was described as "necessary" to prevent slides of the soft rock.

Will our grandchildren gasp with amazement that we tolerated some of the hazards we still tolerate?

4.2 Choosing Less Hazardous Processes

Choosing less hazardous processes is not as simple as it might seem at first sight. Even if an alternative process using less hazardous raw materials or intermediates is available, it may not be economic or may be harder to control. In addition to considering choice of reactants and stability of intermediates, compatibility of materials and choice of solvents and catalysts are important.[59]

4.2.1 Bhopal

The product made at Bhopal (Section 3.6), the insecticide carbaryl, is made from α-naphthol, methylamine, and phosgene. At Bhopal, the first two of these were reacted together to make methyl isocyanate (MIC), the compound that leaked and killed over 2000 people. The MIC was then reacted with α-naphthol to make carbaryl (Figure 4.3).

In the alternative process used by the Israeli company Makhteshim, the same three raw materials are used, but they are reacted in a different order. α-Naphthol and phosgene are reacted together to give a chloroformate ester that is then reacted with methylamine. No MIC is produced.[60]

Neither process is ideal, for both involve the use of phosgene, a gas many times more toxic than chlorine, but the second process at least avoids production of MIC. More fundamentally, instead of carbaryl, could we make an alternative insecticide that is safer to produce, develop pest-resistant plants, or make use of natural predators? I am not suggesting we should (these ideas have disadvantages), but only that the questions should be asked.

α-Naphthol, required for the production of carbaryl, was made at one time by the nitration and reduction of naphthalene, but this process produces traces of carcinogenic α-naphthylamine as a by-product. α-Naphthol is therefore made in a safer manner by the alternative process shown in Figure 4.4.[60]

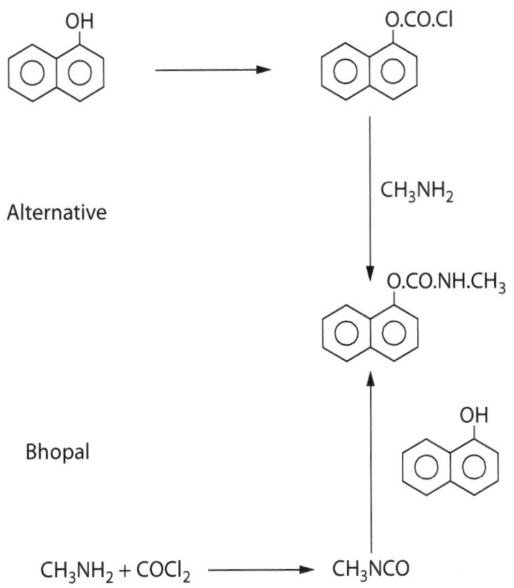

Figure 4.3   Routes to carbaryl.

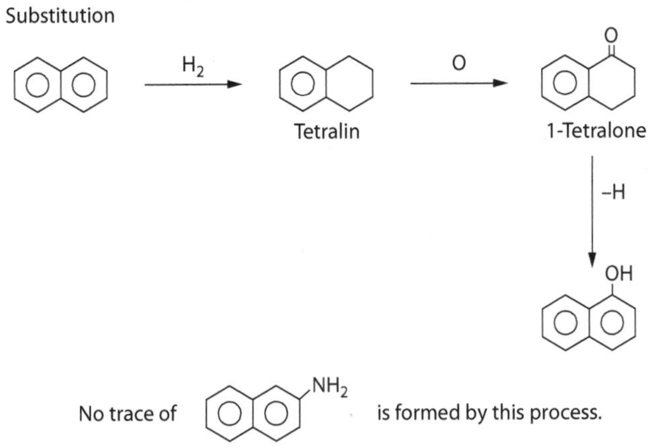

Figure 4.4   α-Naphthol from naphthalene.

4.2.2 DMCC Production

The structural unit

$$
\begin{array}{c}
CH_3 \qquad O \\
\diagdown \qquad \| \\
\qquad N\text{--}C\text{--}O\text{--}R \\
\diagup \\
CH_3
\end{array}
$$

is used in a crop protection chemical. It could be synthesized easily from the alcohol ROH and dimethyl carbamoyl chloride (DMCC). However, DMCC is carcinogenic to animals, possibly to humans, and volatile, and so it was desirable to avoid its use.

An alternative route to the product was developed in which a chloroformate was reacted with dimethylamine to produce a different intermediate:

$$(CH_3)_2\ NH + COClR = (CH_3)_2\ NCOR$$

This alternative route involved the use of phosgene as an intermediate in the preparation of the chloroformate, but this was considered to be a lesser hazard than DMCC.[61]

4.2.3 MDI and TDI Production

The compounds 4,4-diphenylmethane diisocyanate (MDI) and toluene diisocyanate (TDI) have been widely used in the manufacture of synthetic foams, both rigid and flexible. Most of the tonnage required is made using phosgene (Figure 4.5[a]). Much research has been carried out on the development of alternative processes, but without success as of a decade ago. There was much interest several years ago in an alternative process using aromatic nitro-compounds that were carbonylated to alkylurethanes and then pyrolyzed to isocyanates (Figure 4.5[b]), but the promise

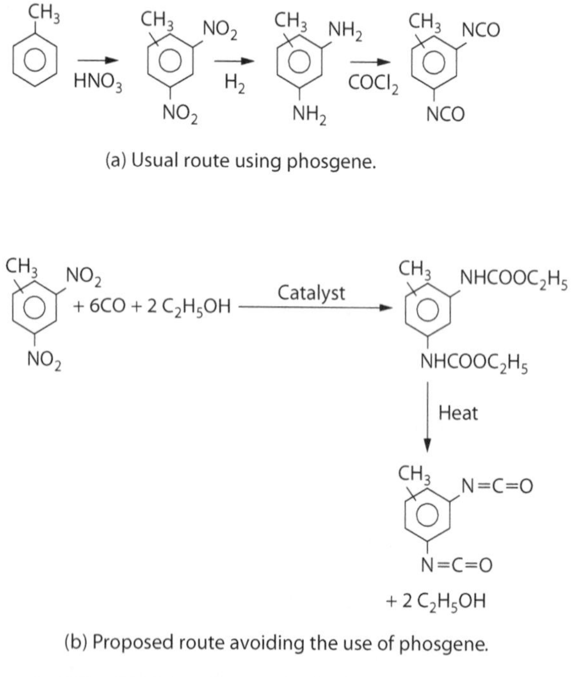

(a) Usual route using phosgene.

(b) Proposed route avoiding the use of phosgene.

Figure 4.5   Routes to toluene diisocyanate.

shown when the process was announced by Atlantic Richfield[62] and Mitsui[63] has not been fulfilled.

However, as mentioned in Section 3.6, manufacturers have considerably reduced or completely eliminated the amounts of phosgene in stock. It can be made immediately before use at the point of use, and it is no longer necessary to store or transport it.

Although the conventional route to MDI and TDI involves the use of a very toxic intermediate, the reaction is only slightly exothermic, and there is little or no danger of a reactor runaway. The carbonylation route, in contrast, is carried out at temperatures at which dinitrotoluene may react violently; it requires the use of carbon monoxide, which is toxic; and the production of nitro-compounds is hazardous (Section 4.2.5). Even if the new route were economic, the choice between the old and new routes would thus not be as clear-cut as it seems at first sight.

Dimethylcarbonate has been proposed as an alternative to phosgene in some syntheses.[64]

4.2.4 KA Production

At Flixborough (Section 3.2.4) a mixture of cyclohexanone and cyclohexanol (usually known as KA or ketone–alcohol mixture) was produced by the oxidation of cyclohexane (Figure 4.6[a]). When the plant was rebuilt after the leak and explosion that killed 28 people, the KA was manufactured by an alternative route, the hydrogenation of phenol. This was widely quoted as a change to an inherently safer route. However, the phenol has to be manufactured, and this is usually done by the

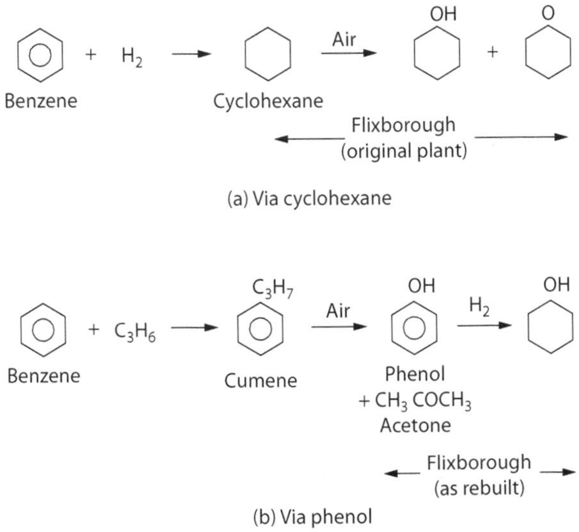

Figure 4.6   Routes to cyclohexanol from benzene.

oxidation of cumene to its hydroperoxide and its "cleavage" to phenol and acetone (Figure 4.6[b]). This process is as hazardous, perhaps more hazardous, than the oxidation of cyclohexane[42] (Section 5.1.1). It was not carried out at Flixborough but elsewhere. There was less hazard on the Flixborough site, but no reduction in the total hazard. The hazard was exported. The rebuilt plant had a short life. It was closed after a few years for commercial reasons.

It is worth commenting at this point on the notion of exporting hazards and risk. Transfer of risk is an accepted business practice in terms of limitation of liability. However, as eloquently expressed by Edwards,[65] it is not acceptable to transfer risk by shifting the production of bulk chemicals in large-scale plants with complex reaction schemes to locales less equipped to deal with the attendant and likely increased hazards. He cites the specific case of such production being transferred from developed to developing nations and uses Bhopal (Section 4.2.1) as an example. We should export inherent safety, not risk.[65]

4.2.5 Other Processes

At one time, benzidine and β-naphthylamine were widely used in the manufacture of dyes and rubber chemicals. Their use was stopped when it was realized that they are carcinogenic. No other routes to the final products were known, and there was risk that they might be metabolized to the carcinogens. Alternative products were therefore developed.[61]

Acrylonitrile, widely used in the manufacture of synthetic fibers, was made by reacting two hazardous materials: hydrogen cyanide and acetylene. It can now be made from propylene, ammonia, and air, which are all considerably less hazardous.[66]

Attempts have been made to produce ethylene glycol directly from ethylene, thus avoiding the need to produce ethylene oxide (Section 4.1.2) as an intermediate. Although a plant was built, it was not an economic success.[66]

Edwards and Lawrence have reviewed and compared six processes for the manufacture of methyl methacrylate.[67,68]

At one time, carbon dioxide was removed from ammonia synthesis gas by a process that involved the use of arsenic. Alternative processes are now available.

Benzene, which is highly toxic and flammable, can be replaced by glucose as raw material for a variety of products, including adipic acid. In the long term, it may be possible to make glucose from cellulose residues such as husks and straw.[69,70]

In the manufacture of silicon semiconductors, toxic gases such as phosphine, diborane, and silane can be replaced by less hazardous liquids such as trimethyl phosphite, trimethyl borate, and tetraethyl-*o*-silicate. Phosphorus oxychloride can replace phosgene, and 1,1,1-trichloroethane can replace hydrochloric acid gas. All these changes have technical as well safety advantages.[71] Similar substitutions have been made in processes for the manufacture of photovoltaic cells.[72]

Hydrofluoric acid (HF) is widely used as an alkylation catalyst but is hazardous, causing severe skin and lung burns to anyone who comes into contact with the vapor. In 1987 in Texas City, 1000 people were hospitalized after a leak of HF. Two alternative processes using solid catalysts have been described.[73,74]

According to Bretherick,[75] "Accident statistics reveal nitration as the most widespread and powerfully destructive industrial unit process operation," and "nitroaromatic compounds and more particularly polynitroaromatic compounds may present a severe explosion risk if subjected to shock, or if heated rapidly and uncontrollably." Most nitro-compounds are manufactured as intermediates and are further processed into other compounds such as amines. Alternative routes to the final products are said to be uneconomic, but perhaps they deserve further study.

In one case, the nitration hazard was reduced by adding the nitro group at a different stage in the process.[76] In another, two acid nitration routes were evaluated for thermal process safety: the classical procedure with nitric acid in the presence of concentrated sulfuric acid, and an alternative in which a different proprietary acid was added in solvent before nitrating with nitric acid.[77]

Dinitrogen pentoxide, produced electrochemically, has been suggested as a nitrating agent in place of the usual mixture of nitric and sulfuric acids. Because it is fully hydrolyzed in solution, reaction times and temperatures are lowered.[78]

In the early days of the Solvay process for the manufacture of sodium carbonate, a manhole at the top of each batch distillation column had to be opened and solid lime tipped in every time the column was charged. Because ammonia vapor could escape, this was hazardous. During the 1870s, Ludwig Mond (the founder, with John Brunner, of Brunner Mond, one of the forerunners of ICI) suggested that milk of lime should be pumped into the columns instead.[79]

From 1832 onward, matches were made from white phosphorus. By 1844 it was known to cause a painful and disfiguring disease called "phossy jaw." Despite the precautions taken, cases of the disease continued to occur and, because alternative processes were available, the use of phosphorus was banned, though not until 1910 in Europe and 1931 in the United States.[80] It is the only industrial disease that has been virtually eliminated by international action.

Combustion of waste products can easily get out of control and produce undesirable by-products. Toxic wastes can now be destroyed electrochemically at low temperatures (70°C–96°C).[81–83] Unlike the products of conventional burning, the off-gas is said to contain no residues. Another method is by treatment with supercritical water at high pressure.[84]

More remarkable is a proposed process for burning coal without producing carbon dioxide and adding to the greenhouse effect. The coal is hydrogenated to methane (exothermic), which is decomposed to hydrogen and carbon (endothermic).[85] Some of the hydrogen is recycled to the hydrogenation plant, and the rest is burned. The carbon is buried (or in the current vernacular, sequestered or stored). At the time of writing this second edition, alternative combustion and energy processes, as well as carbon capture and storage (CCS), occupy much of the world's attention.

Many hazardous compounds can be made safer by adding a group that assists deactivation if they are released to the environment. References 76, 86, 87, and 88 describe other processes in which hazardous reactions or reactants have been replaced by safer ones.

4.2.6 The Need for a New Approach by Research Chemists, Process Designers, and Operators

Despite the examples quoted, it is sometimes still necessary to convince research chemists that they should look for inherently safer processes. An experienced process designer writes that[89]

> it still amazes me how the organic chemists insist on using the most hazardous chemicals to do their "elegant" syntheses. Unfortunately they do not confer with the safety department before they go off on their merry way, and then we have a *fait accompli* which is often very hard to reverse as much time and effort has already been spent on the research.

Few research chemists have any production experience and do not realize the problems involved in transferring laboratory processes to full-scale production. Research chemists sometimes move to production, but movement the other way is unfortunately rare. According to Laird, research chemists pay little attention to work-up, i.e., extracting the product from by-products, solvents, catalysts, and unconverted raw material. Having chosen a synthesis route, research chemists assume that work-up problems will somehow be solved and only rarely consider choosing a route that is less satisfactory in some respects but easier to work up.[90]

Levenspiel[91] writes that

> before starting work on a particular process concept one should set out the criteria for the ideal—never mind whether practical or not—and see how close one can get to the ideal. This requires "thinking" research, sitting around in easy chairs, discussing and discussing—all this before building even the smallest of pilot plants.
>
> Maybe the search for a process concept is best done by getting together a group of knowledgeable free thinkers of varied backgrounds and interests, including mavericks, technicians and "wild men," and let them go at it. Ideas need time to ferment, so let them meet again and again until they all enthusiastically agree that they have come up with the very best.
>
> It may be risky to start out with small bench scale pilot plants without thinking through the operation, because as one progresses one has more and more invested in following the path which has already been

chosen—not just money invested, but intellectual effort and reputation. A momentum is developed to follow a given path which makes it more and more difficult to change directions, to admit that some other way may be better and that one should maybe start afresh in a different direction.

Similar thoughts have been expressed for process designers and operators by Hendershot,[92] who comments that once a hazard has been identified, designers should ask the following questions (in order):

1. Can the hazard be eliminated?
2. If the answer to question 1 is no, can the magnitude of the hazard be reduced?
3. If consideration of questions 1 and 2 has identified alternatives, do these alternatives create new hazards or increase the magnitude of other hazards? If so, have all hazards been evaluated in choosing the best alternative?
4. What technical and management systems will best help to manage the residual risk posed by the remaining hazards?

These questions essentially invoke the hierarchy of controls (Section 1.3) and highlight the need for good management of change procedures when considering inherently safer alternatives (Section 11.2.5). In Hendershot's experience, however, designers and operators often bypass questions 1–3 and skip directly to question 4 in attempting to manage the risk arising from accepted hazards believed to be unavoidable.[92]

For further information about questions to answer before the start of design, see Section 10.2.3.

References

1. J. Gloag, *The Englishman's Castle* (Andover, U.K.: Eyre and Spottiswood, 1945), 46.
2. W. S. Sykes, *Essays on the First Hundred Years of Anaesthesia* (Edinburgh, U.K.: Churchill Livingstone, 1960), 3.
3. D. R. Steuart, *J. Soc. Chem. Ind.* (31 March 1896): 173. Quoted in *Chem. Ind.* no. 6 (18 March 1996): 202.
4. G. Sanctuary, *Shakespeare's Globe Theatre* (London: Shakespeare Globe Trust, 1994), 17.
5. *Chem. Age* (18 October 1930). Quoted in "Fifty Years Ago," *Chem. Age* (17 October 1980): 22.
6. S. S. Stewart, *Air Disasters* (London: Ian Allan, 1986), 11–33.
7. J. Singh, *Heat Transfer Fluids and Systems for Process and Energy Applications* (New York: Marcel Dekker, 1985).
8. C. Butcher, "Thermal Fluid Systems," *Chem. Eng. (U.K.)* 467 (1989): 41–43.

9. D. Ballard and P. W. Manning, *Chem. Eng. Prog.* 86, no. 11 (1990): 51–69.

10. J. Cuthbert, "Choose the Right Heat-Transfer Fluid," *Chem. Eng. Prog.* 90, no. 7 (1994): 29–37.

11. H. L. Febo, "Heat Transfer Fluid Mist Explosion Potential: An Important Consideration for Users" (paper presented at the 29th AIChE Annual Loss Prevention Symposium, Boston, 31 July–2 August 1995).

12. D. R. Frikken, K. S. Rosenberg, and D. E. Steinmeyer, "Understanding Vapor-Phase Heat-Transfer Media," *Chem. Eng. (U.S.)* 82 (1975): 86–90.

13. B. W. Hatt and D. H. Kerridge, "Industrial Applications of Molten Salts," *Chem. Brit.* 15 (1979): 78–81.

14. "New Transformer Fluid," *Chem. Brit.* 20, no. 5 (1984): 392.

15. "Harwell Tests World Beater," *Atom* 331 (1984): 20.

16. B. Dobson, private communication, 1 September 2004.

17. "Synthetic Polymer Quenchants Reduce Heat Transfer Hazards," *Fire Prevention* 146 (1982): 21–23.

18. A. Blatchford, "World Plans to Purge HCFCs," *Chronicle Herald* (Halifax, Canada), 23 September 2007.

19. B. Kagensberg, "Precision Cleaning without Ozone Depleting Chemicals," *Chem. Ind.* 20 (1996): 787–791.

20. P. Hook, "CFC Substitute May Fit the Bill," *Daily Telegraph* (London), 13 November 1996.

21. J. Walker, "Think Laterally," *Chem. Eng. (U.K.)* 530 (1992): 6.

22. L. Mathiasson, "Development of an Industrial Degreasing Process Based on Supercritical Carbon Dioxide," *Arbetsmilöjfonden (Newsletter of the Swedish Work Environment Fund)* 2 (1992): 2.

23. "Dry-Ice Bombardment," *Health Safety Direc.* September (1995): 21.

24. G. Parkinson and E. Johnson, "Supercritical Processes Win CPI Acceptance," *Chem. Eng. (U.S.)* 96, no. 7 (1989): 35–39.

25. "Supercritical CO_2 Reduces Solvent Emissions," *Chem. Eng. (U.K.)* 465 (1989): 33.

26. "Water-Based Paints Coming," *Chem. Eng. (U.K.)* 502 (1991): 10.

27. T. Cantell, ed., *Industry: Caring for the Environment* (London: Royal Society of Arts, 1987), 36.

28. "Radiation Curing," *N.W. Chem. News (U.K.)* 1 (1990): 6–7.

29. H. Fierz, "Assessment of Thermal Safety during Distillation of DMSO," in *European Advances in Process Safety*, Symposium Series No. 134 (Rugby, U.K.: Institution of Chemical Engineers, 1994), 563–574.

30. S. E. Evon, S. Chervin, G. T. Bodman, and A. J. Torres, "Can Solvent Choices Enhance Both Process Safety and Efficiency?" *Process Safety Progress* 18, no. 1 (1999): 1–4.

31. A. Boniface, "The Fizz Factor," *Chem. Eng. (U.K.)* 470 (1990): 17–18.

32. W. N. Brooks, "The Chlor-Alkali Cell: From Mercury to Membrane," *Chem. Brit.* 22, no. 12 (1986): 1095–1106.

33. J. W. Hempseed and R. W. Ormsby, "Explosion within a Helium Purifier," *Plant/Operations Prog.* 10, no. 3 (1991): 184–187.

34. B. R. Dunrobbin, B. L. Werley, and J. G. Hansel, "Structured Packings for Use in Oxygen Service: Flammability Considerations" (paper presented at the 24th American Institute of Chemical Engineers Annual Loss Prevention Symposium, San Diego, California, 19–22 August 1990).

35. "Safe Spade Invention Cuts Mining Accidents," *Safety Management (S. Africa)* February (1998): 5.
36. G. R. Astbury and A. J. Harper, "Large Scale Chemical Plants: Eliminating the Electrostatic Hazards," *Journal of Loss Prevention in the Process Industries* 14, no. 2 (2001): 135–137.
37. U.S. Department of Energy, *Ventilation Ducts: An Underrated Fire Hazard*, DOE Bulletin No. 89-1 (Washington, DC: U.S. Government Printing Office, 1989).
38. T. Bovard, "Reduce Risk of Fire, Smoke Damage by Properly Selecting Insulation," *Hydrocarbon Processing* 78, no. 10 (1999): 81–84.
39. S. LeBlanc, "Flight 111 Fire Started in Area without Detectors, Safety Board Says," *Mail Star* (Halifax, Canada), 5 December 2000.
40. "Move Towards Safer Cigarettes," *Fire Int.* 14, no. 123 (1990): 10.
41. "Mine Cyanide Spill a Disaster for Guyana," *Safety Management (S. Africa)* September (1995): 5.
42. T. A. Kletz, "Fires and Explosions of Hydrocarbon Oxidation Plants," *Plant/Operations Prog.* 7, no. 4 (1988): 226–230.
43. N. Piccinini and G. Levy, "Process Safety Analysis for Better Reactor Cooling System Design in the Ethylene Oxide Reactor," *Canadian Journal of Chemical Engineering* 62 (1984): 541–546.
44. R. G. Richard and I. R. Shankland, "Flammability of Alternative Refrigerants," *ASHRAE J.* 34 (1992): 20–24.
45. T. Dierckx and J. Berghmans, "Safety Aspects of Working Fluids" (paper presented at IIR-IIF Conference, Ghent, Belgium, 12–14 May 1993).
46. A. B. Badger, "The Expanded Use of Ammonia as a Refrigerant as a Replacement for Synthetic Refrigerants" (paper presented at the 43rd American Institute of Chemical Engineers Annual Loss Prevention Symposium, Tampa, FL, 26–30 April 2009).
47. W. M. Glenn, "Better the Devil You Know?" *OHS Canada* 22, no. 6 (2006): 52–57.
48. A. G. Duncan and R. H. Phillips, *Crystallization by Direct Contact Cooling*, Report No. R7873 (Harwell, U.K.: Atomic Energy Research Establishment, 1974).
49. T. A. Kletz, "The Unforeseen Side-Effects of Improving the Environment," *Process Safety Progress* 12, no. 3 (1993): 147–150.
50. Factory Mutual Engineering Corp., "K-Mart Warehouse Fire," *Record* 6, no. 3 (1983).
51. Literature available from BOC Ltd., Guildford, U.K.
52. J. A. Senecal, "Halon Replacement: The Law and the Options," *Process Safety Progress* 11, no. 3 (1992): 182–186.
53. J. A. Senecal, "Halon Replacement Chemicals: Perspectives on the Alternatives," *Fire Technol.* November (1992): 332.
54. *Halon Phase-Out: Advice on Alternatives and Guidelines for Users* (London: U.K. Department of Trade and Industry, 1996).
55. J. C. Yang and W. L. Grosshandler, eds., *Solid Propellant Gas Generators: Proceedings of the 1995 Workshop* (Gaithersburg, MD: National Institute of Standards and Technology, 1995).
56. J. V. Bramley, "Some Safety Aspects of Miscellaneous Mining," *Minerals Ind. Intern.* July (1993): 3–6.
57. M. E. G. Schmidt, "Hydraulic Fluids: Helpful but Hazardous," *Sentinel* (published by Industrial Risk Insurers) 53, no. 4 (1997): 3–13.

58. D. B. Osterwald, *Ticket to Toltec*, 2nd ed. (Lakewood, CO: Western Guideways, 1992), 31.
59. R. L. Rogers and S. Hallam, "A Chemical Approach to Inherent Safety," *IChemE Symposium Series* 124 (1991): 235–241.
60. B. M. Reuben, private communication, 1986.
61. T. K. Wright, "Inherent Safety by Choice of Chemistry" (course on inherently safer plants, University of Manchester Institute of Science and Technology, U.K., September 1982).
62. News item, *Chem. Week*, 9 November 1977, 57.
63. News item, *Eur. Chem. News*, 7 October 1977, 45.
64. L. Bolton, "Enichem Synthesis Unpacks DMC Derivatives Potential," *Chem. Eng. (U.K.)* 517 (1992): 13.
65. D. W. Edwards, "Export Inherent Safety NOT Risk," *Journal of Loss Prevention in the Process Industries* 18, no. 4-6 (2005): 254–260.
66. S. E. Dale, "Cost Effective Design Considerations for Safer Chemical Plant," in *Proceedings of the International Symposium on Preventing Major Chemical Accidents*, ed. J. L. Woodward (New York: American Institute of Chemical Engineers, 1987), 3.79.
67. D. W. Edwards and D. Lawrence, "Assessing the Inherent Safety of Chemical Process Routes," *Process Safety Environmental Protection* 71, no. B4 (1993): 252–258.
68. D. W. Edwards, A. G. Rushton, and D. Lawrence, "Quantifying the Inherent Safety of Chemical Process Routes" (paper presented at the 5th World Congress of Chemical Engineering, San Diego, CA, 14–18 July 1996).
69. J. W. Frost and K. M. Draths, "Sweetening Chemical Manufacture," *Chem. Brit.* 31, no. 3 (1995): 206–210.
70. J. Frost, "Renewable Feedstocks," *Chem. Eng. (U.K.)* 611 (1996): 32–35.
71. Leaflets are available from J. C. Schumacher & Co., Oceanside, CA.
72. V. M. Fthenakis and P. D. Moskowitz, "Health and Safety Aspects of Thin Film Photovoltaic Cell Manufacturing Technique," *Plant/Operations Prog.* 7, no. 4 (1988): 236–241.
73. "New Linear Alkylbenzene Process," *Chem. Eng. (U.K.)* 488 (1991): 16.
74. A. J. McCarthy, J. M. Ditz, and P. M. Geren, "Inherently Safer Design Review of a New Alkylation Process" (paper presented at the 5th World Congress of Engineering, San Diego, 14–18 July 1996).
75. P. G. Urben, ed., *Bretherick's Handbook of Reactive Chemical Hazards*, 6th ed., vol. 2 (Oxford, U.K.: Butterworth-Heinemann, 1999), 246, 251.
76. R. L. Rogers and S. Hallam, *Process Safety Environmental Protection* 69, no. B3 (1991): 149–152.
77. B. Venugopal and D. Y. Kohn, "Chemical Reactivity Hazards and Inherently Safer Technology" (paper presented at the 39th American Institute of Chemical Engineers Annual Loss Prevention Symposium, Atlanta, 11–13 April 2005).
78. A. Jacob, "Electrochemistry Branches Out," *Chem. Eng. (U.K.)* 501 (1991): 10–11.
79. *The 50th Anniversary of BM & Co., 1873–1923* (Northwich, U.K.: Brunner Mond, 1929), 30.
80. D. Hunter, *The Diseases of Occupations*, 6th ed. (London: Hodder & Stoughton, 1978), 370.
81. D. F. Steele, D. Richardson, J. D. Campbell, D. R. Craig., and J. D. Quinn, "The Low-Temperature Destruction of Toxic Waste by Electrochemical Oxidation," *Process Safety Environmental Protection* 68, no. B2 (1990): 115–121.

82. "Organic Waste Destruction the Electrochemical Way," *Chem. Eng. (U.K.)* 463 (1989): 28.

83. D. F. Steele, "A Novel Approach to Organic Waste Disposal," *Atom* 393 (1989): 10–13.

84. P. Svensson, "Look, No Stack: Supercritical Water Destroys Organic Waste, *Chem. Technol. Eur.* 2, no. 1 (1995): 16–19.

85. "Smashing the Greenhouse," *Chem. Brit.* 24, no. 5 (1988): 414.

86. A. Mittelman and D. Lin, "Inherently Safer Chemistry," *Chem. Ind.* (1995): 694–696.

87. D. C. Hendershot, "Chemistry: The Key to Inherently Safer Manufacturing Processes" (paper presented at the American Chemical Society Division of Environmental Chemistry Conference, Washington, DC, 21–25 August 1994).

88. D. C. Hendershot, "Inherently Safer Plants," in *Guidelines for Engineering Design for Process Safety* (New York: American Institute of Chemical Engineers, 1993), chap. 2.

89. S. S. Grossel, private communication, 1994.

90. T. Laird, "Working Up to Scratch," *Chem. Brit.* 32, no. 8 (1996): 43–45.

91. O. Levenspiel, "Chemical Engineering's Grand Design," *Chem. Eng. Res. Dev.* 66, no. 5 (1998): 387–395.

92. D. C. Hendershot, "An Overview of Inherently Safer Design," *Process Safety Progress* 25, no. 2 (2006): 98–107.

Chapter 5

Attenuation

Things that are moderate last a long while.

—**Seneca (ca. 4 BC–65 AD)**

If intensification and substitution are not practicable, then a third road to inherently safer plants is attenuation (or moderation), which entails carrying out a hazardous reaction under less hazardous conditions or storing or transporting a hazardous material in a less hazardous form.

5.1 Attenuated Reactions

5.1.1 Phenol Production

The process for the manufacture of phenol from cumene (mentioned in Section 4.2.4) involves a reaction stage that usually operates within 10°C of the temperature at which a runaway reaction can occur. It is (or was at one time) therefore usual to provide a large tank kept half full of water in which the reactor contents can be dumped by operation of a valve at the base of the reactor. Figure 5.1 shows successive stages in the elaboration of the safety equipment. Originally, the dump valve was hand operated (Figure 5.1[a]). Then dumping was made automatic when a preset temperature was reached (Figure 5.1[b]). A study of the reliability of the instrumentation showed that it was inadequate, and a duplicate system was installed in parallel (Figure 5.1[c]). Finally, it was realized that, if the reaction temperature can be lowered, the chance of a runaway is much reduced, and the dump tank may be unnecessary (Figure 5.1[d]). The reaction volume has to be increased

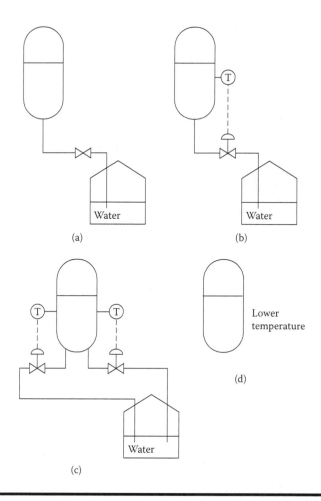

Figure 5.1   Successive stages in the development of the dump system for a reaction that may run away.

slightly to compensate for the lower temperature, but the overall safety is increased. (Attenuation is often the opposite of intensification, as discussed in Section 3.2, but not always; see Section 5.1.2.)

5.1.2 Nitration of Aromatic Hydrocarbons

As mentioned in Section 4.2.5, nitration has been described as the most widespread, powerfully destructive industrial unit process operation.[1] It usually takes place, for economic reasons, in batch reactors at temperatures close to those at which a runaway reaction occurs. (Reference 1 lists accounts of incidents.) The reaction can be carried out safely at these temperatures if the reaction mixture is diluted with a

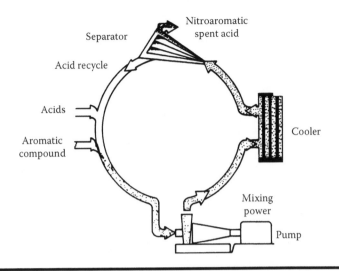

Figure 5.2   The pump nitration circuit: safety by reduction of inventory and dilution. (From "New Continuous Aromatic Nitration Process," *Chem. Eng. (U.K.)* 341 (1979):79. With permission.)

safe solvent and if there is good mixing to compensate for the effects of dilution.[2] Figure 5.2 shows a continuous-loop reactor in which we have both attenuation and intensification. The total inventory is lower than in the traditional batch reactor and, in addition, it is diluted by an excess of sulfuric acid. (Some sulfuric acid is always used.) Vigorous mixing in the circulating pump results in reaction within a few seconds, and the total contact time of the acid and hydrocarbon phases is under a minute. The ratio of sulfuric acid to reactants (nitric acid and hydrocarbon) is 30:1, and thus there are not enough reactants present for a runaway to occur. The maximum possible temperature rise is 15°C.[3,4]

Sulfuric acid might seem a curious choice for a safe solvent, but if it should leak, the incident will be localized. It will not have the devastating effects of a reactor explosion.

5.1.3 Other Reactions

New technology has allowed several processes to operate under less severe conditions than those formerly used. For example, new catalysts have resulted in lower operating pressures in methanol plants and in the Oxo process for producing aldehydes from olefins by carbonylation. Polyethylene and polypropylene can now be produced at lower pressures (Section 3.2.6). Some pyrophoric materials such as butyl lithium are now produced in dilute solution so that they will not ignite on contact with air.[5] A reaction using propylene oxide is now carried out at a reduced pressure at which a runaway decomposition cannot occur.

In some cases, a hazardous reagent or catalyst can be made immobile by attaching an active group to it that binds to a fixed substrate.[6] A recent project has demonstrated how carbon nanotubes can be grown on aluminum oxide carriers to enhance process safety.[7] The general principle espoused by Horng[7] is that production of nanoparticles and nanofibers on microsized carriers or surfaces can help reduce environmental, health, and safety (EHS) impacts.

Similar to the case of nitration (Section 5.1.2), dilution with a safe solvent can help to attenuate the reaction conditions in other processes. Chen[8] describes a cyclohexane oxidation process in which additional water serves to create a minimum boiling azeotrope with the liquid cyclohexane, while the water vapor generated renders the cyclohexane vapor inflammable. Another recent example is the synthesis of an active pharmaceutical ingredient (API) involving the use of highly unstable cyanamide.[9] The use of additional solvent (again, water) was shown by kinetic modeling and adiabatic testing to significantly decrease the likelihood of a runaway reaction.

Other applications of less severe conditions are the use of vibratory feeders instead of screw feeders for moving flammable powders (the vibratory feeders contain no closely spaced moving parts) and the use of gravity or gas pressure instead of pumps for moving unstable liquids (but see Section 8.1.5).

Raw materials often contain impurities that may take part in reactions and ultimately have to be removed from the product. It may be better to remove them from the raw material.

5.2 Attenuated Storage and Transport

The following are examples of attenuated storage and transport:

1. Large quantities of ammonia and chlorine are now usually stored refrigerated at atmospheric pressure and not under pressure at atmospheric temperature. If there is a hole in the tank (below the liquid level) or connecting lines, the liquid flow rate through the hole is less and, because the liquid is cooler, a smaller proportion evaporates. If there is a hole in the tank above the liquid level, then the flow through it is small because there is little or no pressure to drive it.

   Other liquefied flammable gases (LFG) such as propane, propylene, butane, butylene, ethylene oxide, vinyl chloride, and methylamines are often stored refrigerated at low temperature.

   There is no doubt that refrigerated storage is safer than pressurized storage, and if the material is required to be refrigerated, for example, for export by ship, it should certainly be stored in this form. However, if it is required, for use or export, at ambient temperature or under pressure, the whole system (refrigeration, storage, reheating) should be considered. The refrigeration and reheating plants are sources of leakage, and there may be no net increase in safety. Pressurized storage may be as safe or safer if the total quantity stored is no more than several hundred tonnes.

In Japan, following a fire at a plant for filling propane cylinders, regulations required refrigeration of the propane.[10] However, this needs insulated tank trucks, a reduction in payload, and more journeys, which results in more ordinary road accidents. There may have been no increase in safety. An explosion at a propane storage facility in Canada in 2008 has prompted a safety review of the storage and transport of propane (normally under pressure) in the province of Ontario.[11]

2. As an alternative to refrigeration, ammonia can be stored as an aqueous solution. Figure 5.3 shows the reduction in atmospheric concentrations that can be achieved for two accident scenarios: release over 10 minutes of the contents of a vessel holding 250 tons of ammonia (or rather less ammonia as a 28% solution) and fracture of a 2-in. line. The company that made these calculations found that aqueous ammonia could be used for all its processes.[12]

Hendershot[13] has estimated the distance that a leak of monomethylamine vapor will travel for two cases: transfer as a pure liquid and transfer as a 40% solution in water. Fracture of a 1-in. line was considered in each case. The vapor from the solution will be diluted to a nonhazardous level before it reaches the factory fence, whereas the vapor from the anhydrous leak will travel over four times as far and will affect a residential area.

Lacoursiere et al.[14] describe the SO_2SAFE^{TM} technology in which sulfur dioxide gas is dissolved in a high-capacity solvent (diamine absorbent). Thus storage is possible without resorting to potentially hazardous storage as a liquid under pressure. Reference 14 also states that the SO_2SAFE process is less expensive than conventional liquid storage because of the elimination of the equipment required to produce liquid SO_2.

3. For many years, hydrogen has been stored and transported as ammonia, which is "cracked" as required. It can also be transported and stored by hydrogenating toluene to methylcyclohexane and then dehydrogenating it as required.[15]

4. Acetylene has been stored and transported for many years as a solution in acetone.

5. Organic peroxides are liable to decompose explosively and are therefore often stored and transported as solutions despite the increased cost and reduced reactivity. There has to be a balance between reactivity and safety, which, in general, is said to have been achieved.[16] In the United Kingdom, it was recognized that some peroxides had to be diluted before transport, and the maximum permissible container size was set at 1 kg.[17]

6. Some dyestuffs that form explosive powders can be supplied as pastes. When some dyestuffs had to be supplied as a powder, the components were mixed before drying instead of after. Other dusts can be pelletized or suspended in liquid; for example, china clay is now supplied as an aqueous slurry.

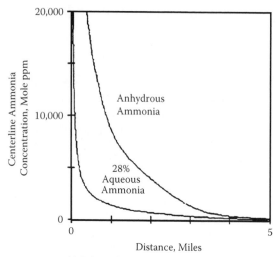

(a) Release Scenario: 10-min. release of storage tank contents.

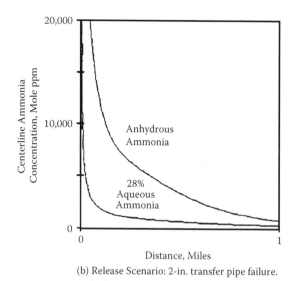

(b) Release Scenario: 2-in. transfer pipe failure.

Figure 5.3 **Comparison of centerline vapor cloud concentration as a function of distance from the release for anhydrous and 28% aqueous ammonia storage for release scenarios of (a) 10-min release of storage tank contents and (b) a 2-in. transfer-pipe failure. (Note: Weather–D stability, 3.4-mph wind speed.) (Copyright 1996 by the American Institute of Chemical Engineers; reproduced by permission of Center for Chemical Process Safety.)**

The hazard in this case is, of course, a health hazard, not an explosive one. Chemicals used in the rubber industry have been incorporated in a rubber premix.[18]

One company overcame a dust hazard by tipping the unopened bags of the dusty reagent into the mixing equipment. The bags were, of course, compatible with the process. Another company found that addition of a little oil to mineral wool reduced the emission of fine respirable fibers.[18]

Other examples of attenuation of dust explosion hazards are given in Section 13.5.

7. Chlorine for use as an antiseptic in swimming pools used to be stored in cylinders, but it is now usually stored as calcium or sodium hypochlorite. In 1983, 79 children were affected by a leak of chlorine at a school swimming pool.[19] However, the suppliers of chlorinating equipment have pointed out that hypochlorites are not entirely nonhazardous and can react with many common materials.[20] As with many other examples in this book, a change for the better in one respect often has accompanying disadvantages. The worst that can happen with hypochlorite is far less than the worst effects of chlorine, and on balance the change seems justified. Another method of avoiding the use of chlorine is to electrolyze brine in situ.

8. The need for high-pressure storage of supercritical fluids (e.g., mixtures of hydrogen and carbon dioxide) in a research laboratory can be eliminated by a novel technology termed "gasless hydrogenation" by its proponents.[21]

5.3 A Railway Analogy

Chapter 5 opened with the statement that attenuation, in part, entails carrying out a hazardous reaction under less hazardous conditions. It is perhaps not too much of a stretch to think of railway track maintenance as a hazardous "reaction" (i.e., activity) that should be conducted under the least hazardous conditions possible. Hopkins[22] describes just this scenario in the background to his analysis of the Glenbrook train accident that occurred in Australia in 1999.

Reference 22 relates the practice of the Queensland rail system management to give track workers exclusive possession of the track in both directions while maintenance work was ongoing. Appropriate time windows facilitated this practice, whereby no trains ran on the track during scheduled track work. This is a form of attenuation achieved in this case by the top element of the hierarchy of controls—elimination of a primary hazard to workers. Other occupational hazards still exist, but the train as a hazard does not.

Compare the above practice to the safety measures in place when it was not possible to eliminate train traffic while track maintenance was carried out. Train drivers were told to slow down (procedural safety—at the very bottom of the hierarchy) and were reminded to slow down by small devices placed on the track and which sent an audible signal when run over by the train (engineered safety—midway in the hierarchy).[22]

Even the elimination of trains during maintenance work affords no absolute guarantee of safety, as evidenced by a train accident in Germany in 2006. In this case, a high-speed magnetic train crashed into a maintenance vehicle on the track, killing 23 people. A newspaper report[23] at the time indicated that maintenance workers were supposed to inform the control room when their vehicle had been taken off the track. Controllers were then required to confirm the presence of the maintenance vehicle in the works garage. Only after this confirmation were the controllers permitted to activate the electrical power allowing the train driver to operate the train. (See also Section 15.3, which discusses operator error.)

References

1. P. G. Urben, ed., *Bretherick's Handbook of Reactive Chemical Hazards,* 6th ed., vol. 2 (Oxford, U.K.: Butterworth-Heinemann, 1999), 246.
2. H. G. Gerrisen and C. M. van't Land, "Intrinsic Continuous Process Safe-Guarding," *Ind. Eng. Chem. Process Des. Dev.* 24 (1985): 893–896.
3. "New Continuous Aromatic Nitration Process," *Chem. Eng. (U.K.)* 341 (1979): 79.
4. Leaflets available from Bofors Nobel Chematur, Bofors, Sweden.
5. S. E. Dale, "Cost Effective Design Considerations for Safer Chemical Plants," in *Proceedings of the International Symposium on Preventing Major Chemical Accidents,* ed. J. L. Woodward (New York: American Institute of Chemical Engineers, 1987), 3.79.
6. D. A. Crowl, ed., *Inherently Safer Chemical Processes* (New York: American Institute of Chemical Engineers, 1996), 37.
7. J.-J. Horng, "Growing Carbon Nanotube on Aluminum Oxides: An Inherently Safe Approach for Environmental Applications," *Process Safety and Environmental Protection* 85(B4), no. 4 (2007): 332–339.
8. J.-R. Chen, "An Inherently Safer Process of Cyclohexane Oxidation Using Pure Oxidation: An Example of How Better Process Safety Leads to Better Productivity," *Process Safety Progress* 23, no. 1 (2004): 72–81.
9. W. Dermaut, C. Fannes, and J. V. Thienen, "Safety Aspects of a Cyanamide Reaction: Inherent Safe Design through Kinetic Modelling and Adiabatic Testing," IChemE Symposium Series No. 153, in *12th International Symposium on Loss Prevention and Safety Promotion in the Process Industries* (Edinburgh, U.K., 2007).
10. "Explosion at Propane Gas Filling Mill," *Safety Eng. News* no. 14, April (Tokyo: Yasuda Fire and Marine Insurance Co., 1989).
11. M. Birk and S. Katz, *Report of the Propane Safety Review* (Toronto, Canada: Queen's Printer for Ontario, 2008).
12. C. G. Carrithers, A. M. Dowell, and D. C. Hendershot, "It's Never Too Late for Inherent Safety," in *International Conference and Workshop on Process Safety Management and Inherently Safer Processes* (New York: American Institute of Chemical Engineers, 1996), 227–241.
13. D. C. Hendershot, "Inherently Safer Plants," in *Guidelines for Engineering Design for Process Safety* (New York: American Institute of Chemical Engineers, 1993), chap. 2.

14. J. P. Lacoursiere, J. N. Sarlis, and P. M. Ravary, "The SO₂SAFE™ Technology for Storage and Transport of Sulfur Dioxide" (paper presented at the 32nd American Institute of Chemical Engineers Annual Loss Prevention Symposium, New Orleans, LA, March 1998).

15. T. Newton, "The Light Stuff," *Chem. Brit.* 33, no. 1 (1997): 29–31.

16. D. J. Lewis, "Explosive Decompositions," *Hazardous Cargo Bulletin* 7, no. 2 (1985): 31–32.

17. *Organic Peroxides (Conveyance by Road) Regulations,* SI No. 2221 (London: Her Majesty's Stationery Office, 1973).

18. British Occupational Hygiene Society Technology Committee, *The Manager's Guide to Control of Hazardous Substances* (Leeds, U.K.: H and H Scientific Consultants, 1996), 12, 13, 54.

19. "Scenario," *Saf. Manage. (South Africa)* 9, no. 12 (1983): 44.

20. Technical Bulletin No. 8405-1, Jensen Beach, FL: Chlorinators Inc., 1984.

21. J. Singh, S. Waldram, J. R. Hyde, and M. Poliakoff, "Supercritical Hydrogenations: Using a Novel, Inherently Safer Approach," IChemE Symposium Series No. 150, in *Hazards XVIII* (Manchester, U.K., 2004).

22. A. Hopkins, *Safety, Culture and Risk: The Organizational Causes of Disasters* (Sydney: CCH Australia Limited, 2005), 67.

23. "Experts Probe Cause of Train Crash," *Chronicle Herald* (Halifax, Canada), 24 September 2006.

Chapter 6

Limitation of Effects

There is always a best way of doing everything, if it be to boil an egg.

—Ralph Waldo Emerson, *Conduct of Life: Behavior*

The previous three chapters have discussed ways of making plants safer by reducing the inventory of hazardous materials and instead using safer materials or hazardous materials in a less hazardous form. This chapter discusses inherently safer ways of limiting the effects of failures (of equipment, control systems, or people) by equipment design or change in reaction conditions rather than by adding protective equipment that may fail or be neglected or may introduce other problems. Public opinion and legislation are making it increasingly difficult to control runaway reactions by venting gas to the atmosphere; instead, we should try to choose reactions and plant designs that make runaway reactions impossible. If we cannot do that, we should at least collect the discharge.

6.1 Limitation of Effects by Equipment Design

The following are examples of limitation effects by equipment design:

1. Spiral-wound gaskets are inherently safer than fiber gaskets because the leak rate is much lower if the bolts work loose or are not tightened correctly.
2. A normal rupture disc is inherently safer than a reverse buckling disc because the latter may slowly "roll over" and rest against the knife blades. The blades then provide support instead of a cutting action, and the disc will not rupture

113

until the pressure is much greater than the normal rupture pressure. Rollover can occur as the result of minor mechanical damage, such as a small indentation, small changes in pressure, or both.[1]

3. A single-side-open explosion barricade (with three walls and a roof) and a freestanding missile wall spaced apart from and parallel to the open side is a common design for such passive barriers.[2] This type of design can be inherently safer than an arrangement that does not incorporate preferential blast wave direction with missile protection.

4. When a leak occurs from a storage tank, if the diked area is small, the evaporation is low and the area of any fire is small. The best design is that often used for tanks containing refrigerated liquefied gases, such as liquefied chlorine or ammonia. The dike is as tall as the tank and about 1 m from it, and only a narrow ring of liquid is exposed to the atmosphere.

   Ferguson[3] describes an interesting application of inherent safety in mitigating off-site impacts of release of a toxic and flammable liquid. The particular chemical site had a large storage tank and a containment dike from which high evaporation rates would occur in the event of a release due to the large pool surface area. Although there is no mention of reducing the size of the storage tank, installation of a revised dike around the tank reduced the surface area available for evaporation by 60%.

5. As discussed in Sections 3.2 and 3.2.8, tubular reactors are safer than pot reactors, for any leak can be stopped by closing a valve, and vapor-phase reactors are safer still because the mass flow rate through a hole of a given size is much less.

6. Do not install larger pipes and valves (especially control valves) than necessary. The leak rate from a fractured 2-in. (50 mm) line is less than half that from a 3-in. (75 mm) line. Drain lines on liquefied petroleum gas tanks should be ¾ in. (19 mm) maximum and sample lines ¼ in. (6 mm) maximum. The fire at Feyzin in France in 1966, which killed 18 people and injured 81, started when a 1½-in. (38 mm) drain valve stuck open.[4]

Decreasing the number of pipelines, flanges, valves, nozzles, and so on, will not affect the size of a leak but will reduce the frequency of leaks.[5] Risk is the combination of size and frequency. Some writers say that it is more effective to reduce the number of leaks than to reduce the inventory and thus the maximum size of any leak. However, large leaks, though infrequent, can cause massive damage and loss of life. We should try to reduce both the number of leaks and their size.

6.2 Limitation of Effects by Changing Reaction Conditions

6.2.1 Different Vessels for Different Stages

A process involved two reaction stages, both of which were originally carried out in one vessel (Figure 6.1[a]). If, by mistake, first-stage reactants (A and B) were added

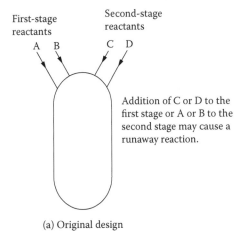

First-stage reactants
A   B

Second-stage reactants
C   D

Addition of C or D to the first stage or A or B to the second stage may cause a runaway reaction.

(a) Original design

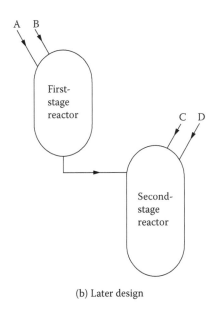

A   B

First-stage reactor

C   D

Second-stage reactor

(b) Later design

Figure 6.1   Old (a) and new (b) designs for a two-batch reaction system.

during the second stage or second-stage reactants (C and D) were added during the first stage, a runaway reaction could occur. Interlocks and training were originally used to prevent such incidents.

A new plant used an inherently safer method: separate vessels for the two stages (Figure 6.1 [b]). Reactants A and B were piped only to the first vessel and C and D only to the second. No extra vessels were needed because the plant contained more than one stream.

Figure 6.2 shows another similar redesign.[6] In Figure 6.2(a), all the raw materials and services are connected to one vessel, and it is possible to introduce one of them during the wrong stage. This is impossible in the design shown in Figure 6.2(b). In general, carrying out different stages in different vessels may allow design parameters such as heating and cooling arrangements, raw material addition, and relief requirements to be tailored more accurately to the needs of each step. But take care that highly toxic or unstable intermediates do not need increased handling.[6]

Overcharging of a reactant can be prevented by charging it via a measuring tank. However, it may still be possible to charge it twice.

6.2.2 Changing the Order of Operations

In a batch process, all the reactants are put into the reactor at once. Reactions are usually under better control in a semibatch process in which we put one reactant in the reactor and add the other(s) gradually; reaction occurs immediately, and unreacted mixtures of reactants cannot accumulate unless mixing fails or catalyst is used up.[6] Unfortunately, mixing has often stopped (or was never started), and the protective system designed to stop addition of further reactant has failed to operate.[7,8] A continuous reactor, especially a tubular one, is better than a semibatch reactor for reasons discussed in Section 3.2 (it has higher integrity, leaks are easily stopped, heat loss is greater, and there is less variation in mixing and thus product quality). Continuous reactors may be practicable for small-scale production, as also discussed in Section 3.2.

Three raw materials (A, B, and C) had to be reacted in a semibatch reactor. Originally A and B were put into the reactor, and C was added slowly. If the temperature control failed, A and B reacted together, and a runaway occurred. If too much C was added, a runaway occurred.

In a newer plant, B and C are put into the reactor, and A is added slowly. If the temperature control fails, B and C do not react together, and there is no runaway. If the A flow controller fails, the rate of addition is limited by the line size. (Note: narrow-bore lines are better than restriction orifice plates because they are less easily removed.) Section 4.2.1 describes a change in the order in which reactions are carried out in a continuous plant.

In another plant, the chance of a runaway was reduced by premixing two reactants and then adding them to a third that was already in the reactor.

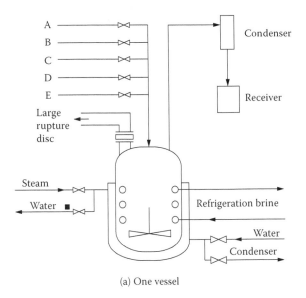

(a) One vessel

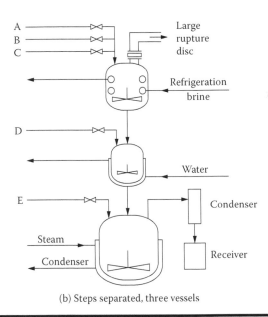

(b) Steps separated, three vessels

Figure 6.2 Alternative designs for a batch process consisting of two reaction steps and solvent strip. (From D. C. Hendershot, "Safety Considerations in the Designs of Batch Processing Plants," in *Proceedings of the International Symposium on Preventing Major Chemical Accidents,* **ed. J. L. Woodward [New York: AIChE, 1987], 3.1–3.16. Reproduced by permission of the American Institute of Chemical Engineers.)**

A concentrated solution of an unstable chemical had to be pumped to the top of a tower for spray drying. The design was changed so that a weak solution was pumped and concentrated immediately before spraying. This reduced the inventory of concentrated solution and avoided the need to pump it. In this case, the inherently safer design cost more to build and operate, but the extra cost was considered worthwhile.

Klein and Balchan suggest the following guidelines for semibatch reactions:[9]

1. Add the final reactant slowly, thus slowing the rate of heat release.
2. Do not use initiators that decompose slowly at reaction temperature or unreacted material may build up in the reactor.
3. If cooling depends on evaporation and reflux of a solvent, establish reflux before reaction starts.
4. Do not mix reactive chemicals in feed tanks.

6.2.3 Changing Temperature, Concentration, or Other Parameters

Section 5.1.1 describes a process that was made safer by operating at a lower temperature farther away from the temperature at which a runaway can occur.

Surprisingly, it is sometimes safer to operate at a higher temperature, for example, when a second nitro group is added to a substituted nitrobenzene by gradual addition of mixed acids (nitric plus sulfuric). Suppose cooling is lost after an equimolar amount of acid has been added. If reaction is carried out at 80°C, the temperature will rise to 190°C, decomposition will start, and a runaway will occur. If, however, the reaction is carried out at 100°C, more reaction takes place while the acid is added and, when cooling is lost, the temperature rises only to 140°C, a temperature at which decomposition cannot occur.[10]

Concentrated sulfuric acid is often used to remove the water produced in a chemical reaction. One such reaction was likely to run away if the temperature was 15°C too high or too much acid was added. The margin of safety was increased by using weaker acid. The rate of reaction was not affected.[11]

Regenass et al.[12] describe how reaction conditions for sulfonation can be chosen so that a runaway is least likely to occur.

Many oxidation processes operate close to the explosive limits, and elaborate protective systems have to be (or should be) installed to prevent the reaction mixture from entering the explosive range. As mentioned in Section 3.2.2, a new catalyst for the oxidation of *o*-xylene to phthalic anhydride allows operation well away from the explosive range.[13] A similar situation was described in Section 3.2.10 for a porous-wall ceramic membrane reactor.[14]

6.2.4 *Limiting the Level of the Energy Available*

Corrosive liquids are often handled in plastic (or plastic-coated) tanks heated by electric immersion heaters. If the liquid level falls, exposing part of the heater, the tank wall may get so hot that it catches fire. One U.S. insurance company reported 36 such fires in 2 years; many of them spread to other parts of the plant. Five were due to failure of a low-level interlock. The inherently safer solution is to use a source of heat that is not hot enough to ignite the plastic: for example, hot water, low-pressure steam, or low-energy electric heaters.[15]

One company thought it had installed a low-temperature electric heater but, after a fire, realized that the temperature was low only when the heating element was covered by liquid.[16]

Above 140°C, ethylene oxide starts to polymerize, and the rise in temperature can lead to explosive decomposition (Section 3.3.1). The risk of overheating ethylene oxide in a distillation column can be avoided by heating with live steam instead of a reboiler.[17]

Many spillages of liquefied petroleum gas are due to overfilling of storage vessels and discharge of liquid from a relief valve that is not connected to a flare system. If the filling pumps can be rated so that their closed-head delivery pressure is below the set point of the relief valves (or the vessels designed so that they can withstand the delivery pressure), then the relief valves will not lift when the vessels are completely filled.

If unstable chemicals have to be kept hot, the heating medium should be incapable of overheating them. Some acidic dinitrotoluene should have been kept at 150°C because it decomposes at higher temperatures. It was heated by steam at 210°C for 10 days in a closed pipeline and decomposed explosively.[18] Excessive heat input can sometimes be prevented by limiting the heat-transfer area.

The use of an unnecessarily hot heating medium led to the runaway reaction at Seveso, Italy, in 1976, which caused a fallout of dioxin over the surrounding countryside, making it unfit for habitation, and more "legislative fallout" than any other chemical plant accident except Bhopal.[19]

A reactor was left for the weekend containing an uncompleted batch of 2,4,5-trichlorophenol (TCP). Its temperature was 158°C, well below the temperature at which a runaway reaction can start (believed at the time to be 230°C, but possibly as low as 185°C). How, then, did the contents get hot enough for a runaway to start?

The reaction was carried out under vacuum at 158°C in a reactor heated by an external steam coil, which was supplied with exhaust steam from a turbine at 190°C and a gauge pressure of 12 bar (Figure 6.3). The turbine was on reduced load because various other plants were also shutting down for the weekend (as required by Italian law), and the temperature of the steam rose to about 300°C. The temperature of the bulk liquid could not get much above 158°C, and so below the liquid level there was a temperature gradient through the walls of the reactor:

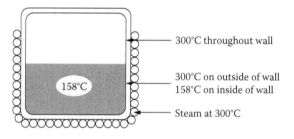

Figure 6.3   The Seveso reactor.

300°C on the outside and 158°C on the inside. Above the liquid level, the walls were at 300°C right through.

When the steam was isolated and, 15 minutes later, the stirrer was switched off, heat passed by conduction and radiation from the hot wall above the liquid to the top 10 cm or so of the liquid, which became hot enough for a runaway to start.[20] If the steam had been cooler, 185°C or less, the runaway could not have occurred.

Another incident due to unforeseen heating had less serious consequences. Various resins were dissolved in a flammable solvent at 150°C in a kettle heated by direct gas firing in a firebox. The mixture was normally allowed to stand for an hour after the heating was turned off. One day, the operator emptied the kettle as soon as the batch was complete. The refractory lining on the inside of the firebox was still hot enough to heat the kettle above the autoignition temperature of the solvent (about 280°C), and 10 minutes later there was an explosion in the kettle. Of course, operators should not take shortcuts and should understand the reasons for waiting periods, but a cooler source of heat would have avoided the hazard.[21]

Some convenience foods come in plastic bags that can be heated in hot water. They cannot be overheated (unless the pan boils dry). In contrast, if the foods are grilled instead, they will be overheated and spoiled if left under the grill for a few minutes too long. The same can also be said of the ubiquitous microwave oven.

6.3 Elimination of Hazards

6.3.1 Elimination of Hazardous Equipment

Once we have equipment, we can always find reasons for retaining it. Intermediate storage was discussed in Section 3.6, and there are other examples in the next chapter on simplification (Section 7.8). The following are examples in which the design is not much simpler, but hazards are nevertheless reduced or removed by the replacement or removal of equipment:

1. Most drains contain a vapor space above the liquid level. If flammable liquid gets into the drain, it may explode. Many companies try to prevent such explosions by avoiding spillages and fitting U-bends to the drain inlets to keep out sources of ignition, but nevertheless explosions occur from time to time. The vapor spaces can be eliminated by using fully flooded (surcharged) drains. These should be installed in all new plants that handle volatile flammable liquids.

2. Many fires and explosions have occurred in the pump rooms of ships carrying oils and chemicals. The best way of preventing such incidents is to eliminate the need for a pump room by using submerged pumps in the tanks.[22]

3. Alcohol vapor and air were removed from extraction ducts by an internal fan. The alcohol concentration was usually about half the lower explosive limit. One day the concentration exceeded the limit, and an explosion occurred. The source of ignition was corrosion of the motor support brackets; the motor fell until it was supported by the cable, which then sparked.

Obviously, the motor should have been properly maintained, but sources of ignition are always likely to turn up, especially when moving machinery is present. It would have been better to have blown air through an extractor that sucked out the vapor.[23]

4. Fires have often occurred when air is compressed, mainly because lubricating oil forms a flammable deposit on the walls of compressor delivery lines. There are ways of preventing these fires,[24] but the most effective way is to avoid the need for compressed air. Instead of using it in air driers, steam can be used instead. It is cheaper than air because the steam is hot enough for recovery of the latent heat.[25]

5. In the past, men installing machinery in warships worked in a confined space exposed to dust, fumes, noise, fire, and the mechanical hazards of restricted access. The solution was to install the deck after the machinery. A mobile tent provided weather protection.[26]

6. Sash windows can trap fingers if the cords holding the windows break. Horizontally sliding windows are safer.

7. When someone bangs his or her head on a scaffold pole projecting through a ladder, the first question asked is usually, "Why weren't you wearing your safety helmet?" rather than, "Why was the scaffold pole sticking through the ladder?"[27] Of course we should all wear our safety helmets, but safety problems can be solved in more than one way. (Section 4.1.1 suggests two ways of preventing furniture fires.)

6.3.2 Elimination of Hazardous Operations

The best way of limiting the effects of hazardous operations is to eliminate (or reduce) the need for them. As with hazardous equipment, once a procedure is well established, we are reluctant to dispense with it. The need for some procedures should, however, be questioned.

If what we don't have can't leak, people who are not there cannot be injured. This is recognized when we locate plants far from centers of population, but this is not so well recognized within the plant. Every time we add another trip or alarm to the plant, we increase the number of hours that someone is exposed to risk because he or she is on the plant testing or maintaining it. Dalzell suggests[28] that we may have gone too far and that, on some plants, adding more protective equipment will increase total risk, not reduce it. He advises, with offshore platforms in mind, that we minimize manning; invest in plant integrity rather than in detection, control, and mitigation systems; and ensure the survival of personnel through an effective mustering and evacuation system.

The following are some operations that might be reduced:

1. *Sampling and analysis.* Many people have been injured while taking samples or carrying them back to the laboratory. Do we need to collect as many samples as we do? Laboratory work is also hazardous. The need for each sample, sample point, and analysis should be questioned. Can we make more use of on-line analysis, preferably noninvasive, so that maintenance of the equipment is less hazardous? For both on- and off-line analysis, intensified methods using very small amounts of material are becoming increasingly common.[29]

Although laboratory work is undoubtedly a key requirement in process development, and not everything can be found in a book, we should all heed the remark of G. J. Quarderer (Dow Chemical Co.) as quoted by Fogler:[30] *Four to six weeks in the lab can save you an hour in the library.*

2. *Maintenance* is a hazardous activity (Section 5.3). Many accidents occur during maintenance, often as a result of poor preparation rather than of the maintenance itself.[31] One way of reducing maintenance is to avoid, when possible, equipment with moving parts (Section 7.9).

The less maintenance we carry out, the fewer accidents we will have. As a first step in reducing maintenance, we need to know how we are spending money at present. Many maintenance cost accounting systems are designed to allocate the costs to the right product and cannot, for example, provide data on the costs of maintaining different sorts of pumps or even the costs of different sorts of work. Table 6.1 shows the breakdown of maintenance costs in one petrochemical company. Each company will differ, but this sort of data should be available.

Table 6.1   Breakdown of Maintenance Costs in a Petrochemical Company

56%	*True maintenance*: repairing or replacing worn-out or corroded equipment, repairing leaks
18%	*Safety work*: routine tests and inspections
11%	*Process work*: cleaning choked and dirty equipment, changing catalysts, fitting and removing hoses and blinds
5%	*Minor modifications* (major ones are charged to capital)
10%	*Overheads*

In the nuclear industry, maintenance is difficult or impossible, and very reliable plants have been developed. For example, centrifuges for concentrating uranium-235 are designed to run for 10 years without maintenance.[32] It would be far too expensive for the process industries generally to copy nuclear designs, but some movement in that direction might be justified.

The nuclear industry does not achieve high reliability by massive duplication, by making everything thicker and stronger, or by using expensive materials of construction or special types of equipment; rather, it achieves high reliability by paying great attention to detail in design and construction. The last item is important. Many failures in the process industries are the result of construction teams not following the design in detail or not carrying out well, in accordance with good engineering practice, details that were left to their discretion.[33]

As a result of the nuclear industry's high-reliability designs, the world's first commercial nuclear reactors at Calder Hall and Chapelcross in the United Kingdom had completed over 40 years of operation by the late 1990s and were licensed to operate for a further 10 years.[34] Both plants continued operation into the twenty-first century, with Calder Hall ceasing power generation in 2003 and Chapelcross in 2004.

3. *Cleaning* can be a hazardous operation. Safer cleaning agents than those sometimes used are described in Section 4.1.1. At one time, polyvinylchloride reactors were entered and cleaned manually after every batch. Other methods were developed when vinyl chloride was found to be carcinogenic. Discharges from water droplets in ships' tanks being cleaned by high-pressure water washing have ignited flammable vapor and caused serious explosions.[35]

4. *Transport* of hazardous intermediates can be dangerous and can be avoided by distributed production at the point of use (Section 3.6). Ponton[36] has suggested that cleaning, transportation, and maintenance could be avoided in the manufacture of pharmaceuticals by using small plastic reactors that are used once and then destroyed. No valves would be needed because lines could be sealed and cut after use. Mixing could be achieved by shaking the

unit. Microwaves could be used for heating. Instruments would be sophisticated, permanent, and noninvasive. At the time of publication of Reference 36, these ideas had not been tested and had been put forward to stimulate discussion.

5. *Unproductive activities.* In the engineering manufacturing industry, when a customer places an order, only 5% of the time that elapses before delivery is spent processing the raw material and thus adding value. The rest of the time is taken by stockholding, inspecting, moving, reworking, and waiting while machinery is repaired or maintained or waiting while administration is completed. It is no good being twice as efficient in the value-adding stage if we lose it all in inefficiency elsewhere. Although U.S. and European industry has concentrated on improving processes, Japanese industry has been more successful in reducing unproductive time and cost. The unproductive activities carry some risk, and eliminating or reducing them will decrease accidents.[37]

6. *Replacing hazardous substances by procedures.* Elimination of hazardous substances is discussed in Section 4.1.1, but here is an example of a hazardous substance replaced by an entirely different type of operation. Instead of using pesticides to protect historical clothing on display, a London museum devised the following procedure: The articles are cooled in a freezer and then returned to a warm room. The eggs hatch. Before the larvae can do any damage, the articles are put back in the freezer. This kills the larvae, and they can be removed by vacuuming.[38]

References

1. T. Duckworth and A. A. McGregor, "The Maintenance of Relief Systems: A Practical Approach," in *Selection and Use of Relief Systems for Process Plants* (Manchester, U.K.: Institution of Chemical Engineers, North West Branch, 1989), 4.1–4.13.
2. D. G. Clark, "Applying the 'Limitation of Effects' Inherently Safer Processing Strategy when Siting and Designing Facilities," *Process Safety Progress* 27, no. 2 (2008): 121–130.
3. D. J. Ferguson, "Applying Inherent Safety to Mitigate Offsite Impact of a Toxic Liquid Release," in *19th Annual International Conference: Emergency Planning Preparedness, Prevention and Response* (New York: Center for Chemical Process Safety, American Institute of Chemical Engineers, 2004), 167–170.
4. T. A. Kletz, *What Went Wrong? Case Histories of Process Plant Disasters*, 5th ed. (Oxford, U.K.: Butterworth-Heinemann, 2009), sec. 6.1.1.
5. L. C. Schaller, "Off Site Risk Assessment and Risk Minimization Guidelines," *Plant/Operations Prog.* 9, no. 1 (1990): 50–51.
6. D. C. Hendershot, "Safety Considerations in the Design of Batch Processing Plants," in *Proceedings of the International Symposium on Preventing Major Chemical Accidents*, ed. J. L. Woodward (New York: American Institute of Chemical Engineers, 1987).
7. T. A. Kletz, *What Went Wrong? Case Histories of Process Plant Disasters*, 5th ed. (Oxford, U.K.: Butterworth-Heinemann, 2009), sec. 3.2.8.

8. T. A. Kletz, *Lessons from Disaster: How Organisations Have No Memory and Accidents Recur* (Rugby, U.K.: Institution of Chemical Engineers, 1993), Sec. 4.2.8.

9. J. A. Klein and A. S. Balchan, "Safe Formulation and Manufacture of Acrylic Resins," in *International Conference and Workshop on Process Safety Management and Inherently Safer Processes* (New York: American Institute of Chemical Engineers, 1996), 329–342.

10. H. Fierz, P. Fink, G. Giger, and R. Gygax, "Thermally Stable Operating Conditions of Chemical Processes," in *Loss Prevention and Safety Promotion in the Process Industries*, vol. 3, Symposium Series No. 82 (Rugby, U.K.: Institution of Chemical Engineers, 1983).

11. H. G. Gerritsen and C. M. van't Land, "Intrinsic Continuous Process Safeguarding," in *Preventing Major Chemical and Related Process Accidents*, Symposium Series No. 110 (Rugby, U.K.: Institution of Chemical Engineers, 1988).

12. W. Regenass, U. Osterwalder, and F. Brogli, "Reactor Engineering for Inherent Safety," in *Eighth International Symposium on Chemical Reactor Engineering*, Symposium Series No. 87 (Rugby, U.K.: Institution of Chemical Engineers, 1984).

13. T. Sato, Y. Nakanishi, and Y. Haruna, "Recycling Vent Gas Improves Phthalic Anhydride Process," *Hydrocarbon Process.* 62, no. 10 (1983): 107–110.

14. J. Coronas, M. Menedez, and J. Santamaria, "The Porous-Wall Ceramic Membrane Reactor: An Inherently Safer Contacting Device for Gas-Phase Oxidation of Hydrocarbons," *Journal of Loss Prevention in the Process Industries* 8, no. 2 (1995): 97–101.

15. International Risk Insurers, *Sentinel* 3rd quarter (1981): 15.

16. *Loss Prevention Bulletin* 131 (1996): 117.

17. J. L. Hawksley and M. L. Preston, "Inherent SHE: 20 Years of Evolution," in *International Conference and Workshop on Process Safety Management and Inherently Safer Processes* (New York: American Institute of Chemical Engineers, 1996).

18. "Explosion in a Dinitrotoluene Pipeline," *Loss Prevention Bulletin* 88 (1989): 13–16.

19. T. A. Kletz, "Seveso," *Learning from Accidents*, 3rd ed. (Oxford, U.K.: Butterworth–Heinemann, 2001), chap. 9.

20. T. G. Theofanous, "The Physicochemical Origins of the Seveso Accident," *Chem. Eng. Sci.* 38 (1983): 1615–1636.

21. "Repeated Explosion of Varnish Kettle," *Loss Prevention Bulletin* 130 (1996): 11–12.

22. K. Fleming, "Safety Takes a Front Seat," *Hazardous Cargo Bull.* 5, no. 6 (1984): 9–11.

23. "Explosion in Storage Facility," *Chem. Saf. Summ.* 55 (1984): 62.

24. T. A. Kletz, *What Went Wrong? Case Histories of Process Plant Disasters*, 5th ed. (Oxford, U.K.: Butterworth-Heinemann, 2009), sec. 12.1.

25. C. Butcher, "Drying Update," *Chem. Eng. (U.K.)* 557 (1994): 15.

26. J. F. Wollaston, "Old Trades/New Trades: Shipbuilding and Ship Repair," *Occup. Med.* 42 (1992): 203–212.

27. A worker quoted by M. A. Needham, "The Relationship between Safety, Quality and Company Culture: A Case Study from the Petrochemical Industry" (paper presented at the IBC Technical Services Conference on the Human Factor in Safety: Implications for the Chemical and Process Industries, Manchester, U.K., 20–21 November 1990).

28. G. Dalzell, "Nothing Is Safety Critical," in *Hazards XIII: Process Safety—The Future*, Symposium Series No. 141 (Rugby, U.K.: Institution of Chemical Engineers, 1997).

29. R. W. Johnson, S. D. Unwin, and T. I. McSweeney, "Inherent Safety: How to Measure It and Why We Need It," in *International Conference and Workshop on Process Safety Management and Inherently Safer Processes* (New York: American Institute of Chemical Engineers, 1996).

30. H. S. Fogler, *Elements of Chemical Reaction Engineering*, 2nd ed. (Upper Saddle River, NJ: Prentice Hall PTR, 1992), 226.

31. T. A. Kletz, *What Went Wrong? Case Histories of Process Plant Disasters*, 5th ed. (Oxford, U.K.: Butterworth-Heinemann, 2009), chap. 1.

32. H. G. Tolland, "Advanced Enrichment Technologies," *Atom* 359 (1996): 2–5.

33. T. A. Kletz, "Pipe Failures," in *Learning from Accidents*, 3rd ed. (Oxford, U.K.: Butterworth–Heinemann, 2001), chap. 16.

34. Health and Safety Executive (United Kingdom), "Calder/Chapelcross: Go ahead for 50 years," *Nuclear Safety Newsletter* no. 11 (1996): 2.

35. T. A. Kletz, *What Went Wrong? Case Histories of Process Plant Disasters*, 5th ed. (Oxford, U.K.: Butterworth-Heinemann, 2009), sec. 15.2.

36. J. W. Ponton, "The Disposable Batch Plant" (paper presented at the 5th World Congress of Chemical Engineering, San Diego, 1996).

37. C. Hope, *Leading through Innovation: Competing with the World* (London: Royal Academy of Engineering, 1996), 13.

38. British Occupational Hygiene Society Technology Committee, *The Manager's Guide to Control of Hazardous Substances* (Leeds, U.K.: H & H Scientific Consultants, 1996), 47.

Chapter 7

Simplification

The machines that are first invented to perform any particular movement are always the most complex, and succeeding artists generally discover that with fewer wheels, with fewer principles of motion than had originally been employed, the same effects may be more easily produced.

—Adam Smith, *The Wealth of Nations*

Complex systems have complex troubles.

—Ernest Siddall, *Complete Programming in English*

Simpler plants are safer than more complex ones because they contain less equipment that can leak and provide fewer opportunities for human error. In that sense, they are inherently safer. However, they may still contain large inventories of hazardous materials at high temperatures and pressures. If we cannot avoid these hazards, we may have to add equipment to control them, even though a more complex plant will result. A truly inherently safer plant would avoid the use of hazardous materials (or contain very little).

7.1 The Reasons for Complexity

Complex designs provide many opportunities for human error or equipment failure. If what you don't have can't leak, equipment you don't install cannot develop faults or be operated at the wrong time or in the wrong way. Simpler plants are therefore, other things being equal, cheaper and safer than complex ones.

127

In 1978, at a conference on loss prevention, a speaker described an explosion in a batch reactor. He showed a diagram of the reactor before the explosion (Figure 7.1[a]) and then one of the reactors as modified after the explosion (Figure 7.1[b]). My neighbor and I looked at each other. Surely there must be a better way of making plants safer. This set me thinking along the lines described in this chapter and the next one. However, I do not intend to criticize the author of the paper. Faced as he was with an existing plant, he did all he could. As we shall see, simplification must start early in plant design.

The reasons for complexity in plant design are as follows:

1. The need to control hazards. If we can design inherently safer plants and avoid hazards, as discussed in Chapters 2–6, then we shall not need to add so much protective equipment to control the hazards. We shall need fewer trips, interlocks, and alarms; fewer leak detectors and emergency isolation valves; less fire protection; and so on. Our plants will be simpler and cheaper.

   Similarly, if we can design plants that are inherently easier to control, as discussed in Section 9.5, we shall need less control equipment.

2. Related to this is our failure to carry out safety studies, and similar studies of operating problems, until late in design. By the time we do carry them out, it is impossible to make basic changes in design, to remove hazards or operating problems, or to simplify the plant, and all we can do is to add equipment to control hazards or overcome operating and control problems. The organizational changes needed are discussed in Chapter 10. The rest of this chapter gives examples of the simplifications that could be made if we looked at designs critically in their early stages. In many cases, the new designs are also inherently safer.

3. Another reason for complexity is following specifications, standards, or custom and practice to the letter when they are no longer appropriate (Section 8.1). Lees[1] displays two diagrams that show how plants get more complex as design proceeds.

4. A fourth reason is a desire for flexibility and availability of spares (Section 8.2).

5. A fifth reason is a desire for technical elegance. To some people, simple designs seem rather crude (Section 8.3.2). However, simpler need not mean primitive and need not be a critical description. It can mean "stripped to bare essentials."

6. Finally, in some cases, we may go too far in trying to remove all risks (see Chapter 15).

Complex procedures and instructions, like complex equipment, can add to costs and provide opportunities for error (Sections 6.3.2 and 7.4).

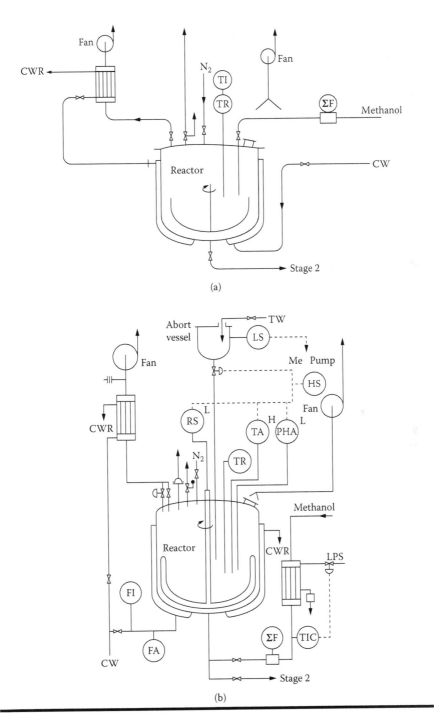

Figure 7.1   A reactor (a) before an explosion and (b) as modified afterward.

7.2 Stronger Equipment Can Replace Relief Systems

When hazardous chemicals are handled, relief devices (that is, relief valves or rupture discs) should not discharge to the atmosphere but to a flare system, for flammable gases; to a scrubbing system, for toxic gases; or to a collection system, for hazardous liquids and solids. Often these are combined: Liquids are collected in a catch pot from which gases pass to flare. These systems are expensive; in addition, flare systems may sterilize large areas of land and produce complaints about noise and light.

It is sometimes possible to dispense with relief devices and all that comes after them by using stronger equipment—strong enough to withstand the highest pressure that can be reached. Zhu and Shah[2] describe a burst-resistant design for pressure vessels over a wide range of service at varying pressure, temperature, and corrosiveness.

Section 13.6.1 explains the use of pressure-resistant vessels for dust explosion mitigation; the concept is also important for other applications. For example, if we have a series of vessels with a pressure drop between them (Figure 7.2[a]), large relief devices are needed on the later vessels in case the letdown valves between the vessels fail open and subject the downstream vessels to the full upstream pressure. (Absence of adequate relief valves has caused some spectacular failures.[3]) If the vessels are made strong enough to withstand the full upstream pressure, the relief system is not needed. Fire relief valves will still be needed if we are handling flammable materials, but they can be allowed to discharge to atmosphere; the discharge may ignite, but, if there is a fire beneath a vessel, a fire on the end of a relief valve tail pipe will not matter so long as it does not impinge on other equipment.

Similarly, in Figure 7.2(b), if the vessel can withstand the pump delivery pressure, a relief device is not needed. (Note: If there is a long length of line between the valve and the pump, the vessel and pipeline must be able to withstand the hammer pressure produced when the valve is closed. This may be much greater than the static closed-head delivery pressure.)

It may be possible to avoid the need for a relief valve on a distillation column by making it strong enough to withstand the pressure developed if cooling or reflux is lost but heat input continues (Figure 7.2[c]). This will not be economic on a large column but may be on a small one. As before, a fire relief valve may still be needed.

Instead of installing vacuum relief devices, we can make equipment strong enough to withstand vacuum. If the equipment contains flammable gas or vapor, vacuum relief valves that admit air should be avoided because they can cause an explosion (unless the amount of air is small). If they admit nitrogen, the supply may be limited. Stronger equipment is usually the safest and simplest solution (Section 10.2.4).

Why are stronger vessels not used more often? Sometimes designers think that a relief device is necessary to conform to the codes, but this is not always the case.[4]

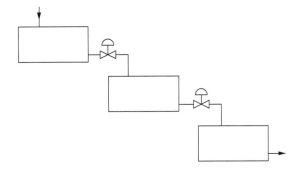

(a) Vessels designed to withstand the full upstream pressure.

(b) Vessel designed to withstand pump closed-head delivery pressure.

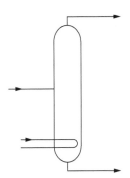

(c) Still designed to withstand maximum possible vapor pressure.

Figure 7.2   Relief valves can be avoided by using stronger vessels.

More often, the reasons are logistic rather than technical. Vessels are ordered early in a project; relief and blowdown reviews come later. By the time we realize that we could simplify the design by ordering stronger vessels, it may be too late to do so; the vessels may already be on order.

Design engineers remind us that it costs more to strengthen a vessel than to supply a relief valve. However, if we can avoid the need for relief valves, we also avoid (or reduce) the need for the flares, scrubbers, and catch pots that come after them. We also avoid the need to maintain the relief valve, usually located at the highest point of the structure, where escape is difficult if anything goes wrong. Reference 5 describes some accidents that have occurred during the maintenance of relief valves.

7.3 Resistant Materials of Construction Can Replace Protective Instruments

Just as stronger vessels can avoid the need for relief devices, so materials of construction resistant to low temperatures can avoid the need for low-temperature trips. Figure 7.3 shows a situation that often arises in low-temperature plants. The compressor is made from a grade of steel that will not withstand temperatures below –50°C. The vapor entering the compressor is therefore warmed by passing it through a heat exchanger that raises its temperature from, say, –100°C to –40°C. If the flow through the other side of the heat exchanger falls (or becomes cooler), the suction temperature will fall, and thus warm compressed gas is recycled automatically to maintain the suction temperature at –40°C. The automatic controller may fail; therefore, in addition, a trip is installed to shut down the compressor if the suction temperature approaches –50°C. A simpler, and also inherently safer, system is to construct the compressor out of a grade of steel that will withstand the lowest temperature that can be reached.

In comparing the costs of the two designs, remember that the instrumented protective system requires regular testing and maintenance, and this roughly doubles its installed cost even after discounting. Instruments cost twice what you think they cost! In contrast, though the low-temperature steel costs more initially, it does

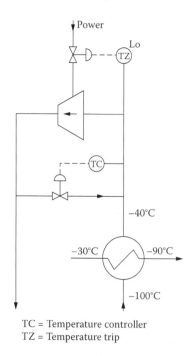

TC = Temperature controller
TZ = Temperature trip

Figure 7.3   The protective instrumentation can be avoided if the compressor can withstand temperatures down to –100°C.

not have any additional running costs attached to it; the compressor maintenance costs will be the same whatever grade of steel is used. Obviously, the decision to use low-temperature steel must be made early in design, before the compressor is ordered, and long before the stage when instrumentation is normally reviewed.

The interaction between low temperatures and process equipment played a key role in the Longford gas plant explosion in Australia in 1998.[6] Catastrophic failure of a heat exchanger that had become abnormally cold released a large quantity of volatile liquefied petroleum gas (LPG). Two plant workers were killed, eight others were injured, and the gas supply to the city of Melbourne was cut for two weeks. Hopkins[6] gives a comprehensive causal analysis of the Longford accident in which cold-metal embrittlement is seen to be a directly contributing cause.

A chlorine-using plant was designed with a chlorine blower made from titanium. This metal is suitable for use with wet chlorine but reacts rapidly with dry chlorine—so rapidly that it is said to burn. The chlorine passed through a water scrubber before reaching the blower, and thus it should normally have been wet, and an elaborate system of trips and controls was designed to make sure that the chance of a water failure was very small.

In this case, an investigation was carried out into the design, and it was decided to scrap the idea of a titanium blower and install a rubber-covered blower instead. This rather old-fashioned piece of equipment is less reliable than a titanium blower, but it does not matter if a rubber-covered blower comes into contact with dry chlorine. It was possible to simplify the design by dispensing with the control and trip system on the scrubber.

7.4 Designs Free from Opportunities for Human Error

Designs free from opportunities for human error are discussed in detail in Reference 7, and other examples are described in Sections 6.2.1 and 9.9. The following are three more examples:

1. To save cost, three waste-heat boilers shared a common steam drum. Each boiler had to be taken off-line from time to time for cleaning, and thus isolation valves were installed in the steam and condensate lines between the boilers and the steam drum (Figure 7.4). On two occasions, valve D3 was closed instead of valve D2, and boiler No. 3, which was on-line, was starved of water and damaged. On the first occasion damage was serious, and afterward, high-temperature alarms were fitted on the boilers. On the second occasion on which the wrong valve was operated, the high-temperature alarm prevented serious damage, but some tubes had to be changed. A series of interlocks were then fitted so that the fuel-gas supply to each unit had to be isolated before a key could be removed from the fuel-gas isolation valve; this key is needed to isolate the corresponding valves on the steam drum.

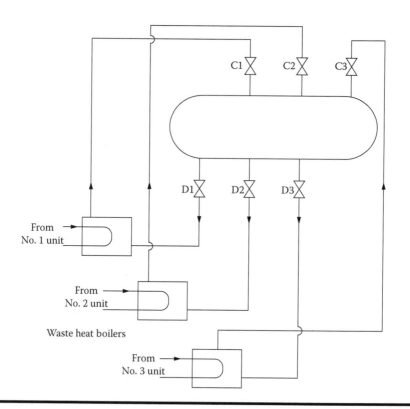

Figure 7.4 Waste-heat boilers sharing a common steam drum.

The chance of an error was increased by the lack of labels and by the arrangement of the valves: D3 was below C2.

A better design, used on later plants, is to have a separate steam drum for each boiler (or group of boilers if several can be taken off-line together). There is then no need for valves between the boiler and the steam drum. This is a decision that must be made early in design.

In this case, the simpler solution is more expensive but safer. We are often willing to spend large sums to achieve safety by complexity, but we are reluctant to spend money on simplicity. Why? Do we feel we are getting something for our money if we see a lot of valves and instrumentation? To quote a newspaper, "Invent a better mousetrap and the world will beat a path to your door. But design a really complicated mousetrap—one that takes four AA-sized batteries and comes with an instruction book half an inch thick—and you'll be crushed in the stampede."[8]

2. If a nitrogen supply is connected to a vessel, it should be disconnected (or blinded) before the vessel is entered. Failure to do so has caused some serious accidents.[9] If the nitrogen supply line is routed across the manhole cover, then the line has to be removed before the vessel can be entered.

3. A worker was seriously injured in an explosion and fire involving ice-resurfacing equipment.[10] The explosion occurred after he mistakenly placed a hot-water hose in the machine's gas tank, vaporizing some of the gasoline, which then came into contact with an ignition source. Was this simply a case of operator error? Hardly, when one considers that the machine's water and gas tanks were painted the same color and were located side by side.[10] The design was clearly flawed: In addition to the color and location of the tanks, why was the gasoline tank receptacle designed such that it could accept a water hose?

As already stated, simple instructions are better than complex ones that people may not fully understand or do not have the time or inclination to read. A simple instruction that covers 99% of the situations that could arise and is read is more effective than a complex one, covering every conceivable eventuality, that few will read and understand. Complex instructions are often written to protect the writer rather than to help the reader (Section 9.7).

7.5 Design Change Can Avoid the Need for Better Instrumentation

Figure 7.5 shows part of a liquid-phase oxidation plant. The liquid inlet and exit lines are not shown. The off gas leaving the reactor contained flammable vapor and about 3% oxygen. Air was added automatically to bring the oxygen content up to

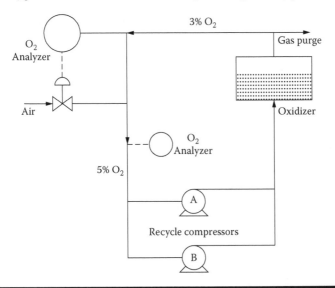

Figure 7.5   Part of a liquid-phase oxidation plant in which an explosion occurred.

5% before the gas was recycled back to the reactor through two recycle compressors. An oxygen content of 10% is needed for an explosion, and so there was a large safety margin. Nevertheless, an explosion occurred, and four men were killed.

Normally, both recycle compressors were on line. If either stopped, the air was isolated automatically by an interlock, which is not shown. One compressor had to be shut down for a few minutes. The operator disarmed the interlock to prevent it from isolating the air supply, but he forgot to reduce the air rate. The oxygen content of the recycle gas rose, the oxygen detector was too slow acting to detect the rise in oxygen concentration in time, and an explosion occurred. As so often happens, the source of ignition was never found.

The protective system was poor. The oxygen detectors should have been quicker acting (Section 7.7), and the operator should not have been able to disarm the interlock. However, instead of relying on instruments and operators to *control* the hazard, it was possible to *remove* it by a change in design: If the air is added directly to the reactor, the recycle gas cannot get into the flammable range. If the air is added this way, there is no need for a recycle system, and the compressors can be dispensed with. Many plants operate without recycle. The plant has other disadvantages, and its advantages are said to be illusory.[11]

The plant was owned by a U.S. company, designed by a Swiss one, and operated by a Spanish one. Perhaps those involved did not understand each other's problems and the environment in which they worked.

7.6 Relocation Can Avoid the Need for Complication

A control unit had to contain sparking electric equipment. The location chosen for it was classified Division (Zone) 2 (that is, flammable vapor was not often present in the area but could be present if a leak occurred nearby; see Section 8.1.3). The control unit was therefore protected by pressurization with nitrogen to prevent any vapor present in the surrounding atmosphere from entering the unit. If the nitrogen pressure fell, a low-pressure switch isolated the power supply.

Nevertheless, an explosion occurred in the control unit. Flammable vapor entered with the nitrogen, which was contaminated. The nitrogen pressure then fell, allowing air to leak in. The low-pressure switch had been made inoperative, and thus when the electricity supply was switched on, an explosion occurred (Figure 7.6).

The recommendations made after the explosion included the following:

1. Prevent contamination of the nitrogen.
2. Do not allow protective equipment to be made inoperative unless authorized by a responsible person and signaled in some way so that everyone knows it is inoperative.
3. Test all protective equipment regularly.

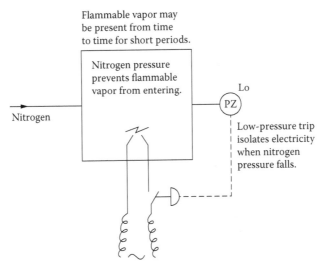

Figure 7.6 **This elaborate system for preventing an explosion inside a box containing electrical equipment in fact caused an explosion. It could have been avoided by moving the equipment.**

4. Use compressed air instead of nitrogen, for the compressed air supply is more reliable.

However, there was a simple way of removing the hazard. If the control unit had been moved a few meters, it would have been outside the Division 2 area, and pressurization with nitrogen (or air) would have been unnecessary. The explosion was the result of installing a poorly designed protective system and operating it incorrectly to guard against an unlikely hazard that could easily have been removed.

Why was the easy way of removing the hazard not foreseen during design? The most likely explanation is that different functional groups often work in isolation and do not come together to discuss hazards and alternative designs. Thus one person or group fixed a location for the control unit, and it happened to be in a Division 2 area. Normally, this would not have caused any problems, but in this case, it meant that the electrical designer had to ask for pressurization with nitrogen. He did not ask whether the control unit had to be in a Zone 2 area, and he did not point out the savings that would result if it could be moved. That was not his job. His job was to supply electrical equipment suitable for the area classification that had already been decided.

This incident happened some years ago, and today many design organizations do try to improve coordination between groups. For more details, see Reference 12.

7.7 Simple Technology Can Replace High Technology

The following are some examples of the replacement of high technology by simpler designs:

1. Many types of fire detectors are available based on the infrared or ultraviolet radiation from a fire, the light absorbed by smoke, the effect of smoke on ions, and so on.

   For many applications, a simpler fire detector can be made by allowing a fire to burn through a plastic tube and release the air pressure inside or burn though a thin wire and break an electric circuit.

2. Many buildings such as compressor houses contain equipment from which leaks of flammable gas may occur, and there have been many explosions in such buildings.[13] Obviously, we should use nonflammable materials when we can (Chapter 4), but this cannot always be done. We should try to prevent leaks, but we will not always be successful. Companies have often installed forced ventilation, but capital and running costs are high. Forced ventilation is usually designed for the comfort of the operators rather than the dispersion of leaks, and it is less effective than natural ventilation. Other companies have installed lightweight wall panels that blow off as soon as an explosion occurs and prevent a large rise in pressure. They protect the equipment but not the operators.

   A better solution is to leave off the walls and allow small leaks to disperse by natural ventilation.[14,15] Even on a still day, natural ventilation will give more air movement than most forced-ventilation systems. Although several tonnes of flammable gas are necessary for an explosion in the open air, a few tens of kilograms are sufficient to destroy an enclosed building.

   In recent years, many enclosed compressor houses have been built to meet new noise regulations. In New Zealand in 1986, I saw two plants that were outstanding by world standards. No expense had been spared to make sure that the most up-to-date safety features were installed, with one exception: Both had completely enclosed compressor houses. The plants had to meet stringent noise regulations, and enclosing the compressors was the simplest way of complying. One of the plants was out in the country without a house in sight, and it seemed that the regulations had been made for the benefit of the sheep. I saw a similar compressor house in the United Kingdom in 1991.

   Perhaps modern compressors leak less often than those built 30 or 40 years ago, and there is less need for good ventilation. However, the leaks that lead to explosions in a compressor house are often not from a compressor but from other equipment such as pipe joints. One such leak occurred because a spiral-wound gasket had been replaced by a compressed asbestos fiber one, probably as a temporary measure, seven years earlier. Once installed, the fiber gasket was replaced by a similar one during subsequent maintenance.[16]

An alternative way to reduce the noise radiation from a compressor is to surround it with a housing made from sound insulation and purge the gap between the compressor and the insulation with a continuous stream of compressed air. This, of course, is more complex than leaving off the walls, but in this case, it is safer.

3. In oxidation plants, it is necessary to measure the oxygen content of some of the plant streams. If too much oxygen is present, an explosion can occur (Section 7.5). Measuring oxygen in a stream of dry gas is straightforward, but if the gas stream contains liquid or vapor, it has to be removed, and this is often difficult.

   One plant was supplied with an elaborate and expensive system for scrubbing the gas with water. It did not work, and the oxygen analyzers fell into disuse. The system was replaced with one for scrubbing the gas with cold hydrocarbon. It worked, but it was expensive to maintain. Finally, the system shown in Figure 7.7 was installed. The vapor condenses in the uninsulated pipe and runs back into the vessel. The system is cheap and simple and has a much shorter response time than more elaborate systems. (Engineering was once defined as "the art of doing that well with one dollar, which any bungler can do with two after a fashion."[17] Is this still true?)

4. Flare and vent stacks and overflow lines should be kept as simple as possible and, when possible, obstructions such as water seals, filters, flame arresters, and molecular seals should be avoided, particularly when there is a continuous or frequent flow through the line. All these devices are likely to choke or freeze.

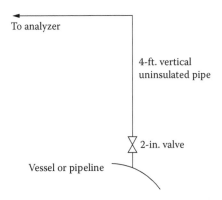

To analyzer

4-ft. vertical
uninsulated pipe

2-in. valve

Vessel or pipeline

Figure 7.7　A simple but effective method of removing liquid from a gas stream.

5. If a vessel has to be protected against fire with water, drenching it with a jet from a monitor is simpler than spraying on water through complex pipework, which is easily damaged and requires regular maintenance if the nozzles are not to choke. However, monitors do require much larger amounts of water and corresponding drainage. Insulation can reduce (but not replace entirely) the demand for water and is passive; that is, it is immediately available and does not have to be commissioned by somebody or something, who or which might fail (Section 9.9).

6. Highly sophisticated methods are used for measuring ventilation rates in houses. In the 1930s, J. B. S. Haldane burned candles, weighed them to measure the weight loss, and measured the concentration of carbon dioxide in the rooms.

7. According to a press report, an American firm spent much time, money, and effort developing a space pen that could work upside down in zero gravity. The Russians came up with a simpler solution. They used pencils.[9]

7.8 Leaving Things Out

Designs can sometimes be simplified by leaving out a vessel and making another carry out two functions. For example, a design showed a reactor overflowing into a catch pot that was fitted with a high-level alarm. Following a hazard and operability study (Hazop) of the flowsheet (Section 10.2.4), the catch pot was left out and the high-level alarm fitted to the reactor.

Some other examples are shown in Figures 7.8 and 7.9. Figure 7.8 shows how a storage tank can be used as a blowdown drum. The tank must, of course, be strong enough to withstand any pressure that might be developed. Figure 7.9(a) shows a compressor for a gaseous refrigerant. It is fitted with a suction catch pot to collect liquid carryover from the heat exchanger in which the liquefied gas is vaporized. In Figure 7.9(b), a disengagement space has been added to the heat exchanger and the catch pot eliminated. A typical vessel is fitted with about five control loops and a relief valve. We save the cost of supplying and maintaining all these if we can leave out the vessel.

Sometimes simpler and cheaper equipment can be used instead of the conventional equipment. For example, on less critical duties than those shown in Figure 7.9, a length of vertical pipe can be used as a catch pot. It should, of course, be fitted with a level controller and a high-level alarm. On pneumatic systems, lengths of, say, 6-in. pipe can be used as compressed-air reservoirs instead of small vessels.

Section 3.2.8 showed how plants can be simplified by combining reaction and mixing in a single operation. Mixers can also be combined with driers, granulators, and heat exchangers.[18] Book and Challagulla[19] describe a method for the adiabatic oxidation of sulfur dioxide to sulfur trioxide in which there are few moving parts and several operations can be performed in a single equipment unit.

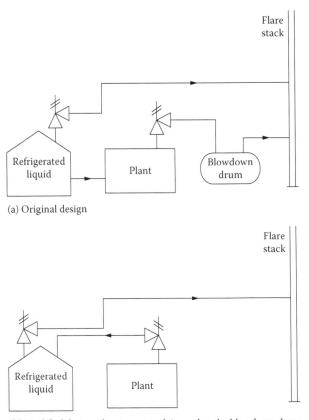

(a) Original design

(b) Modified design; the storage tank is used as the blowdown drum.

Figure 7.8   Relief and flare system on a liquefaction plant.

7.9 Power Fluidics: Avoiding Moving Parts

Power fluidics are devices that carry out operations such as distilling (Section 3.3.2), scrubbing, mixing, and pumping in equipment that has no seals and no moving parts in contact with the process fluids—sometimes no moving parts at all. Power fluidics therefore require little maintenance and have been much used by the nuclear industry.[20,21]

The fluidic check valve or diode is shown in Figure 7.10. There is low resistance to flow out of the tangential opening but high resistance to flow, 200 times higher, in the other direction. There is thus a small flow in the wrong direction, but if this can be tolerated, the valves are very reliable and very good at stopping a pressure surge that might damage upstream equipment.

Two diodes can be combined to produce a fluidic pump (Figure 7.11). The gas supply is cyclic: when it is high, it pushes liquid into the discharge pipe; when it is low, liquid flows out of the supply tank. Figure 7.12 shows another type of fluidic pump.

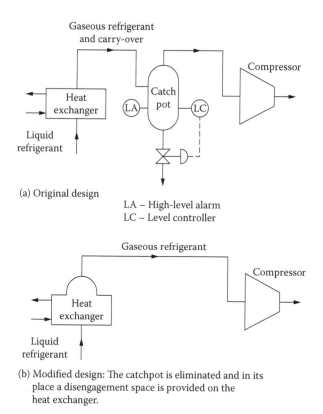

(a) Original design

LA – High-level alarm
LC – Level controller

(b) Modified design: The catchpot is eliminated and in its place a disengagement space is provided on the heat exchanger.

Figure 7.9   Off-gas compressor and catch pot: (a) original and (b) modified designs.

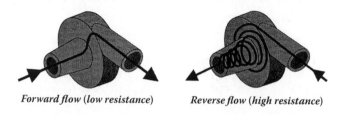

Forward flow (low resistance)   *Reverse flow (high resistance)*

Figure 7.10   Fluidic check valves or diodes. (Reproduced by permission of Accentus plc.)

These pumps are very reliable but expensive to operate and have therefore been used mainly in the nuclear industry and for handling very unstable or toxic materials.

Figure 7.13 shows a fluidic diverter. At first sight, it looks like a simple Y, but by careful design of the shape, all the liquid can be made to flow one way, and can then be diverted by a short pulse of compressed air.

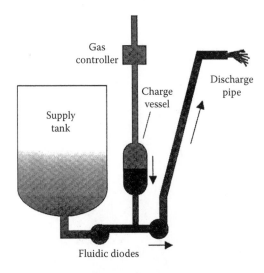

Figure 7.11   Fluidic pump. (Reproduced by permission of Accentus plc.)

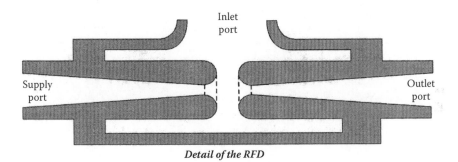

Detail of the RFD

The RFD pump consists of a gas controller, a charge vessel, the RFD, and the discharge pipework. The pump operates by repeatedly supplying compressed air to the charge vessel in two phases: compressed air is passed through the controller to the charge vessel, which forces the liquid through the RFD, where more liquid is sucked from the supply tank, and on to delivery through the discharge pipe. This continues until the charge vessel is empty.

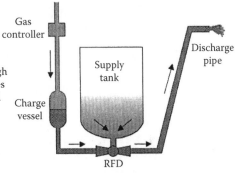

Figure 7.12   Another type of fluidic pump, the reverse flow diverter (RFD). (Reproduced by permission of Accentus plc.)

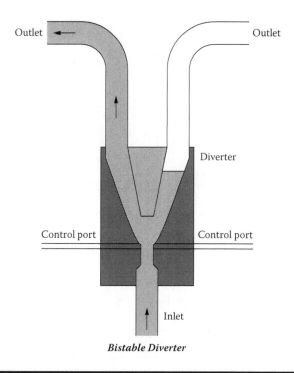

Bistable Diverter

Figure 7.13   Fluidic diverter. (Reproduced by permission of Accentus plc.)

Figure 7.14 shows a fluidic mixer. The ejectors suck liquid into the pulse tubes. When the flow stops, the liquid jets out into the vessel. Section 3.3.2 describes a fluidic gas–liquid contactor.

7.10 Other Examples of Simplification

1. An opportunity missed: two valves next to each other on a line diagram were installed on separate floors. As a result, two operators were needed to operate them.
2. A chlorinated solvent used in the manufacture of a pharmaceutical product gave rise to a chlorinated waste product. A more powerful mixer—a double-helical agitator capable of mixing viscous liquids—made it possible to operate without the solvent.[22]
3. If a liquid has to be extracted from an aqueous solution, it is often necessary to carry out two operations: wash the solution with an organic solvent and then wash the solvent with water to recover the material we want. The two operations can be combined by separating the two aqueous layers by a layer of solvent and then allowing the solute we want to diffuse from one aqueous

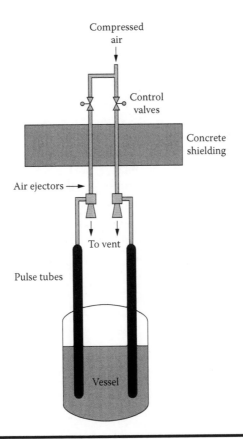

Figure 7.14   Fluidic mixer. (Reproduced by permission of Accentus plc.)

layer to the other. To keep the solvent layer in place, it is absorbed on a rigid membrane.[23]

4. The usual way of making sure that a flowmeter is kept full of liquid is to put it at the bottom of a vertical U-bend. With large pipes, this can be expensive. It may be simpler to use a type of flowmeter, such as a magnetic one, that can measure flow in partly filled pipes. The instrument is somewhat more complex, but the pipework is simple. Make sure that access to the instrument is adequate.[24]

5. Gas compression is expensive and is a source of leaks (Section 7.7). Can we avoid the need for it by generating the gas under pressure? For example could we electrolyze brine and generate chlorine under pressure?

6. Newer designs of pressurized water reactors are much simpler than earlier ones. The Westinghouse design contains 60% fewer valves, 75% less pipework, 80% less control cable, 35% fewer pumps, and 50% less building volume.[25] Nuclear reactors are discussed further in Chapter 12.

7. The loss of the space shuttle *Challenger* in 1986 was due to the failure of the O-rings in the joint between the two halves of the casing. A one-piece design would have been simpler and safer because it would have avoided the need for an extra component that was inevitably weaker than the parts it joined. The original design called for a one-piece casing, but the two-piece design was cheaper.

8. Steam traps are typically mechanical devices and can sometimes become jammed. A version with no moving parts uses the condensate itself to hold back the steam as water slowly drains through a hole in the pipework.[26] Even with these simplified orifice steam traps, caution has been expressed with respect to the possibility of water hammer and the need to fit fine-mesh strainers to prevent blockage (such strainers requiring frequent checking).[27]

9. Magnetic sensors for biomedical and underwater detection applications have been cheaply produced in the form of a ceramic capacitor that harvests the energy of stray magnetic fields. Therefore, no power supply is needed to generate 7 mV of electricity at a cost of one penny.[28]

7.11 Modification Chains

We make a small change to the design of a new or an existing plant. A few weeks or months later, we realize that the change has consequences that we did not foresee, and a further change is needed; later yet, more changes are needed, and we may now wish that we never made the original change, but it is too late to go back.

Here are two examples of changes that produced a chain of subsequent changes and a degree of unwanted complication that was never foreseen. For other examples, see Reference 29.

7.11.1 Example 1: Manhole Covers on Drains

The manhole cover on a plant drain (Figure 7.15) was not airtight, and flammable vapor escaped into the plant atmosphere.

Step 1. To reduce the chance that the vapor will ignite or be breathed by passersby, a vent, 4-m tall, was fitted to the cover.

Step 2. Because lightning might ignite the gas coming out of the vent, a flame arrester was fitted.

Step 3. It was realized that the flame arrester would choke unless it was cleaned regularly, and thus a simple access platform was provided.

Step 4. It was realized that the platform was insufficient; therefore, handrails and toeboards were fitted (as required by U.K. legislation).

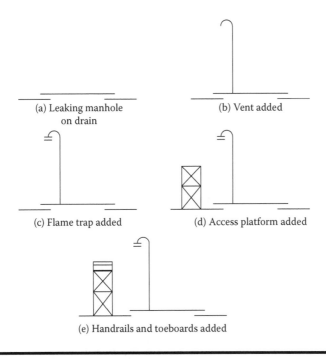

(a) Leaking manhole on drain

(b) Vent added

(c) Flame trap added

(d) Access platform added

(e) Handrails and toeboards added

Figure 7.15 A modification chain: manhole cover.

At one plant, some oil-soaked ground was dug out and replaced by concrete. A scaffold pole standing on the ground was concreted in position and left there when the scaffolding was dismantled. A few years later, someone asked what the pole was for and was told that it was a vent on a drain. It was then fitted with a flame arrester!

7.11.2 Example 2: Slurry Transfer

A slurry had to be transferred under pressure from one vessel to another. To clear chokes in the transfer line and to empty the line at shutdowns, connections were provided so that the line could be steamed from either end (Figure 7.16[a]). Nevertheless, it was feared that chokes might interrupt production. The earlier plants had been batch ones; the new plant was continuous, and interruptions would be more serious.

Step 1. It was therefore decided, after some hesitation, to install a second transfer line for use during the start-up, with the intention that it would be removed when sufficient operating experience had been gained. This transfer line also had to have steam connections so that it could be steamed from either end (Figure 7.16[b]).

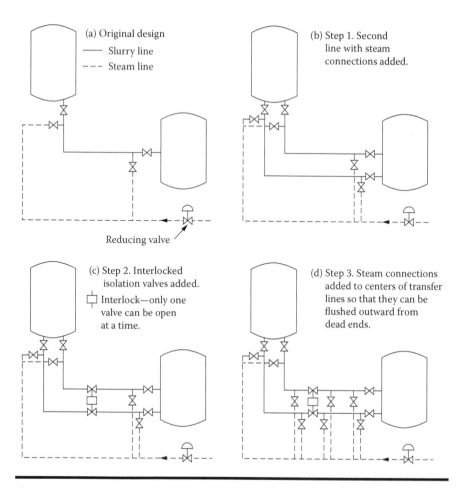

Figure 7.16a-d A modification chain: slurry transfer.

Step 2. The relief and blowdown study drew attention to the fact that both lines could be used at the same time, even though this was not the intention. The downstream vessel needed twice the relief capacity provided. To avoid having to install extra relief valves, interlocked isolation valves were provided (Figure 7.16[c]).

Step 3. The isolation valves introduced two more dead ends in each line; therefore, four more steam connections were provided so that the lines could be steamed from every dead end (Figure 7.16[d]).

Step 4. It was then realized that, to keep the spare line always ready for use, steam had to be kept flowing all the time. The transfer line was designed to withstand the process pressure but not the full pressure of the steam supply. This was acceptable for an occasional flush but was not acceptable if the steam was flowing continuously. A relief valve was therefore fitted to the

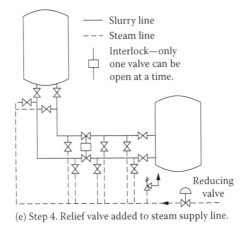

(e) Step 4. Relief valve added to steam supply line.

Figure 7.16e   A modification chain: slurry transfer.

common steam line, downstream of the pressure-reducing valve, to make sure that the steam pressure did not exceed the design pressure of the process equipment (Figure 7.16[e]).

These various steps took place over a period of about 6 months. By the time Step 4 was reached, the designers wished they had not agreed to Step 1, but it was too late to go back. The changes in the design were not all that expensive, but they made the plant more complicated and interrupted the design process. However, in any event, the design team was grateful for the spare line. It allowed the on-line time of the plant to be increased, and 2 years after start-up it had still not been removed.

7.11.3 Prevention

How can we avoid complicating the design and disrupting the design process by the gradual unfolding of modification chains? We can do this by, first, being aware that this can occur and, second, following a procedure that will help us to see the whole chain at the start.

Modification chains are most likely to start when a new line is added to the design (to connect new equipment or to provide new connections to existing equipment) or a new valve is added. New lines may result in contamination, reverse flow, and undersizing of relief valves. New valves may isolate equipment from its relief protection. New lines and valves should be assumed to be guilty of these offenses until proved innocent.

Hazard and operability studies (Hazops: see Chapter 10 and Reference 30) are powerful tools for identifying the consequences of modifications, but those made after a Hazop has been carried out may escape examination. Simpler procedures have been devised for studying modifications that are rather small to justify assembling a full Hazop team.[31,32] Major modifications should always be subjected to a full Hazop. Management of change (MOC) and inherent-safety considerations are discussed in Chapter 11 on Process Safety Management.

Finally, an example from everyday life. I used to ask whether we needed to take so much luggage on holiday. My wife would point out that summer dresses weigh very little. It was years before I realized that every dress was accompanied by a matching handbag and pair of shoes.

References and Notes

1. F. P. Lees, *Loss Prevention in the Process Industries,* 2nd ed. (Oxford, U.K.: Butterworth-Heinemann, 1996), chap. 11, figs. 11.2 and 11.3.
2. G. Zhu and M. Shah, "Steel Composite Structural Pressure Vessel Technology: Future Development Analysis of Worldwide Important Pressure Vessel Technology," *Process Safety Progress* 23, no. 1 (2004): 65–71.
3. Health and Safety Executive, *The Fires and Explosion at BP Oil (Grangemouth) Ltd.* (London: Her Majesty's Stationery Office, 1989).
4. T. A. Kletz, *Dispelling Chemical Engineering Myths,* 3rd ed. (Washington, DC: Taylor & Francis, 1996), item 1.
5. T. A. Kletz, *What Went Wrong: Case Histories of Process Plant Disasters,* 5th ed. (Oxford, U.K.: Gulf Publishing, 2009), secs. 6.1(d) and 10.4.3.
6. A. Hopkins, *Lessons from Longford: The Esso Gas Plant Explosion* (Sydney: CCH Australia Ltd., 2000).
7. T. A. Kletz, *An Engineer's View of Human Error,* 3rd ed. (Rugby, U.K.: Institution of Chemical Engineers, 2001).
8. R. Maybury, "The Siren Song of the Sledgehammer Solution," *The Daily Telegraph* (London), supplement, 14 January 1997, p. 8.
9. T. A. Kletz, *What Went Wrong: Case Histories of Process Plant Disasters,* 5th ed. (Oxford, U.K.: Gulf Publishing, 2009), sec. 12.3.
10. A. Stelmakowich, "Editorial: Rolling the Dice," *OHS Canada* 22, no. 2 (2006): 6.
11. A. Krysztoforski, S. Ciborowski, R. Pohorecki, and Z. Wojcik, "Chemical Reaction Engineering as a Tool in Process Safety Considerations," vol. 2, in *Fifth International Symposium: Loss Prevention in the Process Industries* (Paris: Société de Chimie Industrielle, 1986), 45-1.
12. T. A. Kletz, "Protective System Failure," *Learning from Accidents,* 3rd ed. (Oxford, U.K.: Butterworth-Heinemann, 2001), chap. 2.
13. T. A. Kletz, "A Gas Leak and Explosion: The Hazards of Insularity," *Learning from Accidents,* 3rd ed. (Oxford, U.K.: Butterworth-Heinemann, 2001), chap. 4.
14. D. H. A. Morris, in *Loss Prevention and Safety Promotion in the Process Industries* (Amsterdam: Elsevier, 1974), 317.
15. W. B. Howard, "Interpretation of a Building Accident," *Loss Prevention* 6 (1972): 68–73.

16. J. A. MacDiarmid and G. J. T. North, "Lessons Learnt from a Hydrogen Explosion in a Process Unit," *Plant/Operations Prog.* 8, no. 2 (1989): 96–99.

17. A. M. Wellington, introduction to *The Economic Theory of Railway Location*, 6th ed. (New York: John Wiley & Sons, 1900).

18. "Multi-purpose mixers," *Chem. Eng.* 481 (1990): 39–44.

19. N. L. Book and V. U. B. Challagulla, "Inherently Safer Analysis of the Attainable Region Process for the Adiabatic Oxidation of Sulfur Dioxide," *Computers and Chemical Engineering* 24, no. 2-7 (2000): 1421–1427.

20. N. Hanigan, "Solvent Recovery: Try Power Fluidics," *Chem. Eng. (U.K.)* 555 (1993): 19–20.

21. Search engines (e.g., Google) list many companies that now supply power fluidics. The illustrations were supplied by Accentus plc, a former subsidiary of the U.K. company AEA Technology.

22. "Eliminating Chlorinated Reaction Solvents," *Chem. Technol. Eur.* 3, no. 3 (1996): 22.

23. C. Butcher, "Liquid Membranes: The Time Is Ripe," *Chem. Eng. (U.K.)* 557 (1994): 21–22.

24. I. Robertson, "Magnetic Flowmeters: The Whole Story," *Chem. Eng. (U.K.)* 560 (1994): 17–18.

25. A. Cottrell, "Harwell: The First Fifty Years" (lecture commemorating the 50th anniversary of the foundation of the Harwell Atomic Energy Research Establishment, Harwell, U.K., May 1996).

26. L. Bond, "Steamy Business," *Chemistry & Industry*, 16 November 1998.

27. M. MacDonald, "Steamed Up," *Chemistry & Industry*, 7 December 1998.

28. "Magnetic Field Detectors for Less than a Penny," *Chemistry World* 5, no. 1 (2008): 10.

29. T. A. Kletz, "Modification Chains," *Plant/Oper. Prog.* 5, no. 3 (1986): 136–141.

30. T. A. Kletz, *Hazop and Hazan: Identifying and Assessing of Hazards*, 4th ed. (Rugby, U.K.: Institution of Chemical Engineers, 1999).

31. T. A. Kletz, "A Three-Pronged Approach to Plant Modifications," *Chem. Eng. Prog.* 72, no. 11 (1976): 48–55.

32. S. Mannan, S., ed., *Lees' Loss Prevention in the Process Industries*, 3rd ed. (Oxford, U.K.: Elsevier Butterworth-Heinemann, 2005), chap. 21.

Chapter 8

Simplification: Specifications and Flexibility

No rule is so general that admits not some exception.

—Robert Burton, *Anatomy of Melancholy*

Once simplicity became replaced by complexity, as in Stonehenge III,
one can be virtually certain that science has been replaced by ritual.

—Fred Hoyle, *On Stonehenge*

In Chapter 7, we saw that much complication occurs because we fail to recognize a hazard or an operating problem until late in design, and then all we can do is add something to *control* it. It is too late to change the design and *avoid* the hazard or problem.

This chapter discusses other reasons for complication: (a) adherence to specifications, standards, and customs too closely without asking why they were adopted in the first place and (b) a desire for flexibility in plant operation.

8.1 Following Rules to the Letter

Some unnecessary expense and complication occur because codes, standards, specifications, and custom are applied unthinkingly to circumstances not foreseen by their writers, as the examples in the following subsections show.

8.1.1 Isolation for Entry

It is a well-established principle that before a vessel is entered it should be isolated by physically disconnecting or blinding (slip-plating) all inlet and exit lines as close to the vessel as possible. Two or more vessels should not be isolated as a unit if liquid may be trapped in connecting pipework or valves.

If a distillation column has to be entered, a very big blind is needed in the overhead vapor line. If we put the blind next to the column, access is difficult, and thus it is often put immediately above the condenser (Figure 8.1). Access is then easier, but the plate is still big, expensive, and difficult to handle. Figure 8.2 shows a figure-eight (spectacle) plate for such a duty. It was made from stainless steel, 60-mm (2.4 in.) thick, weighed about 0.7 tonne, and cost about $7000 plus $4000 for the flanges at 1984 prices.

If the blind (or figure-eight plate) is put in the liquid line *below* the condenser, only a small one is needed. We have isolated two vessels—the still and the condenser—as one, but there are no valves or depressions in the overhead line in which liquid can collect. The design of the condenser should be checked to make sure that liquid cannot collect there.

Blinds should normally be designed to withstand the same pressure as the pipeline, and they should normally be made from the same grade of steel. However, a big blind in a column overhead line will be used only when the plant is shut down and freed from process materials. It will not have to withstand pressure

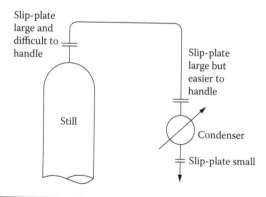

Figure 8.1 Possible slip-plate positions on the overhead line from a distillation column.

Figure 8.2 This spectacle plate, designed for an overhead line, is made from stainless steel 60-mm thick, weighs 0.7 tonne, and costs $7000 plus $4000 for the flanges (1984 prices).

or corrosive chemicals. If the decision is to install a blind, despite what was said above, it can be made thinner than required by the line standard and made from carbon steel.

Obviously, designers (and operators) should not take codes, standards, and specifications lightly, but they should be encouraged to ask themselves the purpose of the rule and whether this purpose can be achieved with equal safety in another way. There should be a system for getting exceptions approved at an appropriate level of management and recording what has been approved and the reasons why.

8.1.2 Fire Protection

A company had long experience in the design of plants for the processing of hydrocarbons and had built up standards for the degree of fire protection it considered necessary. When the company started to design plants for handling chemicals with a much lower heat of combustion, about half that of hydrocarbons, it applied the same standards without asking whether lower standards would be adequate.

8.1.3 Electrical Equipment for Flammable Atmospheres

Areas where flammable atmospheres may be present are classified in a particular scheme as Division (Zone) 1 if a flammable atmosphere is likely to be present during normal operation and Division (Zone) 2 if a flammable atmosphere is not likely to be present during normal operation and, if it does occur, will be present for only a short time. A useful rule of general guidance is to say that an area is Division 1 if a flammable atmosphere is present for more than 10 hours per year. Different types of equipment are used in Division 1 and 2 areas. Division 2 equipment will not spark in normal use but may spark if it develops a fault, typically once in 100 years.

Someone who carries a radio or personal monitor may occasionally enter a Division 1 area. Does the radio or monitor have to be a type suitable for use in Division 1 areas, thus increasing cost and restricting the choice available?

If we interpret the rules literally, the answer is yes, but a little thought tells us that the answer is no. No one should enter a cloud of flammable gas or vapor, except in the most exceptional circumstances (for example, to rescue someone), and only rarely will someone be caught in an unexpected leak. Occasionally, someone may be tempted to enter a cloud to isolate a leak, but the amount of time one spends inside a flammable cloud will still be no more than a few minutes every few years—far less than 10 hours per year. The chance of this coinciding with a fault on the radio or monitor is thus very small, and the device can therefore be a type suitable for Division 2 areas.

Taking Division 2 radios or monitors into Division 1 areas is thus less hazardous than using fixed Division 2 equipment in Division 2 areas. However, when the codes were written, no one considered the special problems of equipment attached to people, and the codes make no provision for them. On the other hand, if people are regularly entering flammable atmospheres, then there is something seriously wrong with the plant equipment and the company policy.

A somewhat similar situation can arise in standardizing the designs of electrical apparatus for explosive atmospheres of gases and of dusts. Eckhoff[1] describes his concerns with respect to the alignment of standards for combustible dusts with those for combustible gases. He emphasizes that such an exercise must differentiate clearly between the ways in which explosive gas and vapor clouds, and explosive dust clouds, are generated and sustained in industrial practice. Further, dust standards must account for open and smoldering dust fires as hazards of concern in addition to dust clouds. Eckhoff[1] suggests that the term *source of release*, while relevant for standards dealing with accidental gas and vapor leaks, is inappropriate and misleading (i.e., introducing unnecessary complexity) in the case of dusts, for which the term *area of dust cloud generation* is better suited.

8.1.4 Spare Equipment

In many companies, it is the normal practice to install a spare for every pump. Spares are then installed for pumps that operate for only a part of the day, for example, pumps used to fill tank trucks or charge batch reactors. At most, an uninstalled spare is all that is usually necessary. An installed spare costs five or six times as much as an uninstalled one because most of the cost is in the pipework and installation.

The need for all installed spare equipment should be examined critically. More reliable equipment can reduce the need for spares (Section 7.9).

Two pumps on different duties sometimes share a common spare. This saves a pump, but the pipework is complex, contamination through leaking valves is possible, and the results of incorrect valve settings can be serious. The practice is not recommended.

8.1.5 Gravity Flow or Pumped Flow?

Gravity flow can attenuate the pressure (Section 5.1.3) and may save the cost of a pump, but the cost of any necessary structures and extra vessels should be taken into account. Also, gravity can only be turned off by a valve, which may leak. Isolating the power to a pump is a more positive form of isolation (when the suction pressure is low).[2]

In a series of plants making the same product, the reactor was fed by gravity from the feed heater. When a new plant, ten times bigger than the existing ones, was designed, the traditional design was followed and the reactor again fed by gravity. A structure was provided for the feed heater, and because the structure was there, other equipment was added to it. It ended up three floors high and was the most expensive item in the design. When someone asked why a structure was needed, the design was changed so that the reactor was fed by a pump, and the structure was eliminated.

Convective flow is discussed in Section 9.11.

8.1.6 Obsolete Rules

Specifications sometimes continue in use after the need for them has passed. At one time, when a rupture disc blew on a high-pressure polyethylene reactor, the discharged gas was liable to explode in the open air. At a neighboring plant, it was found that the vibration caused instruments to trip out, and equipment was installed to prevent this from happening. Designers were still installing this equipment on new units long after the polyethylene plant had found a way of preventing the aerial explosions.

In an interesting article with a most appropriate title, "Tell Me Why," Hendershot[3] discusses the importance of good documentation of the reasons for plant design and plant operating procedures. This applies equally to safety features, especially those involving inherent safety. Inherently safer design measures are particularly susceptible to lapses in corporate memory, given that they are such a fundamental part of the design that their purpose may not be obvious.

8.1.7 An Example from Another Industry

During the nineteenth century, various rules were drawn up for the safe operation of the railways, and they continued in force into the twentieth century. In the

United Kingdom, the Board of Trade and, to some extent, the railway companies themselves, applied these rules to all railways, even little-used country branch lines, and the cost of building and operating them therefore became prohibitive. Lives would have been saved if the railways had *reduced* standards, for then more people would have been able to use them instead of traveling by road.[4]

8.1.8 An Example from History

The Roman emperor Hadrian built a wall to separate England from the wild and less civilized region to the north, now known as Scotland. It was not a defensive wall like that surrounding a castle (the Romans fought their battles in the open), but an administrative boundary to control movement between the two countries.

Gateways were constructed at intervals of a mile along the length of the wall, even when there was a nearly precipitous drop outside the gateway and it could not be used. Presumably, the design rules were written in Rome by someone who did not foresee this possibility, perhaps by the Emperor Hadrian himself, and the men on the job felt they had to follow them to the letter.[5]

Officials do not change. During the 1970s, a replica of a short section of the wall, with gateway and milecastle, was constructed. When the archaeologists applied for planning permission, it was at first refused because the Roman design did not comply with the U.K. building regulations: there was no damp course, and the doorways were too low.

8.1.9 A Traveler's Story

An air-weary traveler arrived at an airport and walked through the sliding glass doors as he had done at scores of similar airports in the past. Expecting the second set of glass doors in front of him to slide open in the usual fashion, he merrily proceeded forward only to be brought up short when no opening appeared. He then realized that he was facing a glass wall, not more sliding doors. There on the glass partition in front of him was a thin line of yellow caution tape with small arrows directing him to turn to his left. He did so, and there indeed was the second set of glass doors which opened so he could walk through them, turn to his right, and then finally approach the check-in counter. The graphic for 'Simplify' in Appendix A does a nice job of depicting this situation.

Intrigued, and a little embarrassed, our intrepid traveler stood back and watched to see how others would cope with this seemingly unnecessary complication. Person after person fared no better, and some even worse, in navigating the glass maze.

So why did the airport management decide to challenge its customers in this manner? The apparent reason is that had the two sets of sliding doors been arranged in the normal straight-through design, a blast of cold air would have reached the ticket agents at their stations during the winter months. One can surmise that it was the custom at this particular airport to locate the agents close to the doors for the convenience of all concerned. But was this really an appropriate specification for a location with a somewhat brisk winter climate?

The solution to this issue lies, of course, in the building design, which once implemented, affords little opportunity for the use of revolving doors, automatic interlocks, and the like (or even better, relocation of the check-in counters). But even when procedural safety fails, as in the case of people not paying attention to directional signage, other solutions exist. On another visit to the same airport a few months later, the same traveler was prepared to follow the arrows and avoid a collision with the glass wall. He need not have worried; there, in front of the glass wall, was an enormous orange traffic cone—the largest he had ever seen!

8.1.10 Conclusions

As stated above, standards and specifications should not be brushed aside, but people should be encouraged to ask whether the writer intended them to be applied in the circumstances in which they are applied, whether changes in technology have produced a better solution, and so on. Often, people do not know whether there is a procedure for authorizing departures from the rules. There should be such a procedure; the reasons for the departure should be documented; and they should be authorized at an appropriate level.

Sometimes, excessive complication is the result not so much of following all the rules, but of striving for a perfect solution. Solutions should not be perfect. They should be adequate (see Chapter 15).

The following remarks were made at a symposium on simplification (see appendix to Chapter 8):

> We have, over the last few years, codified a great number of views about design, views about how we should operate, which in previous eras were left to experience. That codification is very valuable because it puts experience into written codes so that other people can pick up that experiential value very easily. Also,... it's easier to demonstrate what your codes are if you've got something written.
>
> However, a number of us feel that these systems are leading us into styles of design and operation that are less than the best from the point of view of the imaginative use of resources....
>
> [We should] feed back into the interpretation of codes that flavour of human ability to take decisions which is the essence of good design and good operation, rather than the bureaucratic attitude of taking decisions by the book.

8.2 Asking for Too Much Flexibility

Some complication comes from a desire for flexibility in plant operation. Suppose a plant consists of five parallel streams, each of three stages, as shown in Figure 8.3. An example is an early high-pressure polyethylene plant in which the three stages are primary compression, secondary compression, and reaction.

If one stage has to be shut down for repair, the whole stream has to be shut down. Crossovers and valves are therefore installed so that every item can be used in any stream. The resulting "spaghetti bowl" (try to draw it) is expensive, contains many joints and valves that are sources of leaks, and provides many opportunities for error. Was it worth the cost? The money spent on spaghetti might have been better spent on improving the reliability of the components.

Similarly, in a tank farm, the operating team likes to be able to connect any tank to any line. Again, the result is an expensive, error-prone complication of pipes and valves. A better way of providing flexibility is the "snake pit" shown in Figure 8.4. Each line ends in a high-quality stainless steel hose that can be connected to any tank.

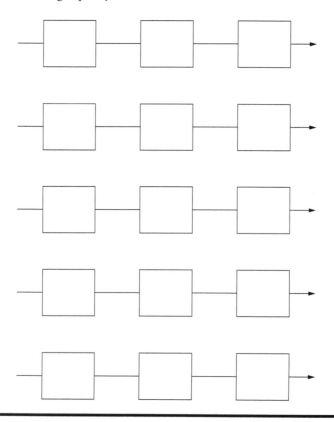

Figure 8.3 **Five parallel streams of three stages. Draw the cross-connections and valves necessary so that any unit can be used in any stream.**

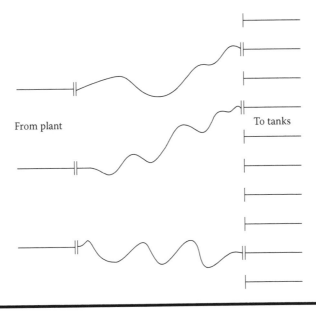

From plant

To tanks

Figure 8.4   A snake pit.

Valves are not shown. Hoses are usually unfriendly equipment (Section 9.4), but in this case they provide a simple solution if the materials handled are nonhazardous.

Spares are discussed in Section 8.1.4.

8.3 Three Problems

Readers are invited to consider the following three problems. There is no known solution to the first one, but solutions are suggested for the other two.

8.3.1 Preventing Reverse Flow

Figure 8.5 shows part of a protective system used to prevent backflow of reactants from a reactor to the storage tank containing one of the raw materials, ethylene oxide. Such backflow has caused serious explosions.[6] Check valves are not nearly reliable enough, and therefore a high-integrity protective system is used to measure the pressure drop in the transfer line and close valves in the line when the pressure drop falls below a preset value. The full system is described by Lawler,[7] and another aspect of the process is discussed in Section 3.2.7.

The liquid is normally added to the vapor space of the reactor, so backflow can occur only when the reactor is overfilled. The problem is to devise a simpler but equally reliable system based on physical principles rather than instrumentation.

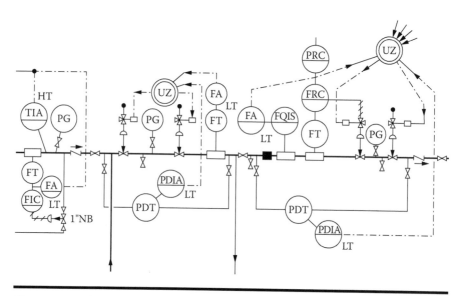

Figure 8.5   Plants as complicated as this are not unusual.

If the liquid is safe and below its boiling point, the pipe and funnel shown in Figure 8.6(a) can be used. Another method is shown in Figure 8.6(b). The height of the inverted U is chosen so that the maximum pressure attainable in the reactor cannot push liquid back over it. The vent at the top of the U acts as a syphon breaker and prevents liquid from siphoning over. This method cannot be used with liquefied gases, and the height required makes it impracticable in other cases.

Figure 8.6(c) shows a variation that could, in theory, be used for liquefied gases, but it is hardly practicable. The raw material is stored on a hill or tall structure and flows by gravity to the reactor.

A break tank (Figure 8.7) is sometimes used to minimize the consequences of backflow. If it does occur, only the amount of raw material in the tank can react, not the contents of the main stock tank. The overflow on the break tank should not go to the main stock tank. This system can also be used to make overcharging of a reactor less likely[8] (but it is still possible to charge it twice).

8.3.2 Dissolving a Solid in Acid

The problem is to simplify the line diagram shown in Figure 8.8. A solid has to be dissolved in acid. There is no use for the hydrogen produced, and it is vented to atmosphere.

Drums of solid are elevated by the hoist, tipped into the hopper, and flow by gravity into the dissolving vessel through a manhole fitted with a quick-release cover that can be opened and closed by the operators. The vessel is then swept out with nitrogen; water and acid are then added. A small flow of nitrogen is kept

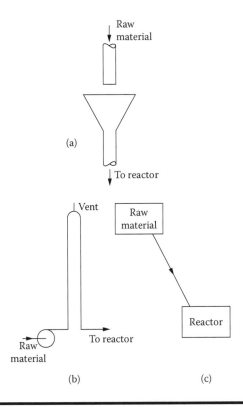

Figure 8.6 Possible methods of preventing backflow.

flowing continuously to make sure no air enters and to agitate the mixture. When the hydrogen flow falls to zero, the batch is complete and can be pumped out.

Figure 8.9 shows how the operation is actually carried out. The reaction "vessel" is a brick-lined pit, about 2.1 × 1.2 × 1.5-m deep (7 × 4 × 5-ft deep) surrounded by standard handrails. Water is put into the pit with a hose until it is 4 bricks (about 0.3 m or 1 ft) deep. The solid, which arrives in 2-m³ drums, is tipped by hand into a large hopper, rather like a wheelbarrow, and then into the pit. Acid is then added until the depth of the liquid is eight bricks.

The contents are raked to make sure that the solid is all below the liquid surface. Reaction occurs with evolution of hydrogen. The acid strength is checked from time to time, and more acid is added if necessary. In cold weather, steam is blown into the liquid to keep the temperature at 25°C–30°C, but the value is not critical. The reaction is complete when hydrogen evolution ceases. The liquid is then pumped out through a hose, the end of which is tied to a ballfloat. The sludge remaining is dug out for further processing.

On several occasions, the hydrogen has caught fire, probably because the solid, which is pyrophoric, was above the surface of the liquid. The fires were soon

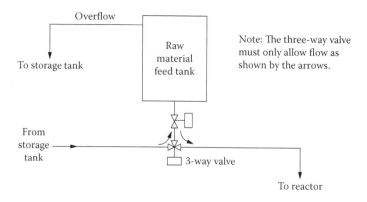

Figure 8.7 A break tank for preventing backflow from a reactor to the main storage tank. (From D. C. Hendershot, "Safety Considerations in the Design of Batch Processing Plants," in *Proceedings of the International Symposium on Preventing Major Chemical Accidents*, ed. J. L. Woodward [New York: AIChE, 1987], 3.1–3.16. Reproduced by permission of the American Institute of Chemical Engineers.)

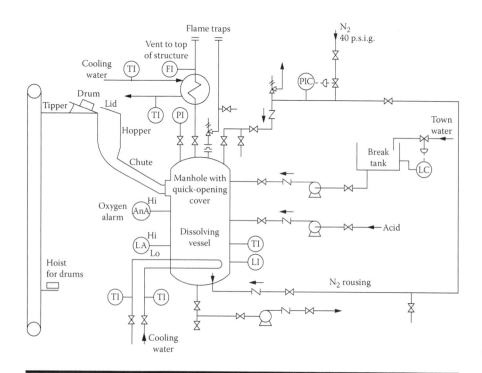

Figure 8.8 Line diagram. Can you simplify this?

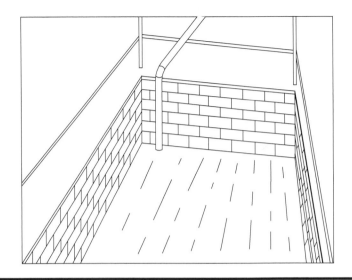

Figure 8.9   How the operation shown in Figure 8.8 is actually carried out.

extinguished with foam. They were confined to the pit and were not dangerous. They could not spread to other equipment.

During the investigation of the first fire, it was suggested that the pit should be enclosed and the hydrogen vented at a safe height. In general, it is not good practice to allow flammable gases or vapors to escape to atmosphere at ground level. However, in this case, the quantity is small (an average of 20 m³/hr over the life of a batch, but at least 10 times higher during the early stages), and the gas is hydrogen, which disperses readily by buoyancy. Any fire is confined to the pit. We could not use this method if, say, propylene were given off.

If the pit is enclosed, most of the features shown on the line diagram follow inevitably, given the usual procedures and standards. The reaction space must be inerted to prevent an internal explosion; an oxygen analyzer is needed so that we know that the inerting is in operation; level measurement is required, and so on.

It is not a satisfactory simplification to neglect the inerting or oxygen measurement and accept an occasional explosion, which could be serious. The acceptable simplification is to redesign the whole process and remove the possibility of an internal explosion by dispensing with the closed vessel.

This example shows that, although sophisticated methods are often necessary, they do not have to be used on every occasion. Sometimes simple methods are adequate. Fingers were made before forks.

On several occasions a group of engineers has been asked to simplify the line diagram. They usually make a few minor changes. Sometimes they make the design more complex. Only rarely do they think of the method actually used, usually when the group includes someone having experience with a similar process.

A young engineer visited a phosphoric acid plant; he saw the palatial offices and the old but well-built plant. He asked whether the plant also made phosphates, and the manager took him to see the food-grade phosphate unit. It resembled the one in Figure 8.9. The operator arrived with bags of sodium and potassium carbonate in a wheelbarrow, tipped them into the pit, added acid from a pipe, and stirred the mixture with a paddle. The manager remarked, "A man with a barrow never breaks down." For the rest of his career, the young engineer never forgot this lesson on the virtues of simplicity.[9]

8.3.3 Removing Liquid from a Sump

Liquid above a given level has to be removed intermittently from a sump to a point outside at the same (or a lower) level. A simple overflow cannot be used because of the nature of the intervening barrier.

Many people faced with this problem would install a pump that is switched on and off by a level controller. They might add a high-level alarm in case the level controller fails (Figure 8.10[a]).

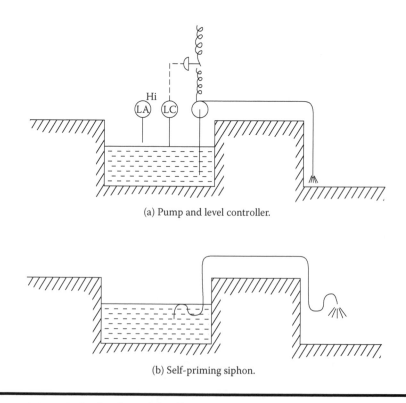

(a) Pump and level controller.

(b) Self-priming siphon.

Figure 8.10   Methods of removing liquid from a sump.

A simpler solution is to use a self-priming siphon (Figure 8.10[b]). So many people asked me if it really works that I made a model out of plastic tubing. We lack an instinctive feel for what can be achieved by gravity and convective flow (Section 9.11).

Appendix: Comments on Simplification

The following comments were made during a symposium on simpler plants. Although made some years ago, they are still relevant. They also show the value of recording discussions, because people are often less inhibited than when writing a paper, more willing to say what they think, and more willing to try out ideas and see how others react to them.

1. "I've been in the company about 18 years, and this is the first time we've had a meeting where people are prepared to stand back and consider the whole scene."
2. "There is danger of confusing innovation with fashion. The minimum capital cost of the mid-60s was fashion, and a whole series of consequences resulted which were not really sensible."
3. "We recently had to modify our ammonia handling system at short notice. We had to use the hardware available, and the system we devised was complex and yet not thermodynamically efficient. In the years to come we may have forgotten the reasons for the design and may copy it on a new plant."
4. "I had an appalling experience in a project. We laid about the design with gay abandon, putting on trip systems as if they were wing mirrors or wheel trims and we finished up with a capital cost twice that which had been sanctioned. We need data so that we can balance the costs of instruments and their maintenance against the cost of thicker vessels, more expensive rotating machinery and so on. These data have to be available to those doing the front-end conceptual design."
5. "The designer needs to know how much it costs to maintain this sort of pump compared to that sort, this sort of heat exchanger compared with that sort, and this information is not available."
6. "When I was a plant engineer I had piles of data locked away in my drawer. Attempts to move the data by mechanistic means haven't been entirely successful. I favour shifting the guy."
7. "Sometimes we install a large number of trips, not in order to solve a safety problem, but just to avoid operating inconvenience or for ease of start-up. That's fine, you can put a value on it, but you can also go too far."
8. "Since hazard and operability are dealt with together, we tend to apply the same standard to both. These different aspects require different standards."

9. "When we take a design from a contractor we tend to add on to it the things necessary for our needs. Perhaps we should be going back a stage in the thinking, however difficult that may be with a contractor."

10. "It costs twice as much to do a job with a contractor because they do things in a conventional way; they put little or no innovation into the job and just plough on down an age-old route. There's no incentive for them to do otherwise. The other disaster in our project execution is trying to do development work on a live project. So if we are going to be innovative we have to start early, well before sanction." [The leading contractors are now more flexible and the first sentence is no longer true. The last sentence, however, is still true.]

11. "If a plant uses a new process, then we are concerned that it works at all. Just how the valves are laid out [Figure 7.4] and similar matters of detail will not get much consideration because there's far too much else to do."

12. "A thousand alarms is a contradiction in terms. Anyone can go into one of our control rooms and see 20 alarms lit and the operator far from alarmed. We've tried to distinguish between alarms and status signals."

13. "We've had two maintenance engineers full-time in the project design team since its inception and they've had profound influence on what we've done."

14. "Design engineers are being put into an impossible position—heads I win, tails you lose. If they design plants, as they have done, with lots of instruments added on, they get pilloried for designing plants that are too expensive and too complicated. If they simply leave the instruments off, they get pilloried for designing plants that blow up and don't work. The only answer is to go back to the beginning and rethink our processes so that there are process solutions to the problems rather than added-on safety solutions." [Many companies are now trying to do this.]

15. "We have a lot of furnaces the design of which is so tight that we have to fit a large number of thermocouples to measure the tube skin temperature. These are the most difficult temperature measurements we have to make. The thermocouples fail within the normal turnaround period of 2 years and create a lot of work. If the heaters had been designed with an adequate margin, a lot of these measurements would be unnecessary."

16. "It's perfectly reasonable that there should be one or two Achilles' heels on a plant. If we get everything right in the design, then it's overdesigned. If we have one or two areas where we have to modify the design, we have opportunities to do so if we build our plants on time as we will not require the full output of the plant immediately."

17. "If there is an involvement of the instrument engineer in the process design, it's not in order to get the instrument engineering right (that will come later); it's in order to get the process engineering right."

18. "As long as we argue about whether we need two, four, six, or eight trips we'll never get anywhere. We'll always prove we need eight, while X [a company

that had had several explosions on similar plants] will install two and go on blowing their plants up. The answer is to use a different design for that unit operation that eliminates the problem. Any arguments about how many trips we need will be sterile. Within the context in which we operate, and with the constraints imposed on us by the Health and Safety at Work Act, there isn't a lot of room for maneuver; we are only going to argue whether we need seven or eight trips, not whether we need two or eight." [We should, of course, try to remove the need for any trips by redesigning the process.]

19. "There needs to be a cultural ambience which is supportive. It is not good enough, if we're trying to innovate and simplify, if people are motivated to say, 'Has this operated before?'"

20. "The production department has, over the last few years, systematized the logging and investigation of incidents and the philosophy of putting in more hardware to ensure the incidents don't recur. This tends to feed back to the design department as 'Process workers are zombies.' How can we avoid giving that impression?"

21. From the production director's summing up:

> It's a tribute to the quality of the people in this room that I've sat here riveted to my chair all day, listening with great attention and interest.
>
> A lot of you are saying you are trapped by the systems we have created. These systems have made us design better plants, but now we are trapped in them and they're making plants which are getting more and more convoluted. How do we break through the system without smashing it?
>
> I will take away the question of the continuity of advice from the production function. Without being too dogmatic, my feeling is that we may have to use more older, wiser, more senior men and less of them rather than younger, keener and less experienced men. The older, wiser man is more inclined not to interfere with the details of design, but to look at the broad outline, to look at the airplane as whole and not argue about how we are riveting the wing spars. [Despite these words, almost everyone over 57, including the speaker, and many younger, retired early during the next few years.]

References

1. R. K. Eckhoff, "Differences and Similarities of Gas and Dust Explosions: A Critical Evaluation of the European 'ATEX' Directives in Relation to Dusts," *Journal of Loss Prevention in the Process Industries* 19, no. 6 (2006): 553–560.
2. C. G. Carrithers, A. M. Dowell, and D. C. Hendershot, "It's Never Too Late for Inherent Safety," in *International Conference and Workshop on Process Safety Management and Inherently Safer Processes* (New York: American Institute of Chemical Engineers, 1996).

3. D. C. Hendershot, "Tell Me Why," *Journal of Hazardous Materials* 142, no. 3 (2007): 582–588.

4. D. St. J. Thomas, *The Country Railway* (Exeter, U.K.: David & Charles, 1976), 120.

5. D. J. Breeze and B. Dobson, *Hadrian's Wall* (London: Penguin Books, 1978), 38.

6. T. A. Kletz, "Accidents Caused by Reverse Flow," *Hydrocarbon Process.* 55, no. 3 (1976): 187–194.

7. H. G. Lawley, "Size Up Plant Hazards This Way," *Hydrocarbon Process.* 55, no. 4 (1976): 247–257.

8. D. C. Hendershot, "Safety Considerations in the Design of Batch Processing Plants," in *International Symposium on Preventing Major Chemical Accidents*, ed. J. L. Woodward (New York: American Institute of Chemical Engineers, 1987).

9. S. Horwich, private communication, 28 August 1995.

Chapter 9

Other Ways of Making Plants Friendlier

> The inventor of spectacles most likely resided in the Italian town of Pisa during the 1280s.... The innovation caught on quickly.... One early problem with eyeglasses was how to keep them on, for rigid arms looping over the ears were not invented until the 18th century.
>
> —**C. Panati,** *Extraordinary Origins of Everyday Things*

Readers may feel that many of the ideas discussed in this and earlier chapters are obvious—so obvious that they are hardly worth spelling out in detail. However, as the quotation above shows, many ideas that appear obvious in retrospect are not thought of for a long time.

9.1 Avoiding Knock-On Effects

Friendly plants are designed so that incidents do not produce knock-on or domino effects. Since the BP Texas City incident in 2005, comprehensive facility siting and plant layout considerations for avoidance of domino effects have received considerable attention in industry. A series of explosions that occurred during restart of a hydrocarbon isomerization unit killed 15 people and injured 180 others. Many of the victims were working inside or in the vicinity of trailers located near a vent stack that released hydrocarbons from an overpressurized distillation column.[1,2]

Domino avoidance is an application that somewhat blurs the boundaries within inherently safer design schemes by drawing close parallels with the use of passive safety barriers (Section 9.11). Additionally, it is not uncommon to see references to the principle of *limitation of effects* when speaking of methods to address domino escalation.[3–7] There is a distinction to be made, though, with the use of limitation of effects as described in the current book. The methods presented in Chapter 6 limit the effects of the hazard at its source. When we think of avoiding domino effects here in Chapter 9, we have accepted that the hazardous event has occurred (fire, explosion, gas release, etc.), and we are attempting to minimize the consequences. Detailed consequence analysis as a component of quantitative risk assessment (QRA) has been demonstrated to be beneficial in this regard for fires and explosions[8] and for toxic gas release scenarios.[9]

Some examples of avoiding knock-on effects are listed here:

1. Friendly plants are provided with corridors, about 15-m wide, between units or between the sections of a large unit, like the firebreaks in a forest, to restrict the spread of fire and reduce explosion damage. If we do have a fire, one unit or section may be damaged or destroyed, but not the whole plant.[10]
2. In designing equipment, we should consider the way in which it is most likely to fail and, when possible, locate or design the equipment so as to minimize the consequences. For example, storage tanks are usually built so that the roof-wall weld will fail before the base-wall weld. If the tank is overpressured, by an explosion or in other ways, the contents of the tank are not spilled, and any fire is confined within the tank walls.
3. When flammable materials are handled, friendly plants should be built outdoors (as discussed in Section 7.7) so that leaks can be dispersed by natural ventilation. Indoors, a few tens of kilograms are sufficient for an explosion that can destroy the building. Outdoors, a few tonnes are necessary for serious damage. A roof over equipment such as compressors is acceptable, but walls should be avoided. Walls are sometimes installed to reduce the noise level outside a compressor house, but the noise problem can be tackled in other ways.

9.2 ■ Making Incorrect Assembly Impossible

Friendly plants are designed so that incorrect assembly is difficult or impossible. For example, compressor valves should be designed so that inlet and exit valves cannot be interchanged. To prevent relief valves from being installed the wrong way round, their inlet and exit lines should have flanges of different sizes. Recall also the water-hose incident described in Section 7.4.

A word of caution: human beings have demonstrated an ability to defeat even the most inherent of safety features. A safety bulletin[11] on the hazards of nitrogen asphyxiation describes an incident in which a cylinder of pure nitrogen was mistakenly delivered to a nursing home along with the correct shipment of pure oxygen cylinders. The nitrogen cylinder had a nitrogen label partially covering an oxygen label and was fitted with nitrogen-compatible couplings. A maintenance employee removed a fitting from an empty oxygen cylinder and used it as an adapter to connect the nitrogen cylinder to the oxygen system. Pure nitrogen was thus delivered to nursing home residents, resulting in four deaths and six injuires.[11]

A different example is shown in Figure 9.1. An aqueous stream was added to an oil stream using the simple T-piece shown in Figure 9.1(a). The water did not mix with the oil, and severe corrosion caused a leak and fire. The device shown in Figure 9.1(b) was therefore designed. It was assembled as shown in Figure 9.1(c), and corrosion was worse. Once it was assembled, it was impossible to see that it had been assembled incorrectly. It should have been designed so that it could not be assembled wrongly, or at least so that any wrong assembly was apparent.

A distance piece was left out during the assembly of a machine that handled explosives. As a result, frictional heating ignited the explosive, and a man was killed. The distance piece was necessary only when another component, a worm wheel, was new and was removed when the worm wheel became worn. This is not a friendly design, and the official report recommended a change to the worm wheel so that the distance piece would never be needed.[12]

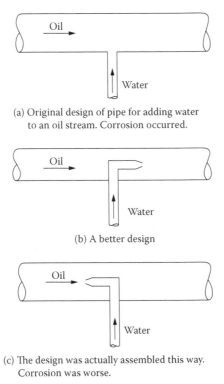

(a) Original design of pipe for adding water
to an oil stream. Corrosion occurred.

(b) A better design

(c) The design was actually assembled this way.
Corrosion was worse.

Figure 9.1   Methods of adding water to an oil stream.

9.3 Making Status Clear

With friendly equipment, it is possible to see at a glance whether it has been assembled or installed incorrectly or whether it is in the open or shut position. One example is quoted in Section 9.2. Other examples include the following:

1. Check (nonreturn) valves should be marked so that installation the wrong way around is obvious. It should not be necessary to look for a faint arrow hardly visible beneath the dirt.

2. Gate valves with rising spindles are friendlier than valves with nonrising spindles because it is easy to see whether they are open or shut. (However, on one occasion, a spindle was knocked off, and the valve was assumed to be shut.) On some types of rising-spindle valves, the gate may work loose. These types should be avoided or at least used in the horizontal position. Ball valves are friendly if the handles cannot be replaced in the wrong position.

3. Figure-eight (spectacle) plates are friendlier than blinds (also known as spades or slip plates) because their position is apparent at a glance. (If blinds are used, their projecting tags should be readily visible, even when the line is insulated.) In addition, figure-eight plates are easier to fit than blinds, if the piping is rigid, and are always available on the job. It is not necessary to search for one.

4. A storage tank made from translucent plastic (Figure 9.2) is friendly because the level in it is never in doubt. However, the materials available at present are suitable only for safe aqueous liquids and are used mainly on farms.

Figure 9.2   A translucent tank—the level is never in doubt.

9.4 Tolerance of Misuse

Friendly equipment will tolerate poor installation or operation without failure. Thus, spiral-wound gaskets (Section 6.1) are friendlier than fiber gaskets because, if the bolts work loose or are not tightened correctly, the leak rate is much less.

Expansion loops in pipework are more tolerant of poor installation than bellows. A bellows failure was the immediate cause of the explosion at Flixborough in 1974 (Section 3.2.4), which is the incident that first stimulated interest in inherently safer designs. The bellows had been used in a way specifically forbidden in the manufacturer's literature; nevertheless, if friendlier equipment had been used, the explosion would not have occurred.

Fixed pipes—or articulated arms, where flexibility is necessary—are friendlier than hoses. For most applications, metal is friendlier than glass or plastic.

Level glasses (except the magnetic type) are unfriendly because the glass is likely to break. They should not be used for hazardous liquids or for high-pressure liquids such as liquefied gases.

Bolted joints are friendlier than quick-release couplings. The former are usually dismantled by a fitter after issue of a permit to work. One worker prepares the equipment, and another opens it up; the issuance of the permit provides an opportunity to check that the correct precautions have been taken. In addition, if the joints are unbolted correctly, any trapped pressure is immediately apparent, and the joint can be remade or the pressure allowed to blow off. In contrast, many accidents have occurred because operators opened up equipment that was under pressure, without independent consideration of the hazards, using quick-release couplings. There are, however, designs of quick-release coupling that give the operator a second chance.[13]

A pump manufacturer has said, "We are being dragged kicking and screaming, by both market forces and users' requirements, to produce pumps that are more tolerant. So in future I think you'll hear less and less of us moaning about the way people use pumps." Some of his company's pumps can run dry without damage.[14] Air-operated diaphragm pumps are friendly in some respects (they can run dry and run against a closed head without damage), but on the other hand, flexible hoses have to be used for suction and discharge. The plus and minus points have to be balanced, and the friendliest pump for one duty may not be suitable for another.[15] Fluidic pumps and valves (Section 7.9) contain no moving parts and therefore cannot wear out.

During the early part of my career with ICI, I was designing instruments for use at the plants. I soon realized that every one of these instruments had to pass a simple test: can someone stand on the instrument without damaging it?

Friendly equipment is designed so that it is not easy to misuse. People are unlikely to borrow, for other purposes, the design of the fire bucket shown in Figure 9.3. A plant installed washbasins so that chemical splashes could be washed off without delay. The basins were covered to keep the water clean, but some people used them as tables (Figure 9.4[a]). Large notices were fixed to the covers, saying,

Figure 9.3   There is little incentive to steal this bucket.

"THIS COVER MUST NOT BE USED AS A TABLE." Figure 9.4(b) shows a more effective solution.

9.5 Ease of Control

If a process is difficult to control, we should look for ways of changing the process before we invest in complex control equipment.[16] Control engineers should be involved early in design, not to design the control system—that will come later—but so that they can influence the choice of process toward one that is easy to control because it has as many as possible of the following features:

1. It is based on the use of physical principles rather than added control equipment that may fail or may be neglected. An example was described in Section 3.2.1.
2. It is robust; that is, it will still operate even though there are approximations in the model of the process on which the control system is based.

Figure 9.4(a) The emergency washbasins were covered to keep the water clean. Because they were used as tables, notices were fixed to them saying, "This cover must not be used as a table."

Figure 9.4(b) A better solution.

3. It is resilient; that is, it can be tested and maintained without interfering with the operation of the plant.
4. It has a flat and slow response to changes in temperature, concentration, and so on, rather than a steep and rapid response, and the safe operating limits are wide.
5. Control limits are not close to maximum safe operating limits, and the margin between them is wide enough to provide scope for recovery if the control system does not operate normally. This applies to composition and its effects on corrosion as well as to temperature and pressure.
6. Very accurate control is not needed. Gearing can sometimes reduce the need for it. If a small change in one variable is accompanied by a large change in another variable, control will be easier if we measure the second variable.

For example, formaldehyde is manufactured by the vapor-phase oxidation of methanol. The methanol vaporizer has to operate close to the flammable limit but should not enter it. Accurate control of the vaporizer temperature is difficult. However, a change of 1°C in the vaporizer temperature causes a change of 30°C in the temperature of the catalyst bed. By measuring the latter and using it to control the steam flow to the vaporizer, very good temperature control can be obtained without the use of sophisticated equipment (Figure 9.5).[17]

7. Positive feedback is avoided; that is, a rise of temperature decreases the rate of reaction. This is a difficult ideal to achieve in the chemical industry. However, there are a few examples of processes in which a rise in temperature reduces the rate of reaction. For example, in the manufacture of peroxides, water is removed by a dehydrating agent. If magnesium sulfate is used as the agent, a rise in temperature causes release of water by the agent, diluting the reactants and stopping the reaction.[18]

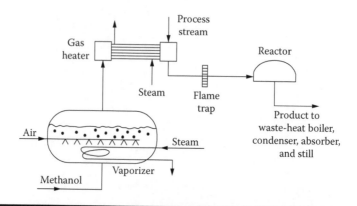

Figure 9.5 Formaldehyde plant. The vaporizer temperature can be controlled accurately by measuring the catalyst temperature.

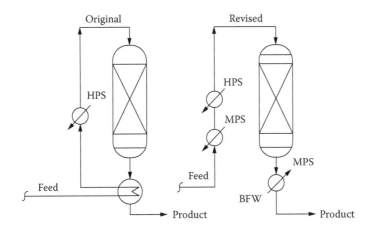

Figure 9.6 Methanation reactor system (exothermic). The original design had positive feedback; if the reactor temperature rose, the feed temperature rose, and the reactor temperature rose further. This is avoided in the revised design. Abbreviations: HPS, high-pressure steam; MPS, medium-pressure steam; BFW, boiler feed water. (From R. J. Caputo, "Engineering for Safer Plants," in *Proceedings of the International Symposium on Preventing Major Chemical Accidents*, ed. J. L. Woodward [New York: AIChE, 1987], 3.43–3.46. Reproduced by permission of the American Institute of Chemical Engineers.)

It is easier to avoid positive feedback in the chemical engineering than in the chemistry. In the original methanation reactor system shown on the left of Figure 9.6, any rise in the reactor temperature causes a rise in the feed temperature, and the reactor temperature rises further. This will not occur with the revised system shown on the right.[19]

The classic case of positive feedback occurred in the design of the Chernobyl nuclear reactor. At low outputs, if it got too hot, the rate of heat production increased, and the reactor got even hotter, so that the rate of heat production increased further, and so on (Section 12.1). In contrast, in every other commercial design, if the temperature rises, the rate of heat production falls, and in some designs it stops completely. The only preventive measure at Chernobyl was an instruction not to operate below 20% output.[20]

Negative feedback provides built-in braking, and positive feedback built-in acceleration.

9.6 Computer Control

The following list describes ways of making computer control more friendly:

1. The software could be made more scrutable. A process engineer who asks for a certain method of control will look at the process and instrumentation diagram to check that the instructions have been understood and carried out. Every process engineer can read these diagrams; the "language" is soon learned. In contrast, if a process engineer asks a software engineer to write a computer program, the result is no more readable than if it were written in Mongolian. Computer languages are difficult or impossible to read unless one is thoroughly familiar with them and uses them frequently; they are not scrutable and cannot readily be checked. Even the specification, the instructions to the programmer, may not be easy to check.

If we want to see that something will do what we want, we normally look at the final product, not the instructions for making it. I want to see food in the supermarket before I buy it, not recipes. In the long term, perhaps we shall see the development of plain English languages that anyone can read.[21] Until this time comes, we will have to make do with the specifications, but at least we should subject them to a thorough examination, using a systematic technique similar to a hazard and operability study (Hazop).[22] It is interesting to note the recent advances in software for performing Hazop studies (e.g., Rahman et al.[23]). It must always be borne in mind that such packages are tools to aid the study team, not to replace them.

Operators may realize that their instructions are in error (either fundamentally wrong, ambiguous, or ungrammatical) and may ignore or correct them, but a computer cannot do so. (In fact, unknown to supervisors and managers, operators sometimes ignore instructions for years.) People, unlike computers, can detect the meaning of imprecise verbiage; we know what is meant if we are asked to save soap and waste paper, but computers do not.

2. Problems arise because software engineers (systems analysts) do not fully understand the process. For example, at one plant, the operating team told the software engineer that when an alarm sounded, the computer should hold everything steady until an operator told it to proceed. The software engineer took this to mean that a water flow rate should be held steady, but the operating team really wanted the controlled variable, a temperature, to be held steady. A human operator could have realized this, whatever the instructions said, and acted accordingly, but a computer could not. People can do what we want them to do; a computer can do only what it is told to do.

3. Studies of the Hazop type, in which the applications engineers take part, can help us overcome the communication problem between process and software engineers. Another way is to combine the jobs. There is a need for people who are equally at home in the fields of process engineering and computer control.

4. Accidents have occurred because old pipe was reused. It looked okay (there was no obvious corrosion), but its previous history had weakened it, and it

failed in later use.[24] Accidents have also occurred because old software was reused; it functioned without problems on the old equipment, but on a new design it failed disastrously. Unforeseen weaknesses on old software caused cancer patients to receive more radiation than their doctors had prescribed[25] and led to the failure of the European space rocket, Ariane.[26]

5. We should allow for the effects of foreseeable hardware failures. For example, at a computer-controlled plant, a hardware failure caused several electrically operated valves to open at the wrong time so that hot polymer was discharged onto the floor of a building. A watchdog that should have warned of the hardware failure was affected by the fault and failed to respond. The system was unfriendly because the interlocks that should have prevented the valves from opening at the wrong time were not independent of the control system, and the watchdog should not have been affected by failures elsewhere.

The hardware failure was not the real cause of the spillage, but merely a triggering event. It would not have had serious results if the design of the system had been better.[27]

Another example: On June 3, 1980, the screens at U.S. Strategic Air Command showed that missiles were heading toward the United States. The immediate cause of the false alarm was a hardware failure, but the underlying cause was unfriendly software. The system was tested by sending alarm messages in which the number of missiles was shown as zero. When the hardware fault occurred, the system replaced the zero by random numbers.[28]

There are other examples of unfriendly software in the literature (e.g., Kletz et al.[22] and Leveson.[25]) They include the effects of modifications, programs that overload the user with too much information, and the effects of slips in writing programs. Leveson[25] suggests that we should not rely on attempts to make software ultrareliable, but instead try to make it safe even if unreliable. The best way of doing so is, of course, to design inherently safer plants, but if that is not possible we have to add protective equipment such as relief valves.

9.7 Instructions and Other Procedures

Figure 9.7(a) shows an unfriendly instruction and Figure 9.7(b) a friendly one. One reason for complex instructions is a desire to cover every conceivable eventuality. The instruction becomes so long and complex that no one reads it. It is better to write an instruction that covers most of the circumstances that can arise and is short and simple enough to be read, understood, and followed. Another reason for complex instructions is a desire to protect the writer rather than help the reader.

An example of an unfriendly procedure: If many types of gaskets or nuts and bolts are stocked, sooner or later the wrong type will be installed. It is better, and cheaper

INSTRUCTION NO: WG 101

TITLE: How to Lay Out Operating Instructions so That They May Be Readily Digested by Plant Operating Staff.

AUTHOR: East Section Manager

DATE: 1 December 1976

COPIES TO: Uncle Tom Cobbly and All

Firstly, consider whether you have considered every eventuality so that if at any time in the future anyone should make a mistake while operating one of the plants on East Section you will be able to point to a piece of paper that few people will know exists and no one other than yourself will have read or understood. Don't use one word when five will do; be meticulous in your use of the English language; and at all times ensure that you make every endeavor to add to the vocabulary of your operating staff by using words with which they are unfamiliar. For example, never start anything; always initiate it.

Remember that the man reading this has turned to the instructions in desperation, all else having failed, and therefore this is a good time to introduce the maximum amount of new knowledge. Don't use words. Use numbers, being careful to avoid explanations or visual displays that would make their meaning rapidly clear. Make him work at it; it's a good way to learn.

Wherever possible, use the instruction folder as an initiative test; put the last-numbered instruction first; do not use any logic in the indexing system; include as much information as possible on administration, maintenance data, routine tests, plants that are geographically close, and training—and then randomly distribute these throughout the folder so that useful data is well hidden, particularly that which you need when the lights have gone out following a power failure.

Figure 9.7(a)   Plant instructions. (a) Memo WG 101.

in the long run, to keep the number of types stocked to a minimum, even though more expensive types than are strictly necessary are used for some applications.

We should aim for safety by design rather than procedures when it is practicable and economic to do so (see Section 9.11 and Table 10.3).

9.8 Life-Cycle Friendliness

We should, when designing a plant, consider the problems of those who have to construct and demolish it, as well as those who have to operate it. Examples of both problems are presented.

Action to Take When a Side ID Fan Trips

1. CHECK THAT A-SIDE FURNACES HAVE TRIPPED
2. ADJUST KICKBACK ON COMPRESSORS TO PREVENT SURGING
3. REDUCE CONVERTER TEMPERATURES
4. CHECK LEVEL IN STEAM DRUMS TO PREVENT CARRYOVER

Panel Operator

1. Shut TRCs on manual
2. Reduce feed rate to affected furnaces
3. Increase feed to Z furnace
4. Check temperature of E-54 column

Furnace Operator

1. Fire up B-side and Z furnaces
2. Isolate liquid fuel to A-side furnaces
3. Change over super heater to B side
4. Check that output from Z furnace goes up to B side

Center Section Operator

1. Change pumps onto electric drive.
2. Shut down J43 pumps.

Distillation Operator

1. Isolate extraction steam on compressor
2. Change pumps onto electric drive

Figure 9.7(b) Plant instructions. (b) Extract from a plant instruction shows the actions that the foreman and four operators should take when the induced-draft fan providing air to a row of furnace trips (known as A side) stops. Compare the layout with that of Figure 9.7(a).

When an ethylene oxide plant was being extended, a new reactor had to be lifted into position over part of the existing plant. The chance of the crane's dropping its load was estimated and found to be rather high. The plant was shut down for several days. If the design team had realized this, they could have laid out the plant so that there would have been no need for the lift. Any extra cost would have been much less than the loss of profit from a three-day shutdown.

The nuclear industry in the United Kingdom grew rapidly from 1946 onward. Little or no thought was given to the problems of dismantling the equipment. If this had been considered, dismantling would have been easier and cheaper.

As we near the end of the first decade of the twenty-first century, life-cycle considerations are clearly front and center in the list of issues being addressed by industry. Ethical concerns, regulations, public opinion, and best business practices have all combined to raise the profile of life-cycle friendliness. Global initiatives such as Responsible Care°, with their emphasis on process and product life cycles, as well as the environment and sustainability, act as active promoters of inherent safety and inherently safer design. Inherent safety has been shown to be applicable through-out a process life cycle[29] and, as discussed in Section 1.1, plays a key integration role with health and environmental protocols. On this latter point, Mannan and Baldwin[30] state:

> Inherently safer technologies implemented in the design phase do not create environmental problems, and inherently cleaner technologies do not create safety problems. Instead of resorting to end-of-pipe solutions, inherently safer and cleaner approaches during design and other phases of a plant's life cycle can eliminate potential conflict between environmental initiatives and safety activities.

9.9 Other Industries

A helicopter crashed because the two rotors got out of phase and touched each other. According to the official report,[31] the cause was corrosion of a gear wheel; monitoring equipment and more thorough testing of design changes were recommended. However, in a friendlier design, the two rotors would have been located so that they could not touch each other.

Monitoring corrosion of the gear wheel will prevent the last accident from happening again, but it will not prevent the next accident because the next time the rotors touch the cause will probably be something else. We should try to remove problems, not tinker with them.

Aircraft with all the engines at the rear are inherently more liable to stall than those with wing engines. A variety of protective equipment is installed to warn the pilots that a stall is imminent and to prompt them to take remedial action automatically. Describing the Trident in 1972, Stewart writes,

> A general and perhaps inaccurate feeling had been left among pilots that the systems were unreliable. False warnings had occurred causing genuine concern.... As a result if the stick push [an automatic stall prevention device] operated in a situation which seemed doubtful to the crew the tendency was to disconnect the stick pusher even though the warning might be genuine.

This caused a Trident to crash with the loss of 118 lives. At the time, it was the worst accident in British aviation history.[32]

The words quoted will sound familiar to all who have worked on process plants! It would be better to use aircraft that are inherently less liable to stall than to add complex stall-prevention equipment.

One would have expected that reciprocating engines, which start and stop twice every cycle, would be less friendly and more troublesome than rotating engines, but this does not seem to be the case. Though reciprocating steam engines have given way to turbines, the reciprocating internal combustion engine still reigns supreme. It seems that the reciprocating engine has been developed to such a peak of perfection that the rotating engine cannot catch up. There is little incentive to develop alternative designs which, if successful, would make so much investment out of date.

In the early days of anesthetics, chloroform was mixed with air and piped to a facemask using the simple apparatus shown in Figure 9.8 that was introduced in 1867. If the two pipes were interchanged, liquid chloroform was supplied to the patient with results that could be fatal. Redesigning the apparatus so that it was friendlier, so that the two pipes could not be interchanged, was easy, but persuading doctors to use the new design was more difficult. They were reluctant to admit that they could make such a simple error. They did not realize that such errors are not due to ignorance or incompetence and that everyone is liable to make simple slips;[33] consequently, as late as 1928, the simple apparatus was still killing people.[34]

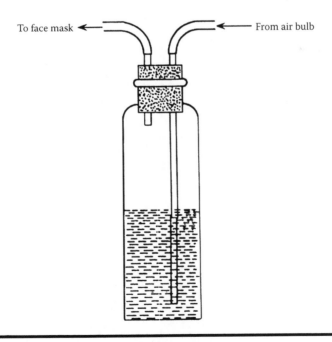

To face mask ← ‖ ‖ ← From air bulb

Figure 9.8 Early chloroform dispenser. If it is connected up the wrong way around, liquid chloroform is blown into the patient.

However, do not assume that doctors are more liable to errors than engineers. In 1989, a fire and explosion in the Phillips polyethylene plant in Houston, Texas, killed 23 people, injured over 130, and caused nearly $750 million worth of damage. Debris was thrown more than 10 km, and the subsequent fire caused two liquefied petroleum gas tanks to rupture. This damaged other units, causing a chain reaction of explosions. The cause of the accident, described in the official report,[35] was similar to that just discussed.

Hot ethylene, dissolved in butane under pressure, was circulated through a reaction loop. Granules of the polymer dropped into settling legs that were emptied by closing a valve at the top and then opening a valve at the bottom. One of the settling legs was choked, and the maintenance department was asked to clear it. They closed the top valve, or so they thought, and started to dismantle the settling leg. The top valve was actually open, and hot ethylene and butane escaped and ignited.

The top valve was operated by compressed air supplied through two hoses; one provided the supply for opening the valve, and the other the supply for closing the valve. The two hoses had been connected the wrong way round, and thus, when the actuator switch said the valve was closed, it was actually open. This is the sort of slip that everyone is likely to make from time to time, and it cannot be prevented by training, by telling people to be more careful, or by threatening them with punishment. The equipment should have been designed so that the hoses could not be connected the wrong way round (Section 9.2). They should have been fitted with different types or sizes of connectors, a change that would have cost little or nothing extra if made during design. We should not expect people to behave in ways that experience shows they do not.

As usual, when there has been a serious accident, more than one thing was wrong. According to the company policy, the settling leg should have been isolated by a blind (slip-plate) or double valve before maintenance was carried out on the settling leg (many companies would ask for both), and the top valve should have been locked shut, but the lock was missing. Procedures had been allowed to lapse because the settling legs had to be maintained frequently, though this should have been a reason for insisting on thorough observance of the safety measures, not relaxing them.

9.10 Analogies

Simple, everyday analogies often help to explain ideas. Two were suggested in Section 2.1. Here are a few more:

1. A tricycle is friendlier than a bicycle.
2. Control: It is easier to keep a marble on a concave-up saucer than on a convex-up saucer. Chernobyl (see Section 9.5 and Chapter 12) was a marble on a convex surface.

3. To prevent a boiled egg from falling over and making a mess we can
   a. Hard boil it.
   b. Put it in the eggcup with the pointed end up so that the center of gravity is lower.
   c. Use the medieval design of eggcup in which the egg lies horizontally.

9.11 Passive Safety

As discussed in Section 1.3, if we cannot use an inherently safer design and have to add safety measures, then passive ones are better than active ones, and active ones are usually better than procedural ones. An example will show why. Water spray turned on by an operator is procedural; the operator may fail to turn it on. Water spray turned on automatically by a fire or heat detector is an active safety measure; the automatic equipment may have been disarmed or may not work. Fire insulation is passive because it does not have to be commissioned, but it is not inherently safe, for it can fall off or be removed for maintenance and not replaced. The inherently safer solution is to use a nonflammable material and remove the possibility of fire, if we can, or use so little flammable material that any fire will be insignificant. Passive safety is thus a sort of halfway house to inherently safer design.

Another disadvantage of active systems is that they may fail safe; that is, they may operate without need and shut down the plant unnecessarily. Though this is preferable to failing to operate when required, it does tempt operators to disarm the systems.

Procedures are usually the weakest form of protection. They corrode more rapidly than steel and can vanish without trace once managers lose interest; a continuous effort is needed to see that they are maintained. Nevertheless, a simple procedure may be more reliable than a cheap or very complex automatic system.[36] Each case should be considered on its merits.

In nature, a hazard is sometimes controlled by a change in procedure instead of a change in design. For example, a newly opened eggshell is bright white and attracts predators. Natural selection might have overcome this by a change in color, but instead most birds that nest in places exposed to predators have developed a routine for removing broken eggshells from their nests.[37] The birds, of course, are compelled by their innate drives to remove the eggshells; they have no choice. The procedure is therefore just as effective as a change in design. With us, this is not the case. Our procedures will lapse unless we work hard to maintain them.

As another example of passive design, consider a reaction that might run away with a rise in pressure that could rupture the reactor.

1. The inherently safer solution, not always possible, is to use a different process that is not liable to run away or, possibly, to use a nonvolatile solvent. A less desirable approach, because the potential for runaway still exists, would be to use a reactor capable of withstanding the highest pressure that might be reached (Section 7.2).

During the design of a new plant it was realized that a flammable mixture might be present at times in the reactor, though the probability was small and the probability that it would explode was smaller. The cost of making the reactor strong enough to withstand an explosion would have made the project uneconomic. The designers therefore suggested that the reactor should be made strong enough to withstand normal operation and strong enough to withstand an explosion without rupturing if it was taken above its design pressure. It was agreed that if this occurred, the reactor would not be used again until it had been examined by a materials engineer and that it might have to be scrapped. The regulator, a member of the U.K. Health and Safety Executive, agreed to this. He insisted on checking the calculation of the rupture pressure, a subject on which he was an expert. No explosion is known to have occurred, but after ten years or so, would the operating and maintenance staff still be aware of the design philosophy? (See the last paragraph of Section 11.2.2, which discusses the need for documentation to preserve institutional memory.)

2. The passive safety solution is to use a rupture disc that requires no initiation beyond a pressure rise (i.e., the initiating event itself).
3. The active safety solution might be to use a trip that isolates the feed supply when the pressure rises.
4. The procedural safety solution is to ask the operator to vent the reactor and isolate the feed supply when the pressure rises. The active equipment is simple and well proven, and in this case no one would consider the procedural solution acceptable.

There is another example of passive design at the end of Section 10.2.6.

All active safety equipment depends to some extent on procedures, for the equipment has to be tested and maintained. Passive equipment is usually less dependent; fire insulation, for example, should be inspected from time to time and may require occasional repairs, but this requires far less attention than automatic water-spray equipment.

Forsberg et al.[38] have described many passive safety features suggested for water-cooled nuclear reactors. Some have not yet been tried out and may, of course, prove to be impracticable. Some of the suggestions, such as the following, may have possible applications elsewhere.

If circulation is lost, a gas-cooled nuclear reactor can be cooled by convection, but most water-cooled reactors need pumped cooling to prevent overheating, which

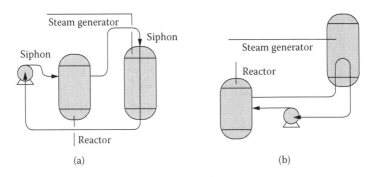

Figure 9.9 **(a) In the Three Mile Island reactor design, siphons and small differences in height prevent circulation by natural convection. (b) A modern design overcomes these difficulties by using a larger height difference and avoiding siphons.**

accounts for the proliferation of emergency cooling systems characteristic of this design. A weakness of the Three Mile Island design, not mentioned in most descriptions, is that the design actually prevented convection cooling. The heat source (the core) was only a little below the heat sink (the steam generator), and there were siphon loops in the water circulation lines that filled with steam and formed vapor locks (Figure 9.9[a]). The reactor may have been designed in this way because the first water-cooled reactors were intended for use in submarines, where a difference in height is impracticable (see Appendix to Chapter 8, Item 3). In recent designs, better layout can give convection cooling equal to 25% of full load, which is enough to remove the residual heat produced by radioactive decay after a shutdown—usually about 6% of full load (Figure 9.9[b]).

Passive safety in this instance requires an instinctive understanding of what (heat and fluid) will move where without being pumped, which is an understanding that the Three Mile Island designers seem to have lacked. How many chemical engineers have this understanding? Safety apart, there may be money to be saved if convection can assist pumped flow; 25% of pumping costs may be worth having. It is easy to draw a pump on a line diagram; it is much harder to jiggle layouts and pipe configurations to maximize convection.

A new plant was a carbon copy of an original one, except for one minor change: The floors were 10 ft (3 m) apart instead of 8 ft (2.5 m). The increase was just enough to prevent convective flow, and efficiency was lost.[39]

References

1. CSB, *Investigation Report, Refinery Explosion and Fire (15 Killed, 180 Injured), BP Texas City, Texas, March 23, 2005*, Report No. 2005-04-I-TX (Washington, DC: U.S. Chemical Safety Board, March 2007).

2. J. A. Baker III and panel, *The Report of the BP U.S. Refineries Independent Safety Review Panel*, January 2007.

3. D. G. Clark, "Applying the 'Limitation of Effects' Inherently Safer Processing Strategy When Siting and Designing Facilities," *Process Safety Progress* 27, no. 2 (2008): 121–130.

4. V. Cozzani, A. Tugnoli, and E. Salzano, "Prevention of Domino Effect: From Active and Passive Strategies to Inherently Safer Design," *Journal of Hazardous Materials* A139, no. 2 (2007): 209–219.

5. V. Cozzani, A. Tugnoli, and E. Salzano, "The Development of an Inherent Safety Approach to the Prevention of Domino Accidents," *Accident Analysis and Prevention* 41, no. 6 (2009): 1216–1227.

6. A. Tugnoli, F. Khan, P. Amyotte, and V. Cozzani, "Safety Assessment in Plant Layout Design Using Indexing Approach: Implementing Inherent Safety Perspective; Part 1: Guideword Applicability and Method Description," *Journal of Hazardous Materials* 160, no. 1 (2008): 100–109.

7. A. Tugnoli, F. Khan, P. Amyotte, and V. Cozzani, "Safety Assessment in Plant Layout Design Using Indexing Approach: Implementing Inherent Safety Perspective; Part 2: Domino Hazard Index and Case Study," *Journal of Hazardous Materials* 160, no. 1 (2008): 110–121.

8. F. I. Khan and P. R. Amyotte, "Modeling of BP Texas City Refinery Incident," *Journal of Loss Prevention in the Process Industries* 20, no. 4-6 (2007): 387–395.

9. S. Jung, D. Ng, J.-H. Lee, R. Vazquez-Roman, and M. S. Mannan, "An Approach for Risk Reduction (Methodology) Based on Optimizing the Facility Layout and Siting in Toxic Gas Release Scenarios," *Journal of Loss Prevention in the Process Industries* 23, no. 1 (2010): 139–148.

10. G. F. Barker, T. A. Kletz, and H. A. Knight, "Olefine Plant Safety during the Last Fifteen Years," *Chem. Eng. Prog.* 73, no. 9 (1977): 64–68.

11. CSB, *Hazards of Nitrogen Asphyxiation*, Safety Bulletin No. 2003-10-B (Washington, DC: U.S. Chemical Safety Board, June 2003).

12. G. J. Jeacocke, *Report on a Fatal Accident in an Explosives Factory at South Normanton, Derbyshire, on 23rd March 1971* (London: Her Majesty's Stationery Office, 1974).

13. T. A. Kletz, *What Went Wrong? Case Histories of Process Plant Disasters*, 5th ed. (Oxford, U.K.: Gulf, 2009), secs. 13.5 and 17.1.

14. R. Palgrave, "Totally Tolerant Pumping," *Chem. Eng. (U.K.)* 466 (1994): 37–42.

15. C. Butcher, "Air-Operated Diaphragm Pumps," *Chem. Eng. (U.K.)* 487 (1990): 31–38.

16. C. H. Barkelew, "Stability of Chemical Reactors," *Chem. Eng. Prog.* 25 (1959): 37–46.

17. R. G. Pickles, "Hazard Reduction in the Formaldehyde Process," in *Loss Prevention in the Process Industries*, Symposium Series No. 34 (Rugby, U.K.: Institution of Chemical Engineers, 1971).

18. H. G. Gerrirsen and C. M. van't Land, "Intrinsic Continuous Process Safeguarding," *Ind. Eng. Chem. Process Des. Dev.* 24 (1985): 893–896.

19. R. J. Caputo, "Engineering for Safer Plants," in *International Symposium on Preventing Major Chemical Accidents*, ed. J. L. Woodward (New York: American Institute of Chemical Engineers, 1987).

20. T. A. Kletz, *Learning from Accidents*, 3rd ed. (Oxford, U.K.: Butterworth-Heinemann, 2001), chap. 12.

21. E. Siddall, *Computer Programming in English*, Paper No. 33 (Waterloo, ON, Canada: University of Waterloo Institute for Risk Research, 1993).

22. T. A. Kletz, P. W. H. Chung, E. Broomfield, and C. Shen-Orr, *Computer Control and Human Error* (Rugby, U.K.: Institution of Chemical Engineers, 1995).

23. S. Rahman, F. Khan, B. Veitch, and P. Amyotte, "ExpHAZOP+: Knowledge-Based Expert System to Conduct Automated HAZOP Analysis," *Journal of Loss Prevention in the Process Industries* 22, no. 4 (2009): 373–380. Similar software has been developed by a team at Loughborough University, U.K., led by P. W. H. Chung. It is marketed by a spin-off company, Smart Plant Process Safety (formerly HAZID Technologies).

24. T. A. Kletz, *What Went Wrong: Case Histories of Process Plant Disasters*, 5th ed. (Oxford, U.K.: Gulf, 2009), sec. 9.1.6.

25. N. G. Leveson, *Safeware: System Safety and Computers* (Reading, MA: Addison-Wesley 1995), app. A.

26. J. L. Lions, *Ariane 5: Flight 501 Failure* (Paris: European Space Agency, 1996).

27. B. W. Eddershaw, "Programmable Electronic Systems," *Loss Prevention Bulletin* 8 (1980): 3–8.

28. A. Borning, "Computer System Reliability and Nuclear War," *Commun. ACM* 30 (1987): 920.

29. M. Hurme and M. Rahman, "Implementing Inherent Safety throughout Process Lifecycle," *Journal of Loss Prevention in the Process Industries* 18, no. 4-6 (2005): 238–244.

30. M. S. Mannan and J. T. Baldwin, "Inherently Safer Is Inherently Cleaner: A Comprehensive Design Approach," *Chemical Process Safety Report* March (2000): 125–132.

31. *Report on the Accident to Boeing Vertol 234 LR, G-BWFC 2.5 Miles East of Sumburgh, Shetland Isles on 6 November 1986* (London: Her Majesty's Stationery Office, 1989).

32. S. Stewart, *Air Disasters* (London: Ian Allan, 1986), 91.

33. T. A. Kletz, *An Engineer's View of Human Error*, 4th ed. (Rugby, U.K.: Institution of Chemical Engineers, 1999).

34. W. S. Sykes, *Essays on the First Hundred Years of Anaesthesia*, vol. 2 (Edinburgh, U.K.: Churchill Livingstone 1960), 3.

35. U.S. Department of Labor, *Phillips 66 Company Houston Chemical Complex Explosion and Fire* (Washington, DC: U.S. Government Printing Office, 1990).

36. T. A. Kletz, *Hazop and Hazan: Identifying and Assessing Process Industry Hazards*, 4th ed. (Rugby, U.K.: Institution of Chemical Engineers, 1999), sec. 3.6.6.

37. R. Harre, *Great Scientific Experiments* (Oxford, U.K.: Phaidon, 1981), 68.

38. C. W. Forsberg, D. Moses, E. B. Lewis, R. Gibson, R. Pearson, W. J. Reich, G. A. Murphy, R. H. Stauton, and W. E. Kohn, *Proposed and Existing Passive and Inherent Safety-Related Structures, Systems, and Components (Building Blocks) for Advanced Light-Water Reactors* (Springfield, VA: National Technical Information Services, 1989).

39. D. A. Crowl, private communication, 1990.

Chapter 10

The Road to Friendlier Plants

Every undertaking begins in discussion, and consultation precedes
every action.

> **—Ecclesiasticus (Wisdom of ben Sira),** *Apocrypha*

Three people together have the wisdom of a Buddha.

> **—Japanese proverb**

Before discussing the action needed for the design of friendlier plants, it will be useful to discuss the constraints that hinder their development. Although there has been much progress since the explosion at Flixborough in 1974 (Section 3.2.4) and the toxic gas escape at Bhopal in 1984 (Section 3.6), particularly in the reduction of the quantities of hazardous materials in storage, progress has not been as rapid as was hoped when inherently safer designs were first advocated in the 1970s.

The field has not been stagnant, however, and has grown over the past decade in recognition and application (as discussed further in Chapter 16). Key progress has been made in some needed areas, for example the development of inherently safer design metrics; see Section 10.4 on the measurement of friendliness. A leading proponent of inherently safer design made these remarks near the close of the twentieth century:

> I believe the major contribution of the inherent safety movement is to
> focus the attention of engineers on hazard elimination and minimization,

and, by providing a structure and methodology, to aid in identifying ways of accomplishing this elimination and minimization. When you give an idea a name, and some specific and defined characteristics, you have created something on which people can concentrate their efforts and energy, resulting in significant accomplishment.[1]

The constraints given in Section 10.1 have been addressed by various authors, for example Kletz,[2] Khan and Amyotte,[3] and Edwards.[4] They are as relevant today as they were when first described by the respective authors in 1999, 2003, and 2005. Time works in favor of breaking down all barriers, but so do concerted and collaborative efforts by industry, academia, and government. Much has been done in this regard, but much more is yet to be accomplished.

10.1 Constraints on the Development of Friendlier Plants

10.1.1 Time and Company Procedures

The first constraint is that company procedures do not always ask for safety studies to be carried out early in design. Safety advisers do not usually get involved, and safety studies are not usually carried out until comparatively late in design, when the line diagrams have been drawn. Then a hazard and operability study, a relief and blowdown review, an electrical area classification, and other safety studies may be carried out, but it is too late to make major changes and *avoid* hazards; all we can do is control them by *adding* protective equipment (Section 2.1).

Why are safety studies not carried out earlier? One reason is that some organizations still look upon safety as a coat of paint: something to be added to the finished design to make sure that people will not be injured. The safety advisers in such companies may be incapable of contributing to the earlier stages of design. (At a safety advisers' conference many years ago, when I was describing the work I was doing, an old-timer said, "I don't get involved in things like that. I leave them to the technical people.") However, a more important reason is that there is rarely time to carry out safety studies early in design. If our marketing colleagues recognize the need for more output and a new plant, they want it soon, or other companies will be ready before us, and the marketing window will have passed. We should, therefore, try to recognize the need for new plants sooner than we have done in the past (Figure 10.1) and should try to develop the "plant after next" philosophy (Section 10.2.1). The change from oil to water cooling described in Section 4.1.2 took place because, for once, the marketing department recognized the need for a new plant 5 years ahead and there was time to think through the problems, real and imagined, that were produced by the change.

"YOU SHOULD HAVE CALLED ME EARLIER."

Figure 10.1   Like this plumber, safety advisors are sometimes called in too late. (Reproduced by permission from the Institution of Chemical Engineers.)

Spending more time discussing the early stages of design will not only make it possible to avoid hazards and the need for expensive additional protective equipment; it will also save costs in other ways. We do not spend much money on a new plant until equipment is procured and installed, but we have committed most of the cost (80% is quoted as typical) by the time the process design is complete. The most significant opportunities for inherent-safety implementation occur early in product and process research and development.[5,6]

10.1.2 Resistance to Change

We tend to resist change. We like to follow the procedures—in plant design and everything else—that we have always followed. We like to use tested designs. If we use a new process or new equipment, perhaps there may be unforeseen problems that will delay the start-up or the achievement of flowsheet output; perhaps it will never be achieved. It is better to stick to the designs we know. In the words of Edwards:[4] "Are we too risk-averse for inherent safety?"

These fears are not without foundation. During the 1960s, a new generation of plants was built, larger than those built before, operating under more extreme conditions. At the same time, there was a demand for minimum capital cost (rather than minimum lifetime cost). Many of these new plants required extensive modification and the expenditure of much money and effort before flowsheet output was achieved. The industry burned its fingers and has tended to play it safe ever since so far as innovation is concerned. Some of the junior managers who struggled to bring these plants online became senior managers in the 1980s and 1990s, and have remained suspicious of innovators' claims. They sympathized with Lord Salisbury, prime minister to Queen Victoria, who said to her, "Change—change—who wants change? Things are bad enough as they are."

The experience of Alcoa provides an example. In 1973, the company announced a new smelting process for aluminum, the culmination of 10 years' research and development. The process was said to reduce energy consumption by 40%. Research and development continued, but in 1984 the plant was abandoned owing to "technical limitations that could not be overcome."[7]

In 1980, a senior engineer in an international chemical company was asked to survey attitudes to innovation. He read through about 15 major expenditure proposals submitted to the head office for approval and found that all but a few claimed, as an advantage, that no innovation was involved. (In one or two cases, he suspected that there was some innovation, but the originators wanted to conceal the heresy.) Obviously, we do not want change for its own sake, but there are times when change may be necessary to increase safety, reduce costs, and meet competition.

Allied to an unwillingness to change is a reluctance to admit that there is a problem and that fundamental change is necessary. A nuclear industry publication has noted the following:[8]

> Whatever enhanced safety level may be desired can be obtained simply by developing the types of reactor we already have. Also, building on what is already proven could bring swifter results with greater confidence than launching into radically new methods that purport to offer inherent safety.

This sounds convincing until we realize that similar arguments, 180 years ago, could have been (and probably were) used to advocate the breeding of better horses instead of developing steam locomotives.

In the chemical industry, despite Flixborough and Bhopal (and ensuing incidents in the 1990s and later), many managers still believe that large inventories are not hazardous because we know how to handle them. They do not realize that when we are dealing with hazardous materials, only very low failure rates, of people and equipment, are acceptable today and that these rates are often beyond, or close to, the best that people and equipment can provide. As quoted in Chapter 2, "It is hubris to imagine that we can infallibly prevent a thermodynamically favoured event."[9]

When a fire is reported in another company, we put the mishap down to the company's poor procedures or failure to follow its procedures. When we have a fire, it is an isolated incident, preventable in future by a minor change in design or procedures. If there is a system in place and it is more or less working, we try to ensure that it goes on working, making the minimum changes necessary. Only when we have had several serious accidents are people willing to admit that fundamental change may be necessary. Worsthorne,[10] writing in a different context, says that "the writing on the wall is not difficult to decipher. Where the difficulty arises is in resisting the temptation to close our minds to the full dismaying purport of the message."

In asking for inherently safer and friendlier designs, we are asking for more than better widgets; we are asking for a cultural change, a change in the design process and in the way engineers approach design problems. This is inevitably slower than a change in technology. Many engineers are happier carrying out calculations than handling ideas. "How to" books sell more copies than ideas books. Universities tend to teach calculation methods rather than softer approaches such as those discussed in this book.[11] Accreditation systems for undergraduate engineering curricula world-wide are now calling for a greater emphasis on business-related skills such as project and risk management. This is a welcome development that may spur change.

During the 1980s and 1990s, there were periods of recession. Few new oil and chemical plants were built, and there were fewer opportunities than in earlier decades to try out new designs. Some of the features of inherently safer design cannot be backfitted on old plants. The exception is storage, and in this area progress has been greatest. However, Section 3.2.9 showed how an old plant was made inherently safer.

10.1.3 The Influences of Licensers and Contractors

The influence of licensers and contractors is firmly on the side of tradition. Why develop new processes when there is a market for those we have already? Section 4.1.2 describes the replacement of kerosene by water as the coolant in the manufacture of ethylene oxide. The licenser was reluctant, until after Flixborough, to agree that there was a need for change and, although willing to design a plant, was unwilling to give the usual guarantees as to output and efficiency.

A design contractor who carried out a critical examination of alternatives, as suggested in Section 10.2.1, found that he had less detailed design work to do and made less money. Next time, he said, he would go back to his old ways.

10.1.4 Control

Will intensification (Chapter 3) make control more difficult? Large inventories in a reactor or the base of a distillation column have a damping effect, smoothing out the effects of minor changes in feed composition, heat input, and other variables.

Control engineers are confident that overcoming reductions in inventory is not an insuperable problem, although in some cases faster-responding instruments may be needed. This constraint is not real. By analogy with computers and other electronic devices, substantial reductions in size and cost may make overcapacity economic and simplify control problems.

10.1.5 Company Organization and Culture

Companies today tend to be organized in business areas rather than functional departments such as research and design—functions that are often contracted to outside companies specializing in these activities. With this sort of organization, it may not be clear who, if anyone, is expected to innovate. Those who control expenditure may be unsympathetic to ideas about better ways of carrying out functions. The business is managed, but no one manages the technology.[12] For example, the large oil and chemical companies have not tried the innovative Higee method of distillation (Section 3.3.2). According to Kirkland,[13]

> If the viability of the project is marginal—which seems to cover most of them—the reaction is ... Do what we did before, only more carefully. Don't attempt anything new and challenging—you may not be able to meet the challenge. Thus we frustrate the innovators, discourage those who would bring some flair into engineering, and ensure that we continue to be seen by the greater public as tighteners of nuts and bolts.
>
> This cannot all be the fault of the "money men" and their failure to understand us. It is often much more our inability to express ourselves in language they can understand.

Some companies have recognized the problem of encouraging innovation in a business-centered organization and tried to overcome it. According to ICI's research and technology director at the time, "We have a process by which scientists can talk about science strategy which cuts across business boundaries. We have seven science groups covering the whole of ICI's scientific ambitions. Each of these scientific areas might serve up to five or six business strategies."[14] Another report says that "certain functions will need to remain central.... Among these precious few are research, technology, and manufacturing."[15]

At a lower level, do individual technologists see it as their job to innovate? The research chemists may be expected to find new products or new routes to old products, but once they have developed the new chemistry, the chemical engineers may see their job as drawing up a flowsheet using established techniques. Although the typical research chemist sees himself or herself as an innovator, many chemical engineers see themselves as practitioners of established procedures.

Is this difference in attitude because most research chemists are PhDs, whereas few chemical engineers have done any research? Is industry in the United Kingdom right to discourage chemical engineers from undertaking research before entering industry? In Switzerland and Germany, and to some extent in the United States, many directors have postgraduate degrees in engineering, and in their companies the risks of innovation are said to be accepted more readily than in the United Kingdom.[16]

As we enter the second decade of the twenty-first century, it is encouraging to see the Royal Society of Chemistry in the United Kingdom promoting the use of an inherent-safety, guideword-led, hazard-identification method throughout the process-design life cycle.[17] What is perhaps most interesting about this approach is the clear articulation of the benefits arising from innovations in both engineering and chemistry.

Ramshaw[18] has discussed the qualities necessary for innovation in a large company:

1. Exceptional tenacity. It takes a long time to persuade colleagues to accept innovation when they are not under pressure to improve operations.
2. Allies and collaborators within the company who will receive credit for any success.
3. An ability to spot applications that demonstrate satisfactory and economic performance. Higee (Section 3.3.2) is comparatively expensive and therefore has been slow to become established; the rotating mop (Section 3.3.3) is cheap and has been used much more.

All of this is not to say that there has been no innovation in industry with respect to inherently safer design, but rather that company organization may not encourage it. There are numerous examples of innovative inherent-safety applications throughout this book. Recent cases include those described for a range of process-related companies,[19] from an EPC (engineering procurement and construction) perspective,[20] and for a specialty chemicals company.[21]

10.1.6 Lack of Knowledge, Tools, and Data

A study carried out in the early 1990s for the U.K. regulatory organization, the Health and Safety Executive, found dearths of knowledge, tools, and data to be major constraints.[22] The author found that safety advisers were familiar with the concept of inherently safer design, but there was still a lack of general awareness in design organizations and among senior managers. A later study (published in 1997) of the extent to which inherently safer designs were used or could be used offshore[23] was similar. It found that many designers and safety professionals were aware of the concept, but few of them had a clear understanding of its possible applications. They may have applied some of its principles but not in a systematic or methodical way. There was not a realization that inherently safer designs can be both cheaper *and* safer because there is less need for added protective equipment and, if we can intensify, the plant will be smaller. A survey of U.S. mechanical engineers came to similar conclusions.[24]

At the close of the 1990s, there was thus a need to raise awareness and to develop techniques and tools that make it easier for designers to develop inherently safer designs. A project[25] to provide these tools was in progress at that time and has now been completed. The INSIDE (*IN*herent *S*HE [Safety, Health, Environment] *In DE*sign) project was a European government/industry project sponsored by the Commission of the European Community to encourage and promote inherently safer chemical processes and plants.[3] The project has developed a set of tools, the INSET (*IN*herent *S*HE *E*valuation *T*ool) toolkit, to identify inherently safer design options throughout the life of a process and to evaluate the options. Table 10.1 summarizes these tools, which are described in detail in Reference 26.

It may be argued that, at the time of writing this second edition, inherent safety has increasingly become common knowledge in the process industries. But what of the status of the tools and methodologies that would facilitate the wide application of the principles of inherent safety? As discussed further in Section 10.4, this has been an area of considerable activity in recent years. Edwards[4] notes that the issue now may not be the availability of inherent-safety assessment tools, but rather the limited use of these tools by industry (for reasons such as subjectivity).[27]

There also continues to be a need for more investigative tools, similar to hazard and operability studies, for examining designs and uncovering ways of avoiding hazards (see Section 10.2 and Chapter 11). Hendershot[28] provides a summary of available tools at the close of the last century, and specifically draws attention to the usefulness of inherent-safety-based checklists.

Turning to data, information on the surface compactness of heat exchangers (Section 3.4) or the holdup of distillation column packings (Section 3.3.1), for example, is rarely quoted in manufacturers' catalogs. Designers do not have the time or patience to collect the data, and the manufacturers may not have them readily available. Lists of such data would be useful. In addition, computer-aided design packages may be unable to calculate plant inventories and cannot readily show the effect of design changes on cost. We can hardly expect designers to minimize the inventory when they cannot easily find out what it is and the effects of changing it. The design packages of the 1990s do not provide easy access to data on flammability and toxicity, which makes it harder for designers to decide which inventories it is most important to reduce.[29] Recent and encouraging developments in this regard are the concept of design support by means of an inherently safer process database[30] and the linking of explosion and other process hazards with design simulation software such as HYSYS.[31,32]

10.1.7 Size

A possible constraint is more subtle and more conjectural: a macho desire to be associated with the biggest. Perhaps we would rather boast that we designed or operate the largest distillation column in the company rather than the smallest. Morris[33] described the Texas philosophy as "if it's big, it's gotta be good; and if it's only good, you gotta make it bigger." A publisher's catalog listed 22 books on dinosaurs. Are they so popular because they were so big?

Table 10.1 Summary of INSET (*IN*herent *SHE* *E*valuation *Tool*) Toolkit

Tool		Description
A		Detailed constraints and objectives analysis
	A.1	Detailed constraints analysis
	A.2	Detailed objectives analysis
B		Process-option generation (including process-waste minimization guide)
C		Preliminary chemistry route options record
D		Preliminary chemistry route rapid ISHE evaluation method
E		Preliminary chemistry route detailed ISHE evaluation method
F		Chemistry route block diagram method
G		Chemical hazards classification method
H		Record of foreseeable hazards
I		ISHE performance indices
	I.1	Fire and explosion hazards index
	I.2	Acute toxic hazards index
	I.3	Health hazards index
	I.4	Acute environmental incident index
	I.5	Transport hazards index
	I.6	Gaseous emissions index
	I.7	Aqueous emissions index
	I.8	Solid wastes index
	I.9	Energy consumption index
	I.10	Reaction hazards index
	I.11	Process-complexity index
J		Multiattribute ISHE comparative evaluation
K		Rapid ISHE screening method

(continued)

Table 10.1   Summary of INSET (*IN*herent *SHE* *E*valuation *Tool*) Toolkit (Continued)

Tool	Description
L	Chemical reaction reactivity-stability evaluation
M	Process SHE analysis/process-hazards analysis and ranking
N	Equipment inventory functional analysis method
O	Equipment simplification guide
P	Hazards range assessment for gaseous releases
Q	Siting and plant layout assessment
R	Designing for operation

Source: INSIDE Project Team Partners, *The INSET Toolkit,* vols. 1 and 2, integrated version, Nov. 2001 (Cheshire, U.K.: AEA Technology, 2001).

10.1.8 False Economies

Jones and Snyder[34] have listed the following "roadblocks and pitfalls" that hinder or prevent the design of inherently safer plants:

1. Cutting design costs, though they are only a fraction of total costs.
2. Saving material costs by cutting down corrosion allowances or using materials that cannot tolerate variations in plant conditions.
3. Reducing the spacing between equipment, thus making operation, maintenance, and firefighting more difficult and increasing the spread of fire (Section 9.1). Modular construction often produces congested plants.
4. Replacing key members of project teams by newcomers who are not committed to previous plans and concepts.

10.2 The Action Needed

Some of the actions needed follow automatically from the descriptions of the constraints, for example, the need for data on the inventory contained in plant equipment. Others merit further discussion.

10.2.1 Recognition of the Need for Friendlier Designs

We will not get inherently safer and friendlier plants unless designers are convinced that such plants are possible and desirable. This book is an attempt to spread that message. We must keep spreading it. One single publication of an example or technique is not enough to catch the eyes of those who might make use of it. We need an orthodontic approach: continuous steady pressure over the years to produce the desired change.[35] The fact that some gains have been made in familiarity with and usage of inherent-safety concepts over the past decade does not mean that we can abandon this approach.

Designers, however, do not live and work in a vacuum, but are influenced by the culture of their organization. If it is unsympathetic to innovation, they will produce traditional designs. Those at the top have to set the tone. Statements of policy carry little weight. The little things count for more. If a designer sees an expenditure proposal supported by a statement that "no innovation is involved," put forward as a reason for satisfaction (Section 10.1.2), the designer assumes that innovation is not wanted.

Senior managers should recognize that there is more to safety than taking an interest in the lost-time accident or injury rates; this should be readily apparent after the 2005 BP Texas City incident and in the current climate of an emphasis on process-safety culture (Chapter 11). As with any other managerial function, they have to recognize the problem and the action required, in this case friendlier designs, and ask for regular reports on progress. They could ask, for example, for an annual report on the progress made in reducing inventories, and before authorizing the construction of a new plant they could ask how its inventory compares with that of the last one. A Japanese national who joined the board of ICI was surprised to find that there was no discussion of new technologies and the progress made in implementing them.[36]

In the United Kingdom, and in some other countries, all chemical engineering undergraduates receive some training in safety and loss prevention, but inherently safer and friendlier design is not always included. It should be. Students should be taught that safety is not (or should not be) a veneer that can be added to their design by a safety expert, but an integral part of the design for which they are responsible. Friendlier design is likely to grow in importance in the years to come, and those students who are not familiar with the concept have not been equipped for their future careers[37] (Section 16.4). Accreditation systems again have a role to play globally in ensuring conformance with meeting the needs of undergraduate engineering students. Why should process-safety audits occur only in industry?

10.2.2 New Design Procedures

Designers, especially process engineers, need some sort of procedure or aide-mémoire to help them consider the various ways of making their designs friendlier. A normal hazard and operability study (Hazop)[38–40] of the line diagrams is not sufficient because it takes place far too late in design for fundamental changes to be made. I suggest that we need the following studies:

1. A study at the conceptual or business-analysis stage, when we are deciding which product to make, by what route, and where the plant will be located. Some questions that should be asked are listed in Section 10.2.3 (see also Section 4.2.6).

2. A study at the flowsheet stage, when we are developing the process design. Some questions to ask are listed in Section 10.2.4. In addition, a hazard and operability (Hazop) or similar study should be carried out on the completed flowsheet. This will not take long, compared with the normal Hazop of the line diagrams, and is well worthwhile.

3. When a Hazop is carried out on the line diagrams, a few additional questions should be asked. These are listed in Section 10.2.5.

The timing of these studies is critical. For each, there is a critical period or window of opportunity. If they are carried out too early, essential information may not be available; if they are carried out too late, the design will be too far advanced, and change will cause expense and delay.

Table 10.2 shows the stages at which each feature of friendly design should be considered, and Figure 10.2 shows how the opportunities for installing inherently safer features decrease as design progresses.[41]

Similar thoughts have been expressed by Tugnoli et al.[42] in their generic concept of the *design space* as a rational representation of the safety choices in design, according to their position with respect to several hierarchies of elements (control levels, inherent-safety principles, and life-cycle stages). Figure 10.3 gives one possible depiction of the design space as an oriented space that illustrates the applicability of inherent safety in process design. A design choice within a given project is represented by a unique point in the space. With this representation, it is clear that inherent safety can be successfully implemented at any single point in the space. As previously illustrated by Table 10.2 and Figure 10.2, though, it is recognized that some zones of the design space will yield better effects (e.g., consideration of the inherent strategy of intensification during process research and development). Any point in the space, however, is available for safety-measure implementations that contribute positively to the overall inherent-safety performance of the project.

In the conceptual stage of the design, a research chemist usually takes the lead; during the flowsheet stage, a chemical engineer takes charge; and during the detailed design stage, a mechanical engineer takes charge. However, all should be involved at all stages. Fulton[43] writes that

> a major hurdle … is research and process engineering being in separate departments.… The corporate structure may cause them, like distant cousins, to meet only rarely at family reunions. Sometimes, research and process engineering meet only when research transmits "the process" to engineering for "final" design. The necessary close cooperation and shared responsibility are squeezed out by organizational rigidity.

Hazard and operability studies of the line diagrams are now common, almost standard, and there is no need to describe them here. Similar studies at earlier stages, particularly at the conceptual stage, are rare. Of course, every company will

Table 10.2   Project Stage at Which Each Feature of Friendly Design Should Be Discussed

	Feature	Conceptual Stage	Flowsheet Stage	Line Diagram Stage
1	Intensification	X	X	...
2	Substitution			
	of chemistry	X	...	...
	of auxiliary materials	...	X	...
3	Attenuation	X	X	...
4	Limitation of effects			
	by equipment design	...	X	X
	by changing reaction conditions	X	X	...
4	Simplification	X	X	...
5	Avoiding knock-on effects			
	by layout	X	X	...
	in other ways	...	X	X
6	Making incorrect assembly impossible	...	...	X
7	Making status clear	...	...	X
8	Tolerance	...	...	X
9	Ease of control	X	X	...
10	Computer control	...	X	X
11	Passive features	...	X	X

assure us that it does not embark on a new design without careful consideration. This is not denied. What is missing in many companies is the formal, structured, systematic questioning procedure characteristic of a Hazop.

The procedures used by some companies to recognize hazards and develop alternatives during the early stages of design are described in the literature.[41,44–48] The previously referenced paper by Hendershot[28] is also helpful in this regard.

The literature[44,45,48] also describes the ICI system of six hazard studies carried out on all projects. The third ICI study is a hazard and operability study, and it

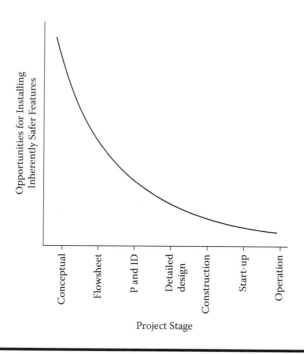

Figure 10.2 Inherently safer features become harder to install as a project progresses.

is therefore tempting to equate ICI Studies 1 and 2 with the first two studies in the three items listed in the opening paragraph of the current section. However, this is not the case. Imperial Chemical Industries Studies 1 and 2 correspond to my second study (and emphasize inherently safer features), but there is nothing similar to my first study. The ICI system is organized and controlled by the engineering design department, and my first study must take place, if it takes place at all, at the very beginning of a project, before the design department is normally involved. My first study asks if we are making the most suitable product, in the most suitable place, in the most suitable way. As of the late 1990s, ICI was trying out earlier studies.

Although many detailed accounts of conventional Hazops have been published,[38–40] little has appeared on the detailed results of studies carried out at earlier stages. One reason for the rapid adoption of hazard and operability studies was the publication of Lawley's 1974 paper,[49] the first published description of the technique, which included a detailed description of an actual study. There remains a need for similar detailed accounts of earlier studies. Some of the recommendations made during a Hazop of a flowsheet study are summarized in Section 10.2.5.

I am not questioning the value of a conventional Hazop of the line diagrams. Hazop is a powerful technique for identifying hazards and operating problems, and

The Design Space

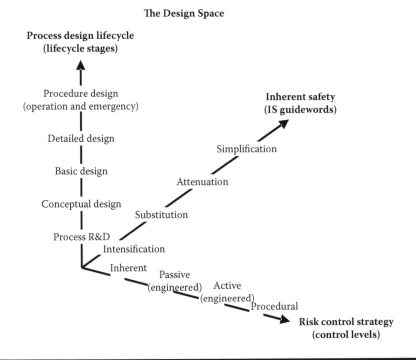

Figure 10.3 A representation of the design space. (Adapted from Tugnoli et al., "Inherent Safety Implementation throughout the Process Design Lifecycle," in *Ninth International Probabilistic Safety Assessment and Management Conference [PSAM 9]* [Hong Kong, 2008]. With permission.)

it should be carried out on all new designs and on existing plants that have been or are to be modified. But alone, it is not sufficient. It comes too late for major changes to be made, and the other studies I have listed should be carried out as well.

There is an important difference between a conventional Hazop of a line diagram and a Hazop of a flowsheet (sometimes called a front-end or coarse-scale Hazop). In a conventional Hazop, we assume that deviations from design conditions are undesirable, and we look for ways of preventing them. In a Hazop of a flowsheet, we also want to generate alternatives. In considering, say, "less of temperature" in a conventional Hazop, we ask if it is possible, if it will be hazardous or prevent efficient operation and, if so, how can it be prevented. In a Hazop of a flowsheet, we also want to know if it will be beneficial and if there is a case for operating at a lower temperature. Clearly, there are strong overtones of inherent safety here.

During the development of a design, even if we do carry out the studies I have suggested, many proposals will be made that cannot be developed or evaluated in the time available. Malpas[50] therefore suggests that we should start work on them so as to be ready for the plant after next. He writes the following:

The technical options available for the next plant are usually limited by time, so if major advances are to be made there has to be thought about the plant after next. Fundamental and significant technical advance takes years to develop, so the time to start thinking about it is precisely when the next plant is being designed, for it is then that the many compromises between what would have been desirable and what is practically acceptable are fresh in the mind. Despite the fact that it is just then that the pressure on everyone to define and execute the next plant is greatest, it is the best time to identify appropriate research targets.

As an example, Malpas[50] quotes ICI's experience with paraquat, the nonpersistent herbicide that made ploughless cultivation possible. An intermediate is 4,4-dipyridyl. The first plant for manufacturing it involved a substantial fire hazard and gave 28% yield. The second was less hazardous and gave 40% yield. While this plant was being built, ICI gave a special team the job of looking further ahead. Within 6 months, it had devised a nonhazardous method with a 96% yield. Design was carried out to the point where the team was confident that the method could be used for the next major expansion.[51]

Glasser and Early[52] suggest that, before we start on the design of a new plant, we should carry out a hazard and operability study on the last plant. If the last plant was not studied, this suggestion has much to commend it. However, if the last plant was studied, the points that could not be incorporated in the design at the time should have been noted then.

The following Sections 10.2.3 to 10.2.6 provide advice on the sort of questions that ought to be asked at critical design stages if we are to achieve inherently safer designs. These sets of questions are essentially checklists for the identification of inherently safer alternatives. (See also Chapter 11.) Appendix A in the recently updated American Institute of Chemical Engineers book on inherently safer chemical processes[53] gives further guidance on the use of inherent-safety checklists.

10.2.3 Some Questions to Answer before the Start of Design

The words *identify*, *prevent*, *control*, and *mitigate*, applied to hazards, are often used to describe the stages to be followed to obtain a safe plant. Inherently safer design could be described as part of prevention, but it is better to add *avoid* or *eliminate* to the list of key words, to emphasize its importance. Designers seem to respond to the word *control* by thinking of added protective equipment; the word does not trigger thoughts of avoidance. Table 10.3 shows an overall approach incorporating this change. Mansfield et al.[25] describe a similar system. See also Section 1.3.

The first stage in the development of a design is the conceptual one in which we decide which product to make, by what route, and where the plant will be located. Some companies call it the *business analysis stage* or *capital analysis stage*, and they may use the term *conceptual stage* to describe what I have called the *flowsheet stage*.

Table 10.3 An Overall Approach to Hazard Management

1.	*Identify* all hazards and their consequences (many will be obvious).
2.	*Assess* all hazards and their consequences; that is, compare them with targets. At this stage, only a very rough assessment is necessary.

Apply Inherent-Safety and User-Friendly Principles

3.	*Avoid* the hazards (or reduce their severity) in one of the following ways:
	Intensify: use less hazardous material (see Chapter 3).
	Substitute: use safer materials or processes (see Chapter 4).
	Attenuate: use hazardous materials in a less hazardous form (see Chapter 7).
	Limit the effects of hazards by equipment design or change of conditions, not by adding protective equipment (see Chapter 6).
4.	*Simplify*: reduce the probability that a hazard will occur by providing less equipment that can fail and fewer opportunities for human error (see Chapters 7 and 8).
5.	*Assess* all hazards and their consequences more thoroughly (see Note 1). Carry on with the following steps unless or until the risks are tolerable.

Apply Additional Safety Principles

As we go down the list, reliability and robustness decrease; that is, the likelihood that the safety features will fail to operate or no longer be present increases.	
6.	*Segregate* people and hazards.
7.	Add *passive safeguards* (see Note 2 and Section 9.11).
8.	Add *active safeguards* (see Note 3).
9.	Apply *procedural safeguards* (see Note 4).

(continued)

Table 10.3 An Overall Approach to Hazard Management (Continued)

Note 1: In the United Kingdom, the regulatory authority—the Health and Safety Executive—has proposed two levels of risk. The upper level is just tolerable and should be exceeded only in exceptional circumstances. Below the lower level, the risk is broadly acceptable. In between, the risk should be reduced if it is reasonably practicable to do so by using the methods of cost-benefit analysis. If the risk is near the upper level, the pressure to reduce it is high and falls as the lower level is approached (Figure 10.4).[128] The HSE goes further and suggests that the upper level should be a risk of death of 10^{-3} per person per year for employees and 10^{-4} for members of the public. For nuclear power generation, the HSE asks for an upper level, for the public, of 10^{-5} per person per year to take account of the fact that the public is particularly concerned about this risk. The lower level proposed is 10^{-6} per person per year for both employees and the public, but the risk should still be reduced below this level when it is reasonably practicable to do so—particularly when accidents could kill many people. It is not essential to quantify the risk to life for every plant, and in any case, this is often impossible at the start of the design process. In many cases, a qualitative estimate, repeated as the design proceeds, is adequate. For major-hazard plants, a quantitative demonstration that the risk is low will make it easier to obtain acceptance by the public and the authorities.

Note 2: Passive safeguards do not need to be switched on by someone or something that might fail. For example, fire insulation is passive, and water spray is active (Section 9.11). Passive safeguards should be inspected regularly.

Note 3: If active safeguards are used, they should be tested regularly.

Note 4: Procedural safeguards are usually the last resort, although they are often necessary. "Prudent designers never use warning, training and personal-protection equipment as substitutes for safe design, guards or safety devices."[24] However, we cannot design our way out of every problem, and simple procedures may be better than cheap or complex active safeguards. All passive and active systems depend, to some extent, on procedures. Procedural safeguards should be audited regularly.

The questions that follow may help design teams at this stage to think of ways of increasing the inherent safety of the design. The questions can be used by individuals as an aide-mémoire, but more will come out of them if they are considered by the design team at a Hazop-style meeting with an independent chairman who will not take *no* or *none* for an answer. If a question is, "Can we … ?" and someone says no, the chairman should ask, "If you had to, how would you?" Even at this early stage, a design team should be involved. We should not let the research chemists choose the process without input from the process, engineering, and control system designers and those who will operate the finished plant (see Section 4.2.6).

Questions should be answered in a brainstorming atmosphere. People should be encouraged to make crazy and impracticable suggestions, if they come to mind, because they may stimulate other people to think of more practicable ones. Let your mind go free.

Remember that if it costs $1 to fix a problem at the conceptual stage, it will cost

about $10 at the flowsheet stage
about $100 at the line diagram stage
about $1000 after the plant is built
over $10,000 to clean up after an accident

1. Is the **product** hazardous? If so, what alternative products have been evaluated? (Cost, convenience, and effectiveness should be considered as well as hazards.)
2. Are the **raw materials, intermediates, solvents, catalysts**, and **other auxiliary materials** hazardous?
3. What **side reactions** are liable to occur and with what results? Will the products be hazardous? Will the raw materials, products, or intermediates react with the air or decompose or polymerize spontaneously?
4. Can **poor mixing** or **distribution** cause undesired reactions or local over- (or under-) heating, and with what results?
5. Will the **impurities** likely to be present in the raw materials have any undesired effects on the reaction, products, or materials of construction? (Section 4.1.5.)

In all these cases, what alternatives are available, and have their cost, convenience, and effectiveness been evaluated?

6. Is the **location** chosen near concentrations of people or other plants? If so, what will be the effects on them? What other locations have been considered? Have they been evaluated? Is there room for firebreaks between the sections of the plant or between units?

The questions listed in the next section should also be asked at the conceptual stage, though a final decision need not be taken until later.

10.2.4 Some Questions to Ask before the Flowsheet Stage

The following questions should be asked in addition to the normal Hazop questions:

Materials: List the inventory of each material in each section of the plant and in storage. As well as process materials, include solvents, catalysts, heat-transfer fluids, and other auxiliary materials. List the following separately:
1. Flashing flammable liquids (that is, flammable liquids under pressure above their atmospheric pressure boiling points)

2. Flashing toxic liquids
3. Refrigerated flammable liquids
4. Refrigerated toxic liquids
5. Unstable materials
6. Flammable gases
7. Toxic gases
8. Flammable liquids below their boiling points
9. Toxic liquids below their boiling points
10. Explosive dusts
11. Corrosive liquids

To help us decide which inventories are so large that we should try to remove or reduce them it may be useful to remember that

a. An explosion in the open air (unconfined vapor-cloud explosion) is unlikely if less than 5 tonnes of vapor and spray are present.
b. Although the vapor produced by a leak of a flashing liquid can be calculated by heat balance, much of the liquid remaining forms a mist or spray that is just as flammable as the vapor and just as toxic. A common general rule is to double the theoretical amount of vapor to allow for this mist and spray.
c. Leaks of corrosive liquids are hazardous only to people who are nearby, whereas leaks of toxic gases and fires and explosions can affect people some distance away.

Reaction

1. If the inventory in the reactor and associated equipment (heaters, coolers, recycle streams, etc.) is large and hazardous, can we change the type of reactor from a pot to a tube, from liquid phase to vapor phase, from externally cooled to internally cooled, or from batch to semibatch or continuous? (See Section 3.2.)
2. Can we reduce the inventory by increasing temperature or pressure, by changing catalyst or catalyst concentration, by better mixing, or by increasing conversion so that there is less recycle? What are the constraints that prevent such action? (See Section 3.2.)
3. Alternatively, can we make the inventory less hazardous by lowering the temperature below the boiling point or by dilution with a solvent? (See Chapter 5.)
4. If the reaction is liable to run away, can we reduce the chance of a runaway by using different vessels for different stages; by changing the order of operations; by changing the temperature, concentration, and so forth; or by limiting the level of the energy available? (See Section 6.2.)
5. If a hazardous solvent is used, can we use a different solvent instead, or no solvent? (See Section 4.1.)

6. Can changes in the reaction section simplify separation or reduce the inventory in the separation section?

Separation
1. If changes in the reaction section cannot reduce the inventory in the separation section, can we reduce the inventory by using an internal dephlegmator instead of an external condenser, by using an internal calandria instead of an external reboiler, by using low-inventory plates or packing, by fitting a narrow base, by using a condenser with a boot instead of a reflux drum, or by combining two columns into one? (See Section 3.3.1.)
2. Can we reduce the inventory of hazardous materials by using a low-inventory distillation method (such as Higee or a thin-film evaporator) or another method of separation instead of distillation? (See Section 3.3.)

Heat transfer (See Sections 3.4 and 4.1.)
1. Can a low-inventory heat exchanger be used?
2. If a shell-and-tube exchanger is used, is the more hazardous fluid in the tubes?
3. Can the inventory be reduced by higher flow rates, extended surfaces, or larger temperature differences?
4. If a heat-transfer medium is used, can we
   a. Use water or other nonflammable medium, a vapor-phase medium, or a medium below its boiling point?
   b. Reduce the inventory by minimizing buffer stock or the inventory in heat exchangers and other devices?
   c. Keep the buffer stock cold?
5. If a refrigerant is used, can we use a nonflammable one or minimize the inventory as above?

Storage (See Section 3.6.): Raw material, intermediates, products, and auxiliary materials such as solvents should be considered in turn.
1. What determines the amount of storage required (shutdowns, planned and unplanned; fluctuations in supply or demand; transport problems)? How else can we overcome these problems?
2. The following should be considered:
   a. Increasing plant capacity
   b. Increasing plant availability
   c. Manufacturing the raw material or using the product on site
   d. Combining reaction steps
   e. Storing at a lower temperature or in a different physical or chemical form

Other questions

1. Can pumps be replaced by gravity or convective flow (to reduce pressure and opportunities for leaks)? (But see Section 8.1.5.)
2. Can flow ratio be controlled by an injector? (See Section 3.2.3.)
3. Can antagonistic materials be segregated? (See Section 6.2.1.)
4. Are any of the materials discharged into vent or drain lines liable to react there?
5. Are any of the liquids discharged into vent or drain lines likely to freeze or polymerize?
6. Can water collect in pockets where it can be heated above its boiling point (or can oil collect in a plant containing an aqueous phase)?
7. How susceptible is the process to human error? Can we reduce the susceptibility by design changes rather than by adding protective equipment?

These questions should be asked during the development of the flowsheet and should be asked again, in addition to the usual Hazop questions, during the Hazop of the flowsheet recommended in Section 10.2.2. Asking these questions at an early stage may save time and expense at the flowsheet stage.

10.2.5 An Example of a Flowsheet Hazop

The following are a few of the 66 changes or queries that were made during the Hazop of a simple flowsheet. The plant consisted of a batch reactor followed by a stripping unit in which an excess of one reactant was removed under vacuum. The study took 9 hours spread over three meetings. Many of the points would have been thought of without a Hazop, but many would have been missed or not thought of until it was too late to make changes.

1. If the reactor is overfilled, it overflows into a catch pot that is fitted with a high-level alarm. Why not fit the alarm on the reactor and dispense with the catch pot?
2. What would it cost to design the reactor to withstand the vacuum produced in the stripper, thus avoiding the need for a vacuum-relief valve that would allow air to be sucked into the reactor with the formation of a flammable mixture?
3. Why do we need two filters? Will a change in type allow us to manage with one?
4. Can we choose a type of bottoms pump for the stripper that will allow us to lower the stripper and reduce the cost of the structure?
5. Can heat exchangers be designed to withstand the maximum pressure possible under all conditions except fire, so that only small fire-relief valves are needed?
6. The material proposed for removing color may be toxic. What other materials are available?

10.2.6 Some Questions That Should Be Asked during Detailed Design

The following questions should be asked during the Hazop of the line diagrams. A negative answer to any question should be justified.

1. Will spiral-wound gaskets be used? (See Sections 6.1 and 9.4.)
2. Will rising spindle valves be used? (See Section 9.3.)
3. If ball valves or cocks are used, can the handles be removed and replaced wrongly? (See Section 9.3.)
4. Will figure-eight plates be fitted at joints that have to be blinded regularly for maintenance or to prevent contamination? (See Section 9.3.)
5. Will expansion loops be used instead of bellows? (See Section 9.4.)
6. If flexible connections are necessary (for example, for filling or emptying tank trucks or cars), will articulated arms be used instead of hoses? (See Section 9.4.)
7. If hoses (or articulated arms) are used, is there a means of venting them before they are disconnected?
8. Will sample points be designed so that samples of hazardous chemicals can be taken without risk of injury to the sampler? Is each sample point essential? (See Section 6.3.2.)
9. If quick-release couplings are fitted to pressure vessels, will they be of a type that gives the operator a second chance if pressure is present? (See Section 9.4.)
10. Have the number of different sorts of gaskets, nuts, bolts, and so on, to be stocked been reduced to the minimum even though more expensive items than necessary are used in some locations?
11. Can any equipment (in the line being Hazoped) be interchanged, installed the wrong way round, or wrongly installed in any other way? (See Sections 9.2 and 9.3.)
12. Will any glass, other brittle material, or plastic be used as material of construction? How can this be avoided?
13. Are pipe sizes restricted as discussed in Section 6.1?
14. Control: See Section 9.5.

As already stated, there is need for published examples of the answers to questions such as these and the way designs have been modified as a result. One example is worth a thousand words.

Mohindra and Stickles[54] have listed many of the deviations that should be considered during detailed design and during a Hazop study, as well as the inherently safer, passive, active, and procedural ways of dealing with them. For example, loss of containment from a vessel through an open drain connection can be prevented by eliminating the connection or not storing the material. If that is not possible, the size of a loss can be reduced by making the connection as small as possible (passive); the probability of a loss can be reduced by fitting self-closing or excess flow valves (active) or by blinding the valve when not in use (procedural). There are other examples in Section 9.11.

10.2.7 The Way Ahead: Conclusions

Despite what has been said in Sections 10.2.3 through 10.2.6, designs will benefit more from tools that help us get the design right the first time than from those, such as Hazop, that find the faults in finished designs. The most useful tools are those that act as advisers rather than monitors or auditors; examples in the former category include the INSET toolkit[26] shown in Table 10.1, as well as some of the methods described in Section 10.4. Nevertheless, the auditing tools are still essential. Being human, we will never get everything right the first time, and it is increasingly necessary to convince the authorities that we have a safe plant and system of working.[55]

10.3 The Influence of the Law

In many countries, the oil and chemical industries are regulated or controlled, to varying extents, by government authorities. In the United Kingdom, for example, before giving planning permission for a hazardous plant, the Planning Authority has to seek the advice of the Health and Safety Executive (HSE), and the advice is usually followed. The HSE is thus in a very powerful position. Should the HSE use this position to insist that, whenever possible, new designs are friendly?

At one time, HSE used to say that, as long as plants are safe, it does not matter how that safety is achieved. Extrinsic safety, obtained by adding protective equipment, is acceptable, provided the equipment is properly designed and maintained. By 1988, the HSE was taking a different view. According to the then-head of the technology division, "I feel the time is ripe for the HSE to communicate and cooperate more widely with chemical plant design offices and I would be interested in feedback. The HSE may have been slow to exploit the full benefits of inherent safety, but I can assure you that … we are now fully alert to its potential and will be following up inherent safety as it affects the design and operation of chemical plant."[56] However, nothing much happened until 1993 when the HSE appointed consultants to advise on what might be done. A first report was issued in 1994,[22] and work has continued.[57] A similar study has been undertaken on the offshore oil industry.[23]

Recent developments in the United Kingdom and the rest of Europe have been summarized by Edwards,[4] who gives a representative list of both Health and Safety Executive (U.K.) and European Union publications relating to the use of inherently safer technologies. He comments that although inherent safety appears time and again as a recommended approach in these guidance documents for safety and environmental regulations, the reference is often implicit and the actual term *inherent safety* may not be used. (A similar observation can be made for reports of investigation of major process incidents; see Chapter 14.) Edwards further comments that perhaps enforcement, not a lack of regulatory requirements, is a key barrier to wide adoption of the principles of inherent safety.[4]

As of the mid-1990s in the United States, the Environmental Protection Agency (EPA) was "pursuing the development of guidance that will help categorize examples of inherently safe [*sic*] chemistry."[58] At that time, the EPA had also started a "Benign by Design Chemistry" initiative seeking to promote pollution prevention in the earliest design stages.[59]

Chapter 10 in Reference 53 describes the current situation in the United States with respect to the development of inherent-safety regulations and the issues resulting from their implementation. Notable in this regard have been the efforts undertaken in Contra Costa County, California (CA)[60] and the state of New Jersey (NJ) to mandate consideration of inherently safer technologies in their chemical security (NJ) and process safety (CA and NJ) programs.[53]

In a world living with the implications of September 11, 2001, security issues have indeed heightened the awareness of the potential benefits of inherent safety,[61–64] especially with respect to intensification and substitution. A recent press release[65] by Edward R. Hamberger, President & CEO of the Association of American Railroads contained these statements:

> It is time for the nation's big chemical companies to stop making the dangerous chemicals that can be replaced by safer substitutes or new technologies currently in the marketplace....

> The most desirable solution to preventing chemical releases is to reduce or eliminate the hazard where possible, not control it. [Quoting the National Research Council, part of the National Academy of Sciences]

It is perhaps telling that this release was issued at the time of a hearing on the Chemical Facility Anti-Terrorism Act of 2008, held by the U.S. House Homeland Security Committee.

In the European Union, including the United Kingdom, under the Seveso Directive, companies with more than defined quantities of hazardous chemicals in their plants have to demonstrate that they can handle them safely. This encourages companies to reduce their stocks. Also, as mentioned in Section 3.6, new Canadian environmental regulations have legislated the requirement for an emergency response plan at facilities storing or using any of a number of listed substances at or above specified threshold quantities.

The attitude of legislators and regulators reflects public opinion. The public, not without reason, doubts the ability of the chemical industry to keep hazardous materials under control. Inherently safer designs should reassure them. Otway and Haastrup[66] doubt this. They write, "Reducing risks by changes in design philosophies and physical plant [is] unlikely to have a large enough effect on public attitudes to create a climate for the acceptance of technologies previously rejected." However, even the public should be able to realize that what you don't have, can't leak. Otway and Haastrup's reservations apply more to the nuclear industry, where the inherently safer designs now under development inevitably still produce radioactive by-products.

One would hope that a global initiative such as Responsible Care®—a clear industry response to Bhopal—with its emphasis on life-cycle stewardship, would help assure the public of a new ethic on the part of the process industries.[3] Unfortunately, one would also have to hope that it will not be the popular media informing the public in matters of process safety. A recent episode of an American television program showed a security breach at a chemical facility resulting in the release of methyl isocyanate (MIC, the Bhopal chemical) directly into an occupied control room. While convenient in terms of establishing a scenario for the hero to "save the day," such portrayals do little to enhance public confidence with respect to the handling of hazardous materials.

If regulators do try to encourage the design of inherently safer and friendlier plants, they will not succeed in doing so only by passing new regulations. In the United Kingdom, there is well-established tradition that companies are required to do only what is reasonably practicable, weighing in the balance the size of a risk against the cost of removing it (see Note 1 in Table 10.3 and Figure 10.4). Under the 1974 Health and Safety at Work Act there is a general obligation to provide a safe plant and method of work and adequate instruction, training, and supervision so far as is reasonably practicable. If there is an inherently safer process than the one proposed, then the HSE can insist that it be used. The company can argue that it is not reasonably practicable, but will have to demonstrate this. In the environment the position is similar: Companies are expected to use the best available technology not entailing excessive cost (BATNEEC).

The flexibility required to encourage inherently safer designs does not come so easily in the United States, where there is a long tradition of governing by regulation and of preferring sticks to carrots. This encourages companies to appoint managers whose strengths are knowing the regulations and how to comply with them rather than those who look for imaginative solutions. However, there are signs of change. A joint study by EPA and DuPont recognized that the best environmental solutions are produced when governments set realistic goals and priorities and industry agrees to meet them in the most efficient way.[67] A potentially significant step has also recently been taken by the American Institute of Chemical Engineers with the publication of a framework for risk-based process-safety management.[68]

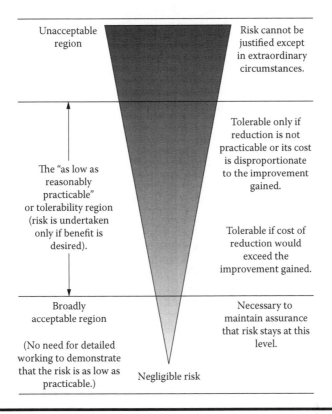

Figure 10.4   An interpretation of "as low as reasonably practicable." (Reproduced by permission of the Health and Safety Executive.)

10.4 The Measurement of Friendliness

If we modify a design so that the plant contains only 10 tonnes of liquefied petroleum gas (LPG) instead of 100 tonnes, it is obvious that we have a safer plant and do not need a special technique to tell us so. But suppose we can replace 100 tonnes of LPG by 10 tonnes of chlorine. Will we have a safer plant? This is not a theoretical question, as is shown by the examples in Section 4.2. Several ways of answering the question have been suggested.

10.4.1 The Mortality Index

A quick method is based on Marshall's mortality index:[69] the average number of people killed by the explosion of a tonne of liquefied flammable gas (LFG) or the release of a tonne of chlorine or ammonia. He shows that the historical record is as follows:

Chlorine	0.30
Ammonia	0.02
Liquefied flammable gas	0.60
Unstable substances	1.50

From these figures, we could deduce that 10 tonnes of chlorine are 20 times safer than 100 tonnes of LFG.

However, it is not quite as simple as this. The probability of a leak may not be the same at the two plants. One may contain more sources of leak, such as pumps or drain points, or leaks may be more likely because operating conditions are more extreme. And not all leaks of LPG ignite. Therefore, more sophisticated indices have been devised.

10.4.2 The Dow, Mond, Loughborough, and Other Indices

The best known of these sophisticated measures of unfriendliness are the Dow Fire and Explosion Index (F&EI),[70] now supplemented by the addition of a Chemical Exposure Index (CEI), and the Mond Index,[71,72] a development of the Dow indices. Numbers are assigned to the properties and inventories of the various materials in the plant, and the operating conditions and these numbers are then combined to give a unified index of woe[73]—a single figure that measures the hazardousness of the whole plant or section of a plant. We can then use this number to compare different designs or to decide how much protective equipment to install. The numbers have no physical meaning, but they do give a rough indication of relative hazard, just as the star rating of a hotel gives a rough indication of its standard.

Section 3.2.9 describes how an existing plant was made inherently safer. The decision to modify it was made when its Dow Chemical Exposure Index was found to be 60% higher than that of any other of the company's European plants.[74]

Changes to the design of latex plants have reduced the Dow Fire and Explosion Index from the severe range to the intermediate. The modifications include a change from batch to semibatch operations and also changes to layout.[75] (A better layout is more user-friendly, but not everyone would call it inherently safer; we make a plant safer by adding distance between equipment.)

A recent effort by Etowa et al.[76] has attempted to quantify the inherent-safety aspects of both Dow indices: the F&EI and CEI. Using the Bhopal incident as a case study, they found that as operating pressure was lowered (attenuation) or inventory reduced (intensification), the F&EI decreased as expected. The hazard reduction was modest, however, because of the high material hazard of methyl isocyanate. This clearly indicates the need for substitution of alternative reaction chemistry.

The relationship between the Chemical Exposure Index and the diameter of the opening for vapor release was also investigated.[76] With an assumption of a standard 5-cm (2 in.) pipe, the CEI was calculated as 750. This value dropped to 540 (a significant decrease) for a pipe diameter of 3 cm. These findings, although perhaps obvious, provide a quantitative measure of the importance of optimum piping designs for processes handling hazardous materials.

In a later study, Suardin et al.[77] integrated the Dow Fire and Explosion Index in their procedure for conceptual process design. An interesting feature of their work was the expression of the F&EI as a function of material inventory and operating pressure, with subsequent use of the index as an optimization constraint for design choice selection.

Edwards and coworkers at Loughborough University, U.K., have described a prototype index of inherent safety (PIIS) based on seven factors: inventory, flammability, width of explosive range, toxicity (threshold limit value), reaction temperature, reaction pressure, and reaction yield. They have used the index to compare six methods for the manufacture of methyl methacrylate (MMA) and have compared the results with assessments made by eight experts. The agreement was good. They have also compared the index values with the costs of making MMA by the six routes to see whether the cost of the inherently safer processes is cheaper (Section 2.4). The agreement was poor; however, as they point out, inherently safer plants are cheaper because they require less added protective equipment, but this was not taken into account in the cost estimates. When alternative processes are compared, the total capital cost is usually estimated by factorial cost accounting, that is, by multiplying the cost of main plant items by a factor, and no allowance is made for different levels of protective equipment.[78,79]

Khan et al.[80] conducted a comparative evaluation of the above indices (Dow, Mond, and Loughborough), along with the Safety Weighted Hazard Index (SWeHI),[81] for the case of a direct oxidation-based ethylene oxide plant. Inherent-safety principles (termed *guidewords*) investigated were: minimization (intensification), substitution, attenuation, simplification, and limitation of effects. No single indexing procedure was found to adequately capture all of these guidewords, although each index did provide some measure of inherent-safety assessment.

Hendershot devised an index at the developmental stage, based on five other indices: the Dow Fire and Explosion Index, the Dow Chemical Exposure Index (index of acute toxicity), an index of chronic toxicity, an index of environmental risk, and an index of transportation risk.[82]

Hawksley and Preston[83] suggest the use of a semiquantitative "target" diagram, such as the one shown in Figure 10.5. The various factors listed around the outer part of the diagram are estimated in any appropriate way and converted into a scale of 1–5. The profiles produced make it easy to compare different processes or plant designs.

Khan and Amyotte[3] have reviewed other inherent-safety indices and assessment approaches that were either completed or in development in 2002. Table 10.4 gives further examples from 2002 to 2009. The references are organized by country of origin of the authors. This is not to suggest that inherent-safety research should be carried out with national borders in mind (it should not), but rather to indicate the global scope of the work in this area.

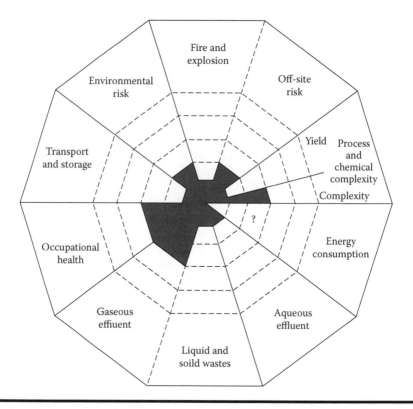

Figure 10.5 Inherent SHE target diagram. (Copyright 1996 by the American Institute of Chemical Engineers; reproduced by permission of Center for Chemical Process Safety.)

Building on the pioneering work of the Loughborough[78,79] and INSIDE[26] teams, the development of inherently safer design metrics has been the subject of considerable interest and activity over the past decade. In addition to the international nature of these activities, the following observations can be made from the references cited in Table 10.4:

1. Inherent-safety researchers can be quite creative in naming their various indices and methodologies—some especially so!
2. Many of the methods deal specifically with the early concept and route-selection stages of the design process.
3. Some approaches use sophisticated mathematical and problem-solving techniques (e.g., fuzzy logic and statistical analysis).
4. There is an increasing tendency to link inherent safety with environmental and health issues in an effort to achieve an integrated approach (see also Section 1.1).

Table 10.4   Examples of Development of Inherent-Safety Assessment Methodologies and Elucidation of Assessment Considerations (2002–2009)

Country	Reference	Contribution
Canada	84	Integrated Inherent Safety Index (I2SI)
	85	Further development of I2SI to include cost model
Denmark and Portugal	86	Method for identifying retrofit design alternatives of chemical processes; uses Inherent Safety Index (ISI) developed by researchers in Finland
Finland	87	Discussion of implementation of inherent safety throughout process life-cycle phases; use of Inherent Safety Index (ISI) developed earlier by these researchers
	88	Comparative evaluation of three inherent-safety indices with expert judgment at process-concept phase
Finland and United Kingdom	89	Process Route Healthiness Index (PRHI) for quantification of health hazards arising from alternative chemical process routes; application is in early stages of chemical plant design
India and United Kingdom	90	Graphical approach for evaluating inherent safety based on earlier developed Loughborough Prototype Index of Inherent Safety (PIIS)
Italy	91	Procedure for assessment of hazards arising from decomposition products formed due to loss of chemical process control; applicable to consideration of substitution principle
	92	Procedure and indices for evaluating inherent safety at preliminary process-flow diagram (PFD) stage for hydrogen storage options
	93	Further development of PFD method[92] by use of quantitative Key Performance Indicators (KPIs) to remove subjective judgment

(continued)

Table 10.4   Examples of Development of Inherent-Safety Assessment Methodologies and Elucidation of Assessment Considerations (2002–2009) (Continued)

Country	Reference	Contribution
	94	Further development of procedure for decomposition product analysis[91] to account for acute and long-term harm to human health, ecosystem damage, and environmental media contamination
Malaysia	31	Integrated Risk Estimation Tool (iRET) for inherent-safety application at preliminary design stage; iRET links the design-simulation software HYSYS with an explosion-consequence model
	32	Further development of iRET[31] to incorporate a quantitative Inherent Safety Level (ISL), thus enabling integration of design simulation software with an Inherent Safety Index Module (ISIM); application is again at the preliminary design stage
	95	Evolution of ISIM[32] to a Process Route Index (PRI) for comparison and ranking of different routes to manufacture the same product based on hazard potential of routes
Russia	96	Use of nonlinear optimization method to select inherently safer operational parameters for given configuration of reactor equipment and materials; primary concern is cooling failure
Switzerland	97	SREST (Substance, Reactivity, Equipment and Safety Technology) layer assessment method for Environment, Health, and Safety (EHS) aspects in early phases of chemical process design
	98	Comparative evaluation of various methods for assessing EHS hazards in early phases of chemical process design; see Section 1.1 (Reference 4 in Chapter 1)

(continued)

Table 10.4 Examples of Development of Inherent-Safety Assessment Methodologies and Elucidation of Assessment Considerations (2002–2009) (Continued)

Country	Reference	Contribution
Singapore	99	Indexing procedure for inherent-safety analysis at process route selection stage
	100	Indexing procedure for inherent-safety analysis at process flowsheet development stage; discussion of *iSafe*, an expert system for automating procedures developed in References 99 and 100
	101	Methodology for integrated inherent-safety and waste-minimization analysis during process design
	102	Inherent Benignness Indicator (IBI), a statistical analysis-based method for comparing alternative chemical process routes
Singapore and Finland	103	Application of TRIZ methodology for creative problem solving to design inherently safer chemical processes
United States	104	Use of game theory to achieve inherently safer operation of chemical reactors
	105	Fuzzy logic-based index for evaluation of inherently safer process alternatives with the aim of linking to process simulation
	106	Linking of inherent safety and environmental concerns with optimization of process scheduling; see Section 1.1 (Reference 5 in Chapter 1)

5. There have been attempts to incorporate inherent-safety assessment into process-design simulators.
6. Some of the indices have been in existence long enough for comparative evaluations to be made among them.

Figure 10.6 shows the framework for the Integrated Inherent Safety Index (I2SI) developed by Khan and Amyotte.[84,85] I2SI is a guideword-based approach to inherent-safety assessment with subindices to account for hazard potential,

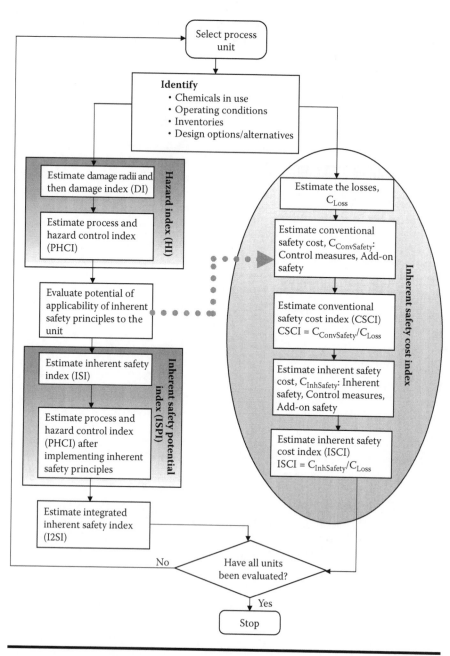

Figure 10.6 Conceptual framework of Integrated Inherent Safety Index (I2SI). (From Khan, F. I. and P. R. Amyotte, *Journal of Loss Prevention in the Process Industries.* 18, no. 4–6 (2005). With permission.)

inherent-safety potential, add-on control requirements, and economic factors. Like many of the other approaches listed in Table 10.4, it is intended for use in a comparative, not absolute, mode for the evaluation of process alternatives.

As mentioned earlier in Section 10.1.6, Edwards[4] has remarked on the fact that inherent-safety assessment tools are now available, but they appear to have seen limited use by industry. This may be caused in part by the subjective judgment required by some of these tools and also by their attendant complexity. Industrial practitioners do indeed want simple-to-use hazard- and risk-assessment tools, but they do not want these tools to be simplistic and lacking in scientific and engineering rigor. This does pose somewhat of a dilemma for researchers, but it is encouraging to see the continuing efforts in this regard.

As a final note on Table 10.4, one can see by the reference citations that virtually all of the work has been conducted in universities and government research institutions. This is merely an observation, not a criticism. Similarly, it would seem that most (not all) of the papers published and presented by industry in this field are examples of successful inherent-safety designs or case studies of unsuccessful designs. It would be interesting to see more industrial design methods and inherent-safety index usage reported in the public domain.

10.4.3 Estimation of Effects

A third method of measuring friendliness is to estimate the probability that certain effects will occur, for example, the death of a person living near the plant. The calculations are complex because we need to estimate the following:

1. The frequency of equipment failure
2. The amount of hazardous material that leaks out
3. The fraction that forms vapor and spray
4. How far the material will spread, on average, in each direction
5. If the material is toxic, the direct effect on people; if it is flammable, the effect on people of (a) the heat radiation if it ignites and (b) the overpressure if it explodes

Several computer packages (see, for example, Reference 107) are now available for carrying out the calculations. Their absolute accuracy may not be high (especially in some of the earlier versions), but their relative accuracy is greater, and alternative designs can be compared. Their advantage, compared with the Dow and Mond Indices, is that the results have a physical meaning.

Updated listings of methods and software for probability analysis and consequence analysis can be found in recent papers by Tixier et al.[108] and Reniers et al.[109] These authors have reviewed 62 and 30 techniques, respectively, many of which are well suited to quantitative risk assessment (QRA).

Johnson et al.[110] suggest calculating the amount of stored energy; this includes chemical energy (assuming complete reaction or combustion) and pressure and thermal energy above ambient levels. This method does not take into account the probability that the energy will be released or its rate of release. As an alternative, they suggest calculation of the area affected by a fire, explosion, or toxic release possibly multiplied by the average number of people in the area.

10.4.4 Insurance Indices

Several insurance companies have developed indices that give a measure of financial risk. Two well-known indices are IFAL (instantaneous fractional annual loss)[111] and the IC Insurance Index.[112] The IFAL is a measure of the average rate of loss, whereas the IC Insurance Index is more like the Dow and Mond indices, for it is based on arbitrary factors that measure the quality of the hardware, the software, and the firefighting facilities. It should also be noted that the Dow Fire and Explosion Index procedure enables the estimation of financial parameters such as actual maximum probable property damage (actual MPPD) and business interruption (BI), measured in dollars.[70]

10.5 Looking for New Knowledge

If companies carry out research on inherently safer design, they have to decide whether it should be project- or equipment-oriented. That is, should they look for better methods of making product A; or better methods of reaction, distillation, heat transfer; and so on?

It is usually easier to get money for project research than for research on new designs of equipment because those who authorize expenditure may feel that the new equipment, even if successful, may never be used or never used sufficiently to recover the development costs (see Section 3.3.2 on Higee). On the other hand, individual projects may not be able to afford research on new equipment; research may be possible only if funded by the company as a whole or by a group of companies.

In deciding whether to fund research on inherently safer design, companies should remember that, as stated in Section 2.4, it is not an isolated but desirable improvement to increase safety, but part of the total package the chemical industry needs in the years to come. Inherently safer plants are simpler, cheaper, and more energy efficient.

Whether we carry out research on inherently safer and friendlier processes, we should remember that lack of knowledge does not prevent us from designing them. Even if we do no research on them at all, we can still make use of the knowledge already available, much of which is not fully used.

This applies to safety and loss prevention generally, not just to friendlier designs. We cannot make gold from lead because there is no known method. But if we want to make our plants safer, we are not hindered by lack of knowledge. Whether we succeed depends on our energy, drive, and commitment. One quickly becomes rather unpopular at meetings called to discuss some aspect of research on safety by asking, "How are you going to persuade people to use the new knowledge you are going to give them when they are not using the knowledge that is available already?" Yet the question is entirely valid and needs to be asked.

One source of knowledge that is not exploited to the full is accident investigation. Most accident reports deal only with the immediate causes of the accident, the triggering events, but not with ways of *avoiding* the hazard.[113] Most systems of accident investigation are designed to find out why control of hazards was lost; they do not ask why the hazard was tolerated and whether it could have been avoided.[114] If there has been a fire, the report usually discusses the reason for the leak of flammable material and the source of ignition, but it does not ask whether it was essential to have so much flammable material present, if a safer material could be used instead, or whether the flammable material could be used in a less hazardous form. Such changes can usually be made only in the storage areas of existing plants, but they should nevertheless be noted and thus be available for the design of the next plant.

The following questions may help accident investigators or investigation teams think of some of the less obvious ways of preventing accidents. It will be more effective if the questions are answered by a team rather than a person working alone. The team leader should be reluctant to take *nothing* or *we can't* as an answer. As with the questions listed earlier, the leader should reply, "If you had to, how would you?" Further discussion of the interaction between inherent safety and accident investigation is given in Chapter 11, dealing with process-safety management.

What is the purpose of the operation involved in the accident?
Why do we do this?
What could we do instead?
How else could we do it?
Who else could do it?
When else could we do it?
Where else could we do it?
What equipment failed?
How can we prevent failure or make it less likely?
How can we detect failure or approaching failure?
How can we control failure (i.e., minimize consequences)?
What does this equipment do?
What other equipment could we use instead?
What could we do instead?
What material leaked (exploded, decomposed, etc.)?
How can we prevent a leak (explosion, decomposition, etc.)?
How can we detect a leak or approaching leak (etc.)?

What does this material do?

What material could we use instead?

What safer form could we use the original material in?

What could we do instead?

Which people could have performed better? (Consider people who might supervise, train, inspect, check, or design better than they did as well as people who might construct, operate, and maintain better than they did.)

What could they have done better?

How can we help them to perform better? (Consider training, instructions, inspections, audits, etc., as well as changes to design.)

What is the purpose of the operation involved in the accident?

Why do we perform this operation?

What could we do instead?

How else could we do it?

Who else could do it?

When else could we do it?

10.6 The Final Decision

What should we do if an inherently safer or friendlier design is possible but is significantly more expensive than a less friendly design made safe by adding protective equipment?

If the traditional or extrinsically safe design really is cheaper, then we should use it. Such plants can be safe if enough protective equipment is added and it is tested and maintained correctly and regularly. However, before comparing the costs, consider the following:

1. Make sure we are comparing lifetime costs, not just capital costs. Dalzell[115] argues that we need to consider the contributions of inherent safety to plant longevity and reliability and the attendant reduction in both incident likelihood and personnel exposure.

2. Make sure that the cost estimates are realistic. Instruments cost twice what we think they do if the costs of testing and maintaining them are taken into account.

3. Remember that we will have to install more protective equipment than on the last plant, for standards are rising, and also that the operating team will probably install yet more during the life of the plant.

4. Make sure that the costs of protective equipment reflect the degree of hazard and are not simply estimated by applying a factor (Section 10.4.2).

5. Remember that, in the long-term, new technologies follow a learning curve and become cheaper. We will rarely innovate if we compare the first costs of a new design with those of an established design.

If the costs of the two designs are about the same, then we should build the friendlier plant because the intangibles are in its favor. If little or no hazardous material is present, we do not have to persuade the public and the authorities that it will not leak out and harm people or the environment.

10.7 Inherently Safer Design Offshore: A Case History

An account of the initially slow progress made in introducing inherently safer design offshore illustrates the problems described in earlier sections.

Following the destruction of the Piper Alpha offshore oil platform in the North Sea in 1988,[116,117] a new regulatory regime was introduced. It was hoped that the new goal-setting regulations would encourage the use of inherently safer designs, but by 1995 little had happened. The Offshore Division of HSE therefore asked AEA Technology to assess the extent to which such designs were used or could be used offshore. Their report[23] shows that, although many designers and safety professionals were aware of the concept, few of them had a clear understanding of its possible applications. They applied some of its principles, but not in a systematic or methodical way. They did not realize that such designs can be both cheaper *and* safer because there is less need for added protective equipment and, if we can intensify, the plant will be smaller. The report concluded that there was a need to raise awareness and to develop techniques and tools that would make it easier for designers to develop inherently safer designs.

Recent reviews of the potential for application of inherent-safety concepts in off-shore oil and gas activities have been published[118] and presented[119] by Khan et al.

Intensification (reducing inventory) is particularly valuable offshore because space and weight are at a premium. Intensified designs of oil–water separators and heat exchangers are available, and equipment such as slug catchers, contactors, and separators could be intensified. On the other hand, designers have typically been reluctant to make use of new designs; their reasons for doing so may be understandable, but if this philosophy had been carried to extremes, iron horses would never have replaced horses.

More fundamentally, could we eliminate the need for offshore or manned platform,[120] by making greater use of multiphase metering and pumping and subsea installations,[121] or by horizontal drilling from onshore?

The principles of inherently safer design can be applied to structures as well as process equipment. For example, designers should, when possible, avoid the use of process materials that can attack steelwork, locate structures to avoid shipping lanes and weak foundations, choose designs that fail progressively rather than suddenly, and employ designs that are easy to inspect. Many of these steps are being taken already, of course, but what appears to be lacking in the design of both structures and processes is a systematic consideration of alternatives; as a result, opportunities will be missed.

Apart from lack of awareness and the consequent lack of a systematic approach, two other constraints were hindering the growth of better designs in the late 1990s. The first was the lack in the major offshore companies of a central resource to support and fund the development of ideas that could benefit the company as a whole but which no single project could afford to sponsor. The second constraint was the need for a change in the design process so that safety studies take place earlier in design. To overcome these constraints, it is not enough to convince designers of the need for change. We have to convince senior managers as well.

Some progress in offshore inherent-safety applications appears to have been made over the past decade or so. Chia et al.[122] describe the incorporation of "inherent-safety challenge" concepts as a design tool during front-end engineering design (FEED) on a major offshore gas project in Southeast Asia. This required the close collaboration of process-safety specialists with design and operations personnel.

Other workers have recently discussed the role of inherent safety in offshore design and operations. Kjellen[123] describes the Norwegian experience in preventing accidents through design: (a) a human-centered, inherently safer approach in which the focus is on workplace design to minimize operator error and (b) an energy-barrier approach. Sklet[124,125] also analyzes the use of barriers offshore, which can cover the full hierarchy of controls (Section 1.3), ranging from inherent to procedural. An increased emphasis on inherently safer design for avoidance and prevention of major offshore hazards, rather than reliance on control and mitigation measures, has also been adopted by some U.K. designers.[126]

Crawley[127] has argued that the holistic approach taken to integrate the assessment of safety, health, and environment (SHE) hazards throughout the life cycle of an offshore rig could be a helpful model for the onshore process industries. Provided that the offshore model recognizes the hierarchical role of inherently safer design, this would be a welcome development.

Appendix: Finding Examples of Inherently Safer Design

At the time of publication of the first edition of this book in the late 1990s, it was difficult to find papers on inherently safer design in databases because the term was not widely used as a keyword or indexing term. A computer search for "inherently safer design" produced few references (mainly papers by T. A. Kletz!). A search for papers indexed under "inherent safety" and also containing the word "design" produced numerous references to the nuclear industry. A search for "intensification" produced numerous references to image intensification and some Russian papers on intensification of heat transfer in the steel industry, but only one related to the intensification discussed in this book. A search for papers in which "intensification" and "safety" both appear reduced the number of references to four, but three still referred to image intensification. A search for "process intensification" yielded several Russian papers (mainly in the *Journal of Applied Chemistry in the USSR*) on heat transfer, absorption, milling, magnetic treatment, and compacting powdered mixtures, as well as references to some of Ramshaw's papers. This seemed to be the most promising indexing term at the time.

"Substitution" is, of course, widely used by chemists in a different sense from that discussed in papers on inherently safer design, but a search for papers including both "substitution" and "safety" did produce a few references to fuel substitution.

In 1998 (first edition of this book), I wrote that those who wished to keep up-to-date with the literature on inherently safer design would have to do so by perusing journals and exchanging information with colleagues.

Now, in 2009, the situation has changed for the better. There is sometimes no substitute for a hard-copy search in the library (if only to relieve the tedium of the keyboard), and it is of course always enjoyable to engage in collaboration and discussion with one's colleagues. But improvements in computer search engines, increased use of electronic reference formats, and wider adoption of "standard" inherent-safety keywords have all helped to facilitate information retrieval. Witness the substantial number of new references in this second edition, particularly in Chapter 3 (about 30% more references) and Chapter 10 (twice as many). And undoubtedly there are more examples of inherent-safety usage out there, even in the search sites mentioned in the preface to this second edition.

Most of the major process-safety journals have featured a special issue on inherent safety some time in the past few years. Most process-safety symposia routinely include a session on inherent safety or list inherent safety/inherently safer design as a conference topic in the call for papers.

It is not just enhanced document search and delivery methods, and publication and presentation venues, that have facilitated inherent-safety awareness over the past decade. Simply put, there are more people conducting research in the field, and perhaps just as important, more researchers primarily interested in process safety and inherent safety rather than other more traditional areas of chemical engineering. Additionally, there are more industrial practitioners presenting and publishing inherent-safety successes and approaches within their engineering projects.

This is not to claim "mission accomplished." There is no evidence to suggest that we should abandon the orthodontic approach of continuous steady pressure over the years (as mentioned in Section 10.2.1) to achieve the desired results. But perhaps there is some cause for cautious optimism—at least with respect to inherent-safety awareness and knowledge, if not application.

References

1. D. C. Hendershot, "Putting Inherent SHE on the Map in the USA," in *Conference on Inherent SHE: The Cost Effective Route to Improved Safety, Health and Environmental Performance* (London, U.K., 1997).

2. T. A. Kletz, "The Constraints on Inherently Safer Design and Other Innovations," *Process Safety Progress* 18, no. 1 (1999): 64–69.

3. F. I. Khan and P. R. Amyotte, "How to Make Inherent Safety Practice a Reality," *Canadian Journal of Chemical Engineering* 81, no. 1 (2003): 2–16.

4. D. W. Edwards, "Are We Too Risk-Averse for Inherent Safety? An Examination of Current Status and Barriers to Adoption," *Process Safety and Environmental Protection* 83, no. B2 (2005): 90–100.

5. D. C. Hendershot, "An Overview of Inherently Safer Design," *Process Safety Progress* 25, no. 2 (2006): 98–107.

6. M. S. Mannan and J. T. Baldwin, "Inherently Safer Is Inherently Cleaner: A Comprehensive Design Approach," *Chemical Process Safety Report* March (2000): 125–132.

7. A. Bisio, "Energy Efficiency: Further and Faster," *Chem. Eng. (U.K.)* 487 (1990): 27–28.

8. "Monitor," Reactions, *Atom* 394 (1989): 35.

9. P. G. Urben, review of "Learning from Accidents in Industry," *J. Loss Prev. Proc. Ind.* 2, no. 1 (1989): 55.

10. P. Worsthorne, "Tell It as It Is," *Sunday Telegraph*, 17 March 1991, p. 24.

11. M. L. Preston and R. D. Turney, "The Process Systems' Contribution to Reliability Engineering and Risk Assessment" (paper presented at the COPE-91 Conference, Barcelona, Spain, 14–16 October 1991).

12. O. A. Asbjornsen, "Technical Management: A Major Challenge in Industrial Competition," *Chem. Eng. Prog.* 84, no. 11 (1988): 27–32.

13. C. J. Kirkland, *The Channel Tunnel: Engineering under Private Finance: Innovation or Frustration* (London: Fellowship of Engineering, 1989).

14. C. Reece, quoted by R. Stevenson, "Charles Reece: Picking Winners for ICI and UK PLC," *Chem. Brit.* 22, no. 8 (1988): 695–696.

15. N. Rothwell, "Pulling Together: The Synergies," *The Roundel* 70, no. 1 (1992): 10–13.

16. G. G. Haselden, "Role of Postgraduate and Research Studies," in *Chemical Engineering Education*, Symposium Series No. 70 (Rugby, U.K.: Institution of Chemical Engineers, 1981).

17. Royal Society of Chemistry, *Environment Health and Safety Committee Note on Inherently Safer Chemical Processes, Appendix 1*, Version 2: 19/07/2007 (2007), http://www.rsc.org.

18. C. Ramshaw, "Problems of the Innovation in a Large Company," in *Management of Change in the Process Industries* (Manchester, U.K.: Institution of Chemical Engineers, North Western Branch, 1989).

19. G. I. J. M. Zwetsloot and N. A. Ashford, "Inherently Safer Production: A Natural Complement to Cleaner Production," *Industry and Environment* 25, no. 3-4 (2002): 84–88.

20. D. Kehn, "Inherently Safe Process Plants: Practical Methods and Examples," in *American Institute of Chemical Engineers 2005 Spring National Meeting, Process Plant Safety Symposium* (Atlanta, 2005).

21. K. Study, "A Real-Life Example of Choosing an Inherently Safer Process Option," *Process Safety Progress* 25, no. 4 (2006): 274–279.

22. D. P. Mansfield, *Inherently Safer Approaches to Plant Design*, Report No. AEA/CS/HSE R1016 (Warrington, U.K.: AEA Technology, 1991).

23. AEA Technology, *Improving Inherent Safety: A Pilot Study* (Sudbury, U.K.: HSE Books, 1997).

24. B. W. Main and A. C. Ward, "What Do Design Engineers Really Know about Safety?" *Mech. Eng.* 112: 44–50.

25. D. Mansfield, Y. Malmen, and E. Suokas, "The Development of an Integrated Toolkit for Inherent SHE," in *International Conference and Workshop on Process Safety Management and Inherently Safer Processes* (New York: American Institute of Chemical Engineers, 1996).

26. INSIDE Project Team Partners, *The INSET Toolkit*, vols. 1 and 2, integrated version, Nov. 2001 (Cheshire, U.K.: AEA Technology, 2001).

27. M. S. Mannan, W. J. Rogers, M. Gentile, and T. M. O'Connor, "Inherently Safer Design: Implementation Challenges Faced by New and Existing Facilities," *Hydrocarbon Processing* March (2003): 59–61.

28. D. C. Hendershot, "Tools for Making Process Safety a Part of the Design of a Chemical Process," in *5th Asia Pacific Responsible Care Conference and Chemical Safety Workshops* (Shanghai, PRC, 1999).

29. A. G. Rushton, D. W. Edwards, and D. Lawrence, D., "Inherent Safety and Computer-Aided Process Design," *Process Safety and Environmental Protection* 72, no. B2 (1994): 83–87.

30. R. Srinivasan, K. C. Chia, A.-M. Heikkila, and J. Schabel, "Supporting Design through a Database of Inherently Safer Processes," *Loss Prevention Bulletin* 177 (2004): 16–25.

31. A. M. Shariff, R. Rusli, C. T. Leong, V. R. Radhakrishnan, and A. Buang, "Inherent Safety Tool for Explosion Consequences Study," *Journal of Loss Prevention in the Process Industries* 19, no. 5 (2006): 409–418.

32. C. T. Leong and A. M. Shariff, "Inherent Safety Index Module (ISIM) to Assess Inherent Safety Level during Preliminary Design Stage," *Process Safety and Environmental Protection* 86, no. 2 (2008): 113–119.

33. J. Morris, *Coast to Coast* (London: Faber, 1956).

34. D. W. Jones and P. G. Snyder, "Project Management Principles for Inherently Safer Design," in *International Conference and Workshop on Process Safety Management and Inherently Safer Processes* (New York: American Institute of Chemical Engineers, 1996).

35. R. J. Marcus, "Refusing to Advertise," *Chemtech* 24 (1994): 56.

36. S. Saba, "The Difference between Japanese and Western Companies," *The Roundel* 67, no. 4 (1989): 84–88.

37. T. A. Kletz, "Should Undergraduates Be Instructed in Loss Prevention?" *Plant/Operations Prog.* 7, no. 2 (1988): 95–98.

38. T. A. Kletz, *Hazop and Hazan: Identifying and Assessing Process Industry Hazards*, 4th ed. (Rugby, U.K.: Institution of Chemical Engineers, 1999).

39. T. A. Kletz, "Eliminating Potential Process Hazards," *Chem. Eng. (U.S.)* 92, no. 7 (1985): 48–68.

40. S. Mannan, ed., *Lees' Loss Prevention in the Process Industries*, 3rd ed. (Oxford, U.K.: Elsevier Butterworth-Heinemann, 2005), chap. 8.

41. R. J. Caputo, "Engineering for Safer Plants," in *International Symposium on Preventing Major Chemical Accidents*, ed. J. L. Woodward (New York: American Institute of Chemical Engineers, 1987).

42. A. Tugnoli, F. Khan, and P. Amyotte, "Inherent Safety Implementation throughout the Process Design Lifecycle," in *Ninth International Probabilistic Safety Assessment and Management Conference (PSAM 9)* (Hong Kong, 2008).

43. J. W. Fulton, "Bring in the Engineer Early," *Chemtech* 14, no. 1 (1984): 40–42.

44. J. L. Hawksley, "Process Safety Management," *Plant/Operations Prog.* 7, no. 4 (1988): 265–269.

45. J. L. Hawksley, "Risk Assessment and Project Development," *The Safety Practitioner* 5, no. 10 (1987): 10–16.

46. J. W. Hempseed, "Hazard Risk Analysis Associated with Production and Storage of Industrial Gases," in *11th International Symposium on the Prevention of Occupational Accidents and Diseases in the Chemical Industry* (Annecy, France: International Social Security Association, 1987).

47. D. E. Wade, "Reduction of Risks by Reduction of Toxic Material Inventories," in *International Symposium on Preventing Major Chemical Accidents*, ed. J. L. Woodward (New York: American Institute of Chemical Engineers, 1987), 21.

48. R. D. Turner, "Designing Plants for 1990 and Beyond," in *Proceedings of the International Conference on Safety and Loss Prevention in the Chemical and Oil Processing Industries, Singapore, Oct. 1989* (Rugby, U.K.: Institution of Chemical Engineers, 1990).

49. H. G. Lawley, "Operability Studies and Hazard Analysis," *Chem. Eng. Prog.* 70, no. 4 (1974): 45–56.

50. R. Malpas, "The Plant after Next," *Engineering* 246 (1978): 563–565.

51. R. Malpas and D. Davies, "The Industry after Next," *Link-Up* 2 (1986): 28–29.

52. M. J. Glasser and W. F. Early, "The Applicability of Inherently Safer Designs to Smaller Facilities" (paper presented at the 23rd Annual Loss Prevention Symposium, Denver, Colorado, 6–10 March 1988).

53. Center for Chemical Process Safety, *Inherently Safer Chemical Processes: A Life Cycle Approach*, 2nd ed. (Hoboken, NJ: John Wiley & Sons, 2009).

54. S. Mohindra and R. P. Stickles, "Selecting the Design Bases for Safety Systems: Inherently Safer Design Concepts," in *International Conference and Workshop on Process Safety Management and Inherently Safer Processes* (New York: American Institute of Chemical Engineers, 1996).

55. M. L. Preston, D. C. Richards, and A. G. Rushton, "CAPE: Crusading for Process Safety," *Comput. Chem. Eng.* 20 supplement (1996): S1533–S1538.

56. A. Barrell, "Inherent Safety: Only by Design," *Chem. Eng. (U.K.)* 451 (1998): 3.

57. D. Mansfield, R. Turney, R. Rogers, M. Verwoerd, and P. Bots, "How to Integrate Inherent SHE in Process Development and Plant Design," in *Major Hazards Onshore and Offshore*, Symposium Series No. 139 (Rugby, U.K.: Institution of Chemical Engineers, 1995).

58. P. S. Tobin, "Risk Management through Inherently Safe Chemistry" (paper presented at the American Chemical Society Conference, Washington, DC, 21–25 August 1994).

59. P. T. Anastas, "Benign by Design Chemistry" (paper presented at the American Chemical Society Conference, Washington, DC, 21–25 August 1994).

60. D. A. Moore, "The Regulation of Inherent Safety," *IChemE Symposium Series No. 149* (2003): 247–256.

61. J. Johnson, "Simply Safer," *Chemical & Engineering News* 81, no. 5 (2003): 23–26.

62. H. H. Lou, R. Muthusamy, and Y. Huang, "Process Security Assessment: Operational Space Classification and Process Security Index," *Process Safety and Environmental Protection* 81, no. 6 (2003): 418–429.

63. D. A. Moore and J. E. L. Rogers, "The Challenge of Inherent Safety," in *18th Annual International Conference Proceedings, Center for Chemical Process Safety* (AIChE, Scottsdale, AZ, 2003), 121–128.

64. D. A. Moore, "Judging Effectiveness of Inherent Safety for Safety and Security of Chemical Facilities," in *20th Annual International Conference Proceedings, Center for Chemical Process Safety* (AIChE, Atlanta, 2005).

65. E. R. Hamberger, "Homeland Security Committee Urged to Consider Safer Chemicals: Chemical Companies Should Stop Manufacturing Extremely Dangerous Chemicals" (press release, Association of American Railroads, Washington, DC, February 27, 2008).

66. H. Otway and P. Haastrup, "On the Social Acceptability of Inherently Safe Technologies," *IEEE Trans. Eng. Manage.* 36, no. 1 (1989): 57–60.

67. P. Tebo, quoted in "Prevention Is Better than Cure," *Financial Times* (London), December 15, 1993.

68. Center for Chemical Process Safety, *Guidelines for Risk Based Process Safety* (Hoboken, NJ: John Wiley & Sons, 2007).

69. V. C. Marshall, "Assessment of Mortality Indices," in *Hazardous Materials Spills Handbook*, ed. G. F. Bennett, F. S. Feates, and I. Wilder (New York: McGraw–Hill, 1982).

70. Dow Chemical Company, *Fire and Explosion Index*, 7th ed., and *Chemical Exposure Index*, 1st ed. (New York: American Institute of Chemical Engineers, 1994).

71. B. J. Tyler, "Using the Mond Index to Measure Inherent Hazards," *Plant/Operations Prog.* 4, no. 3 (1985): 172–175.

72. B. J. Tyler, A. R. Thomas, P. Doran, and T. R. Grieg, "A Toxicity Hazard Index," *Chem. Health Safety* 3 (1996): 19–25.

73. A term suggested by A. V. Cohen and D. K. Pritchard, in *Comparative Risks of Electricity Production Systems* (London: Her Majesty's Stationery Office, 1980).

74. R. Gowland, "Applying Inherently Safer Concepts to a Phosgene Plant Acquisition," *Process Safety Progress* 15, no. 1 (1996): 52–57.

75. N. E. Scheffler, "Inherently Safer Latex Plants," *Process Safety Progress* 15, no. 1 (1996): 11–17.

76. C. B. Etowa, P. R. Amyotte, M. J. Pegg, and F. I. Khan, "Quantification of Inherent Safety Aspects of the Dow Indices," *Journal of Loss Prevention in the Process Industries* 15, no. 6 (2002): 477–487.

77. J. Suardin, M. S. Mannan, and M. El-Halwagi, "The Integration of Dow's Fire and Explosion Index (F&EI) into Process Design and Optimization to Achieve Inherently Safer Design," *Journal of Loss Prevention in the Process Industries* 20, no. 1 (2007): 79–90.

78. D. W. Edwards and D. Lawrence, "Assessing the Inherent Safety of Chemical Process Routes," *Process Safety Environ. Protec.* 71, no. B4 (1993): 252–258.

79. D. W. Edwards, A. G. Rushton, and D. Lawrence, "Quantifying the Inherent Safety of Chemical Process Routes" (paper presented at the 5th World Congress of Chemical Engineering, San Diego, 14–18 July 1996).

80. F. I. Khan, R. Sadiq, and P. R. Amyotte, "Evaluation of Available Indices for Inherently Safer Design Options," *Process Safety Progress* 22, no. 2 (2003): 83–97.

81. F. I. Khan, T. Husain, and S. A. Abbasi, "Safety Weighted Hazard Index (SWeHI): A New User-Friendly Tool for Swift yet Comprehensive Hazard Identification and Safety Evaluation in Chemical Process Industries," *Process Safety and Environmental Protection* 79, no. 2 (2001): 65–80.

82. D. C. Hendershot, "How Do You Measure Inherent Safety Early in Process Development?" (paper presented at the International Conference and Workshop on Process Safety Management and Inherently Safer Processes, Orlando, FL, 8–11 October 1996 and not included in published proceedings).

83. J. L. Hawksley and M. L. Preston, "Inherent SHE: 20 years of Evolution," in *International Conference and Workshop on Process Safety Management and Inherently Safer Processes* (New York: American Institute of Chemical Engineers, 1996).

84. F. I. Khan and P. R. Amyotte, "Integrated Inherent Safety Index (I2SI): A Tool for Inherent Safety Evaluation," *Process Safety Progress* 23, no. 2 (2004): 136–148.

85. F. I. Khan and P. R. Amyotte, "I2SI: A Comprehensive Quantitative Tool for Inherent Safety and Cost Evaluation," *Journal of Loss Prevention in the Process Industries* 18, no. 4-6 (2005): 310–326.

86. A. Carvalho, R. Gani, and H. Matos, "Design of Sustainable Chemical Processes: Systematic Retrofit Analysis Generation and Evaluation of Alternatives," *Process Safety and Environmental Protection* 86, no. 5 (2008): 328–346.

87. M. Hurme and M. Rahman, "Implementing Inherent Safety throughout Process Lifecycle," *Journal of Loss Prevention in the Process Industries* 18, no. 4-6 (2005): 238–244.

88. M. Rahman, A.-M. Heikkila, and M. Hurme, "Comparison of Inherent Safety Indices in Process Concept Evaluation," *Journal of Loss Prevention in the Process Industries* 18, no. 4-6 (2005): 327–334.

89. M. H. Hassim and D. W. Edwards, "Development of a Methodology for Assessing Inherent Occupational Health Hazards," *Process Safety and Environmental Protection* 84, no. 5 (2006): 378–390.

90. J. P. Gupta and D. W. Edwards, "A Simple Graphical Method for Measuring Inherent Safety," *Journal of Hazardous Materials* 104, no. 1-3 (2003): 15–30.

91. V. Cozzani, F. Barontini, and S. Zanelli, "Assessing the Inherent Safety of Substances: Precursors of Hazardous Products in the Loss of Control of Chemical Systems," in *Proceedings of American Institute of Chemical Engineers Spring National Meeting* (New Orleans, 2006).

92. G. Landucci, A. Tugnoli, C. Nicolella, and V. Cozzani, "Assessment of Inherently Safer Technologies for Hydrogen Storage," IChemE Symposium Series No. 153, in *12th International Symposium on Loss Prevention and Safety Promotion in the Process Industries* (Edinburgh, U.K., 2007).

93. G. Landucci, A. Tugnoli, and V. Cozzani, "Inherent Safety Key Performance Indicators for Hydrogen Storage Systems," *Journal of Hazardous Materials* 159, no. 2-3 (2008): 554–566.

94. M. Cordella, A. Tugnoli, F. Barontini, G. Spadoni, and V. Cozzani, "Inherent Safety of Substances: Identification of Accidental Scenarios due to Decomposition Products," *Journal of Loss Prevention in the Process Industries* 22, no. 4 (2009): 455–462.

95. C. T. Leong and A. M. Shariff, "Process Route Index (PRI) to Assess Level of Explosiveness for Inherent Safety Quantification," *Journal of Loss Prevention in the Process Industries* 22, no. 2 (2009): 216–221.

96. A. Kossoy and Yu. Akhmetshin, "Simulation-Based Approach to Design of Inherently Safer Processes," IChemE Symposium Series No. 153, in *12th International Symposium on Loss Prevention and Safety Promotion in the Process Industries* (Edinburgh, U.K., 2007).

97. S. Shah, U. Fischer, and K. Hungerbuhler, "A Hierarchical Approach for the Evaluation of Chemical Process Aspects from the Perspective of Inherent Safety," *Process Safety and Environmental Protection* 81, no. 6 (2003): 430–443.

98. I. K. Adu, H. Sugiyama, U. Fischer, and K. Hungerbuhler, "Comparison of Methods for Assessing Environmental, Health and Safety (EHS) Hazards in Early Phases of Chemical Process Design," *Process Safety and Environmental Protection* 86, no. 3 (2008): 77–93.

99. C. Palaniappan, R. Srinivasan, and R. Tan, "Expert System for the Design of Inherently Safer Processes; 1: Route Selection Stage," *Industrial Engineering Chemistry Research* 41, no. 26 (2002): 6698–6710.

100. C. Palaniappan, R. Srinivasan, and R. Tan, "Expert System for the Design of Inherently Safer Processes; 2: Flowsheet Development Stage," *Industrial Engineering Chemistry Research* 41, no. 26 (2002): 6711–6722.

101. C. Palaniappan, R. Srinivasan, and I. Halim, "A Material-Centric Methodology for Developing Inherently Safer Environmentally Benign Processes," *Computers and Chemical Engineering* 26, no. 4-5 (2002): 757–774.

102. R. Srinivasan and N. T. Nhan, "A Statistical Approach for Evaluating Inherent Benign-ness of Chemical Process Routes in Early Design Stages," *Process Safety and Environmental Protection* 86, no. 3 (2008): 163–174.

103. R. Srinivasan and A. Kraslawski, "Application of the TRIZ Creativity Enhancement Approach to Design of Inherently Safer Chemical Processes," *Chemical Engineering and Processing* 45, no. 6 (2006): 507–514.

104. A. Meel and W. D. Seider, "Dynamic Risk Assessment of Inherently Safe Chemical Processes: Accident Precursor Approach," in *American Institute of Chemical Engineers 2005 Spring National Meeting* (AIChE, Atlanta, 2005).

105. M. Gentile, W. J. Rogers, and M. S. Mannan, "Development of a Fuzzy Logic-Based Inherent Safety Index," *Process Safety and Environmental Protection* 81, no. 6 (2003): 444–456.

106. E. M. Al-Mutairi, J. A. Suardin, S. M. Mannan, and M. M. El-Halwagi, "An Optimization Approach to the Integration of Inherently Safer Design and Process Scheduling," *Journal of Loss Prevention in the Process Industries* 21, no. 5 (2008): 543–549.

107. Center for Chemical Process Safety, *Guidelines for Chemical Process Quantitative Risk Analysis* (New York: American Institute of Chemical Engineers, 1989).

108. J. Tixier, G. Dusserre, O. Salvi, and D. Gaston, "Review of 62 Analysis Methodologies of Industrial Plants," *Journal of Loss Prevention in the Process Industries* 15, no. 4 (2002): 291–303.

109. G. L. L. Reniers, B. J. M. Ale, W. Dullaert, and B. Foubert, "Decision Support Systems for Major Accident Prevention in the Chemical Process Industry: A Developers' Survey," *Journal of Loss Prevention in the Process Industries* 19, no. 6 (2006): 604–620.

110. R. W. Johnson, S. D. Unwin, and T. I. McSweeney, "Inherent Safety: How to Measure It and Why We Need It," in *International Conference and Workshop on Process Safety Management and Inherently Safer Processes* (New York: American Institute of Chemical Engineers, 1996).

111. H. B. Whitehouse, "IFAL: A New Risk Analysis Tool," in *The Assessment and Control of Major Hazards*, Symposium Series No. 93 (Rugby, U.K.: Institution of Chemical Engineers, 1985).

112. F. P. Lees, *Loss Prevention in the Process Industries*, 2nd ed. (Oxford, U.K.: Butterworth-Heinemann, 1996), sec. 5.6.

113. T. A. Kletz, *Learning from Accidents*, 3rd ed. (Oxford, U.K.: Butterworth-Heinemann, 2001).

114. A. Brazier and J. Skilling, "Potential Sources of Data for Use in Human Factor Studies," in *Major Hazards Onshore and Offshore II* (Rugby, U.K.: Institution of Chemical Engineers, 1995).

115. G. Dalzell, "Is Operating Cost a Direct Measure of Inherent Safety?" *Process Safety and Environmental Protection* 81, no. 6 (2003): 414–417.

116. W. D. Cullen, *The Public Inquiry into the Piper Alpha Disaster* (London: Her Majesty's Stationery Office, 1990).

117. B. Appleton, "Piper Alpha," in *Learning from Accidents*, by T. A. Kletz, 2nd ed. (Oxford, U.K.: Butterworth-Heinemann, 1994), chap. 17.

118. F. I. Khan and P. R. Amyotte, "Inherent Safety in Offshore Oil and Gas Activities: A Review of the Present Status and Future Directions," *Journal of Loss Prevention in the Process Industries* 15, no. 4 (2002): 279–289.

119. F. I. Khan, B. Veitch, and P. R. Amyotte, "Evaluation of Inherent Safety Potential in Offshore Oils and Gas Activities," in *23rd International Conference on Offshore Mechanics and Arctic Engineering* (Vancouver, BC, Canada: 2004).

120. G. Dalzell, "What Makes an Inherently Safe Platform?" in *17th International Conference on Offshore Mechanics and Arctic Engineering* (Lisbon, Portugal, 1998).

121. BP Magazine of Technology and Innovation, "All Pumped Up," *Frontiers* April (2008): 36–37.

122. S. Chia, K. Walshe, and E. Corpuz, "Application of Inherent Safety Challenge to an Offshore Platform Design for a New Gas Field Development: Approaches and Experiences," *IChemE Symposium Series* 149 (2003): 257–265.

123. U. Kjellen, "Safety in the Design of Offshore Platforms: Integrated Safety versus Safety as an Add-On Characteristic," *Safety Science* 45, no. 1-2 (2007): 107–127.

124. S. Sklet, "Hydrocarbon Releases on Oil and Gas Production Platforms: Release Scenarios and Safety Barriers," *Journal of Loss Prevention in the Process Industries* 19, no. 5 (2006): 481–493.

125. S. Sklet, "Safety Barriers: Definition, Classification, and Performance," *Journal of Loss Prevention in the Process Industries* 19, no. 5 (2006): 494–506.

126. M. M. Oakden and M. Eades, "The Application of Inherently Safer Design Principles to the Design of a New Offshore Gas Processing Platform," in *9th SPE International Conference on Health, Safety and Environment in Oil and Gas Exploration and Production* (Nice, France, 2008).

127. F. K. Crawley, "Optimizing the Life Cycle Safety, Health and Environment Impact of New Projects," *Process Safety and Environmental Protection* 82, no. 6 (2004): 438–445.

128. Health and Safety Executive, *The Tolerability of Risks from Nuclear Power Stations*, 2nd ed. (Sudbury, U.K.: HSE Books, 1992).

Chapter 11

Inherently Safer Design and Process-Safety Management

Management is doing things right; leadership is doing the right things.

—**Peter F. Drucker**

Given the above quote by Peter Drucker, perhaps *leadership* should be substituted for *management* in the title of this chapter. Without capable and courageous leadership, safety management efforts are effectively wasted. The industry term is, however, process-safety management, and that is the subject of this chapter.

Process-safety management deals with the identification, understanding and control of process hazards to prevent process-related injuries and incidents. Explicit incorporation of inherent-safety concepts in the basic definition and functional operation of the various elements comprising a process-safety management system can help to improve the quality of the safety management effort. Zwetsloot and Ashford[1] remark that inherent safety has the potential to go beyond the technological domain and become a strategic tool for safety management. These same authors also state that adoption of inherent-safety practices requires companies to have the willingness, opportunity, and capability to change.[2] This returns us to the earlier point on leadership, or more broadly, safety culture.

In this chapter, safety management systems are first discussed. This is followed by an examination of the elements that make up a typical process-safety management system. Safety culture is then briefly addressed from the perspective of a commitment to the principles of inherent safety. Several examples, both technical and nontechnical, are provided to reinforce the discussion and also to supplement those given in earlier chapters. Much of the material in the current chapter is drawn from the work of Amyotte et al.[3,4]

11.1 Safety Management Systems

A key engineering tool is a management system appropriate for the risks being addressed (process safety, occupational safety, health, environment, equipment reliability, etc.).[5] Safety management systems are recognized and accepted worldwide as best-practice methods for managing risk. They typically consist of 10–20 program elements that must be effectively carried out to manage the risks in an acceptable way. This need is based on the understanding that, once a risk is accepted, it does not go away; it is there waiting for an opportunity to happen unless the management system is actively monitoring company operations for concerns and taking proactive actions to correct potential problems.[6]

Having an effective management system for process-related hazards (fire, explosion, release of toxic materials, etc.) is therefore a critical corporate objective in the chemical process industries. An approach widely used in Canada is PSM, process safety management (where PSM is the application of management principles and systems to the identification, understanding, and control of process hazards to prevent process-related injuries and incidents). The complete suite of PSM elements is shown in Table 11.1, taken from the *Process Safety Management Guide* of the Canadian Society for Chemical Engineering (CSChE).[7]

This guide was prepared by the Process Safety Working Group of the former Major Industrial Accidents Council of Canada (MIACC) in conjunction with the Process Safety Management Committee of the Canadian Chemical Producers' Association (CCPA). With the dissolution of MIAC in 1999, rights to the guide were transferred to the CSChE. The material in the CSChE PSM guide[7] is based on that developed by the Center for Chemical Process Safety (CCPS)[8] of the American Institute of Chemical Engineers. This route was adopted because the CCPS approach to process-safety management was determined to be comprehensive; well supported by reference materials, tools, and an organizational structure; and based on a benchmark of leading or good industry practice rather than on minimum standards.[7]

Table 11.1 will therefore look familiar to PSM practitioners in the United States. It will also be relevant to those engaged in process-safety efforts in other parts of the world, given the incorporation of best practices in the elements and components shown in Table 11.1. Other systems may have more or fewer elements, the terminology may be somewhat different, or a specific management system may be mandated by regulation, but the underlying concepts are the same. Process safety cannot be left to chance, and there is much to be gained by a systematic and rigorous approach

Table 11.1 Elements and Components of Process-Safety Management (PSM)

No.	Element		Component
1	Accountability: objectives and goals	1.1	Continuity of operations
		1.2	Continuity of systems
		1.3	Continuity of organization
		1.4	Quality process
		1.5	Control of exceptions
		1.6	Alternative methods
		1.7	Management accessibility
		1.8	Communications
		1.9	Company expectations
2	Process knowledge and documentation	2.1	Chemical and occupational health hazards
		2.2	Process definition/design criteria
		2.3	Process and equipment design
		2.4	Protective systems
		2.5	Normal and upset conditions (operating procedures)
		2.6	Process-risk management decisions
		2.7	Company memory (management of information)
3	Capital project review and design procedures	3.1	Appropriation request procedures
		3.2	Hazard reviews
		3.3	Siting
		3.4	Plot plan
		3.5	Process-design and -review procedures
		3.6	Project management procedures and controls
4	Process-risk management	4.1	Hazard identification
		4.2	Risk analysis of operations
		4.3	Reduction of risk
		4.4	Residual-risk management
		4.5	Process management during emergencies

(continued)

Table 11.1   Elements and Components of Process-Safety Management (PSM) (Continued)

No.	Element		Component
		4.6	Encouraging client and supplier companies to adopt similar risk-management practices
		4.7	Selection of businesses with acceptable risk
5	Management of change	5.1	Change of process technology
		5.2	Change of facility
		5.3	Organizational changes
		5.4	Variance procedures
		5.5	Permanent changes
		5.6	Temporary changes
6	Process and equipment integrity	6.1	Reliability engineering
		6.2	Materials of construction
		6.3	Fabrication and inspection procedures
		6.4	Installation procedures
		6.5	Preventive maintenance
		6.6	Process, hardware, and systems inspection and testing
		6.7	Maintenance procedures
		6.8	Alarm and instrument management
		6.9	Decommissioning and demolition procedures
7	Human factors	7.1	Operator-process/equipment interface
		7.2	Administrative control versus hardware
		7.3	Human error assessment
8	Training and performance	8.1	Definition of skills and knowledge
		8.2	Design of operating and maintenance procedures
		8.3	Initial qualifications assessment
		8.4	Selection and development of training programs

(continued)

Table 11.1   Elements and Components of Process-Safety Management (PSM) (Continued)

No.	Element		Component
		8.5	Measuring performance and effectiveness
		8.6	Instructor program
		8.7	Records management
		8.8	Ongoing performance and refresher training
9	Incident investigation	9.1	Major incidents
		9.2	Third-party participation
		9.3	Follow-up and resolution
		9.4	Communication
		9.5	Incident recording, reporting, and analysis
		9.6	Near-miss reporting
10	Company standards, codes, and regulations	10.1	External codes/regulations
		10.2	Internal standards
11	Audits and corrective actions	11.1	Process-safety management systems audits
		11.2	Process-safety audits
		11.3	Compliance reviews
		11.4	Internal/external auditors
12	Enhancement of process-safety knowledge	12.1	Quality control programs and process safety
		12.2	Professional trade and association programs
		12.3	CCPS program
		12.4	Research, development, documentation, and implementation
		12.5	Improved predictive system
		12.6	Process-safety resource centre and reference library

Source: Adapted from CSChE, *Process Safety Management*, 3rd ed. (Ottawa, ON: Canadian Society for Chemical Engineering, 2002).

to managing a company's efforts to reduce its process-related risks. Continuous improvement is a key feature of safety management systems, and in this regard it is important to note the recent CCPS work on developing a framework for risk-based process-safety management (Section 10.3).[9]

Inherent safety is *implicitly* present in PSM. How best then to *explicitly* incorporate inherent safety throughout the management system elements displayed in Table 11.1? One needs to examine both the elements and the method of concept integration.

On the first point, the CCPA Process Safety Management Committee has collected and analyzed data on an annual basis for process-related incidents reported by CCPA member companies using a procedure known as PRIM (Process-Related Incidents Measure). The 2004 PRIM analysis of 89 reported incidents demonstrated that six of the PSM elements in Table 11.1 contributed to 85% of the total incidents (as shown in Table 11.2).[4]

A possible use of the data in Table 11.2 is to prioritize the PSM elements in Table 11.1 for particular attention with respect to inherent-safety consideration. This process is assisted by the breakdown of each PSM element into its components (or subelements), as illustrated in Figure 11.1[4] for element 6, process and equipment integrity. Figure 11.1 gives incident data for 2004 as well as the five reporting periods prior to 2004.

The PRIM methodology uses multiple analysts (i.e., self-reporting by companies) with overall review by a team of process-safety experts. It is therefore best to draw only broad conclusions as to the relative importance of particular PSM elements, especially with respect to trend analysis from year to year (Figure 11.2).[4] For example, it seems reasonable to conclude from Figure 11.1 that *preventive maintenance* and *maintenance procedures* have historically been thought to contribute significantly to incident causation when *process and equipment integrity* has been

Table 11.2   Incident Causation According to PSM Element (2004 PRIM data)

PSM Element No.	PSM Element	% of Incidents
6	Process and equipment integrity	23.8
2	Process knowledge and documentation	21.2
4	Process-risk management	16.8
7	Human factors	8.9
5	Management of change	7.3
3	Capital project review and design procedures	6.5

Source: P. R. Amyotte et al., "Incorporation of Inherent Safety Principles in Process Safety Management," *Process Safety Progress* 26, no. 4 (2007): 333–346.

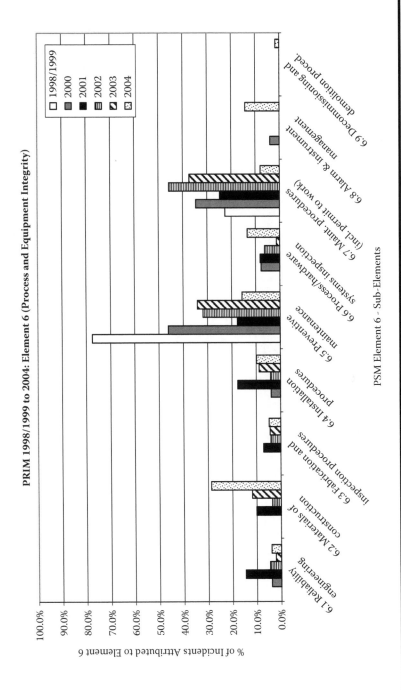

Figure 11.1 Incident causation according to PSM element 6 (PRIM data).

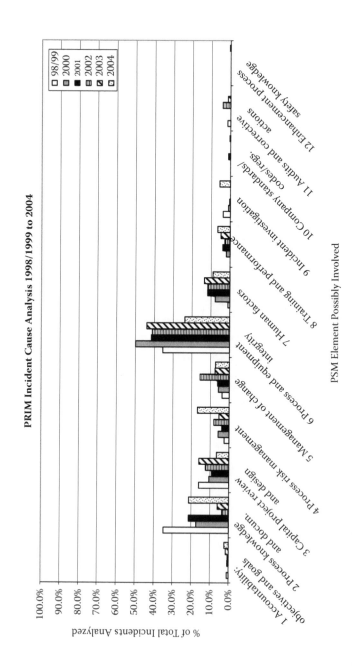

Figure 11.2 Incident causation by PSM element from 1998/99 to 2004 (PRIM data).

Table 11.3   Inherent-Safety Guide Words

Guide Word	Description
Intensify (minimize)	Use smaller quantities of hazardous materials when the use of such materials cannot be avoided
	Perform a hazardous procedure as few times as possible when the procedure is unavoidable
Substitute	Replace a substance with a less hazardous material or processing route with one that does not involve hazardous material
	Replace a hazardous procedure with one that is less hazardous
Attenuate (moderate)	Use hazardous materials in their least hazardous forms or identify processing options that involve less severe processing conditions
Simplify	Design processes, processing equipment, and procedures to eliminate opportunities for errors by eliminating excessive use of add-on (engineered) safety features and protective devices

Source: Adapted from A. Goraya, P. R. Amyotte, and F. I. Khan, "An Inherent Safety-Based Incident Investigation Methodology," *Process Safety Progress* 23, no. 3 (2004): 197–205.

flagged. (Recall the discussion in Section 6.3.2 on maintenance as a hazardous activity.) A defensible conclusion from Figure 11.2 is that, for the seven-year reporting period, deficiencies in PSM elements 2–7 inclusive have been viewed as key contributors to process-related incidents in Canadian chemical companies. The importance attached to each of these six elements has varied over this period, but each has crossed an arbitrary threshold of having contributed to at least 10% of the total incidents during a given year (at least once during the seven-year period).

On the method of inherent-safety concept integration, both qualitative and quantitative techniques are available. Many of these have been described in Sections 10.2 and 10.4, and others are discussed in the remainder of this chapter. The qualitative protocols presented here make use of both guide words and checklists, two well-established and proven tools that are widely used in the field of process safety.[10]

Recommended guide words based on inherent safety and user friendliness, and their descriptions, are shown in Table 11.3. These guide words are simply the four most general and widely applicable principles of inherent-safety/user-friendliness as presented in Chapters 3, 4, 5, and 7, respectively. The description for each is purposely brief and is focused on materials, process routes, equipment, and procedures.

Use of these guide words as "mind triggers" during a particular process-safety activity (e.g., management of change, Section 11.2.5; incident investigation, Section 11.2.9) will help ensure that the concepts of inherent safety are visible within the activity. These guide words are intended as a supplement to existing tools that may already be in use within a specific process-safety protocol. Additional guide words can be added to capture ideas not strictly represented by the set of four in Table 11.3.

Four series of checklist questions built around the guide words in Table 11.3 are given in Table 11.4. The intention would be to use these more detailed indicators of inherent safety and user friendliness at an appropriate time during a particular protocol (Section 11.2.5 and Section 11.2.9). As with the guide words, the aim is not to replace existing checklists, but rather to supplement these perhaps more traditional tools with ones that directly incorporate inherent-safety considerations. The checklist in Table 11.4 is not all inclusive, and is in fact a mix of both process-safety and occupational-safety issues. It is intended as an illustrative example of the type of thinking required to move beyond the usual engineered/procedural form of checklist questions. Other inherent-safety-based checklists can be found in Section 10.2 and Appendix A in *Inherently Safer Chemical Processes: A Life Cycle Approach*.[11]

11.2 Process-Safety Management System Elements

While the PRIM type of analysis (Section 11.1) does offer a method for prioritized resource allocation aimed at improvement, opportunities for inherent-safety application will be missed if all management system elements are not examined. For example, it is not surprising that management accountability (element 1 in Figure 11.2) is rated lower than other elements as an incident causation factor in a system of self-reporting by engineers who may be predisposed to finding technical solutions. Incident investigation (element 9 in Figure 11.2) might be thought of as reactive and, therefore, of limited use as a preventive measure. This is an erroneous conclusion, as shown earlier in Section 10.5 and reinforced here in Section 11.2.9. With respect to process-safety knowledge (element 12 in Figure 11.2), it can be difficult to know what you do not know.

11.2.1 PSM Element 1—Accountability: Objectives and Goals

Management commitment at all levels is necessary for PSM to be effective. The objectives for establishing accountability are to demonstrate the status of process safety compared to other business objectives (e.g., production and cost), to set objectives for safe process operation and to set specific process safety goals. These objectives should be internally consistent.[7]

Table 11.4   Inherent-Safety Checklist

Guide Word	Checklist Question
Intensify (minimize)	Is the storage of all hazardous gases, liquids, and solids minimized?
	Are just-in-time deliveries used when dealing with hazardous materials?
	Are all hazardous materials removed or properly disposed of when they are no longer needed or not needed in the next X days?
	Are all tasks involving working at height done in as few trips as possible?
	Are all tripping hazards minimized and all walking surfaces tractional during all weather conditions?
	Are workplaces designed such that employee seclusion is minimized?
	Is shift rotation optimized to avoid fatigue?
	Are awkward positions and repetitive motions minimized?
	Are attempts made to completely eliminate raw materials, process intermediates, or by-products?
	Are elbows, bends, and joints in piping minimized? (also relates to Simplify)
Substitute	Can a less toxic, less flammable, or less reactive material be substituted for use?
	Is there an alternative way of moving product or equipment so as to eliminate human strain?
	Can a water-based product be used in place of a solvent- or oil-based product?
	Are all allergenic materials, products, and equipment replaced with nonallergenic materials, products, and equipment when possible?
Attenuate (moderate)	Can potential releases be reduced via lower temperatures or pressures, or elimination of equipment?
	Are all hazardous gases, liquids, and solids stored as far away as possible to eliminate disruption to people, property, production, and environment in the event of an incident (avoiding knock-on effects)?

(continued)

Table 11.4   Inherent-Safety Checklist (Continued)

Guide Word	Checklist Question
	When purchasing new equipment, are acceptable models available that operate at lower speeds, pressures, temperatures, or volumes?
	Are low-noise equipment and machinery taken into consideration when making new purchases?
	Are food and beverages offered at the workplace appropriate to minimize negative impact and maximize positive impact?
	Are all power tools deenergized when not in use for extended periods?
	Are all lights, sprinklers, accesses/exits, ventilation ducts, electrical installations, and machines clear of piled materials?
Simplify	Are all manuals, guides, and instructional materials clear and easy to understand, especially those that are used in an emergency situation?
	Are equipment and procedures designed such that they cannot be operated incorrectly or carried out incorrectly?
	Are machine controls located to prevent unintentional activation while allowing easy access for stopping the machine?
	Are all machines, equipment, and electrical installations easily isolated of all sources of power?
	Is the workplace designed for consideration of human factors (i.e., an ergonomically designed workplace)?

Source: Adapted from A. Goraya, P. R. Amyotte, and F. I. Khan, "An Inherent Safety-Based Incident Investigation Methodology," *Process Safety Progress* 23, no. 3 (2004): 197–205.

The essence of this quote was addressed by the vice president and general manager of Petrochemicals, Imperial Oil Limited, in his keynote talk during the 55th Canadian Chemical Engineering Conference.[12] He explained that the fundamental underpinning of Imperial Oil's Operations Integrity Management System is the belief that safety is a corporate value and that performance improvement requires leadership for breakthrough results.

The issue of whether a company believes it is possible to achieve a higher standard of safety has also been addressed in the recent book by Andrew Hopkins.

Hopkins[13] describes three concepts that address a company's cultural approach to safety, and makes the argument that the three are essentially alternative ways of talking about the same phenomena:

Safety Culture: The concept of a safety culture embodies the following subcultures:[13]
1. A *reporting culture* in which people report errors, near-misses, substandard conditions, inappropriate procedures, etc.
2. A *just culture* in which blame and punishment are reserved for behavior involving defiance, recklessness, or malice, such that incident reporting is not discouraged.
3. A *learning culture* in which a company learns from its reported incidents, then processes information in a conscientious manner, and finally makes changes accordingly.
4. A *flexible culture* in which decision-making processes are not so rigid that they cannot be varied according to the urgency of the decision and the expertise of the people involved. Hopkins[13] uses the examples of the *Challenger* and *Columbia* shuttle disasters to explain this point. He comments that decisions in these cases were made by senior NASA officials operating under layers of bureaucracy, rather than by the engineers best equipped to make the decisions.

Collective Mindfulness: The concept of collective mindfulness embodies the principle of *mindful organizing*, which incorporates the following processes:[13]
1. A *preoccupation with failure*, so that a company is not lulled into a false sense of security by periods of success. A company that is preoccupied with failure would have a well-developed reporting culture.
2. A *reluctance to simplify* data that may, at face value, seem unimportant or irrelevant, but which may in fact contain the information needed to reduce the likelihood of a future surprise. (Note that simplification here is not a good thing, unlike the user-friendly principle of simplification.)
3. A *sensitivity to operations*, in which frontline operators and managers strive to remain as aware as possible of the current state of operations and to understand the implications of the present situation for future functioning of the company.
4. A *commitment to resilience*, in which companies respond to errors or crises in a manner appropriate to deal with the difficulty, and a *deference to expertise*, in which decisions are made by the people in the company hierarchy who have the most appropriate knowledge and ability to deal with the difficulty.

Risk Awareness: Hopkins[13] claims that risk awareness is synonymous with collective mindfulness (which is obviously closely related to the concept of a safety culture). He also describes a culture of *risk denial*, in which it is not simply a matter of individuals and companies being unaware of risks, but rather that there are mechanisms that deny the existence of risk.

Hopkins[13] uses these different cultural approaches to demonstrate how a lack of inherent-safety consideration may be interpreted as the placing of production ahead of safety—essentially as a lack of management accountability and commitment to safety. He does this by presenting a case study involving the Royal Australian Air Force and its deseal/reseal program that resulted in numerous maintenance workers suffering severe health problems. The basic case is described as follows:[13]

> Soon after the Air Force took delivery of its F111s in the early 1970s, it became apparent that the chemicals used to seal the insides of the fuel tanks deteriorated rapidly, allowing fuel to leak out. Maintenance workers therefore had to crawl inside the tanks and apply various toxic desealants and then resealants, often by hand, and sometimes by spray. They were supposed to be protected from these chemicals by gloves, respirators, and impermeable suits, but this equipment proved to be ineffective.

Hopkins[13] refers to the Air Force's attitude as "platform-centric" (where platform means aircraft, i.e., equipment or assets). He comments that it is much better to eliminate hazards at the design stage of equipment or processes than to rely on personal protective equipment for worker safety. This is, of course, the hierarchy of controls described in Section 1.3. The overreliance on personnel protective equipment (PPE) in this case meant that alternative nontoxic chemicals were not identified, and the maintenance program was designed with the needs of the physical asset (aircraft) as a priority. Hopkins[13] cites the fact that, in some instances, detergent and hot water were as effective as the toxic solvent actually used to remove old sealant. This is a clear example of a lack of attention to the inherent-safety principle of substitution. One would also have to question the appropriateness of a procedure that does not minimize the need for personnel to repeatedly enter what is essentially a hazardous confined space.

Of direct relevance to PSM element 1 (accountability: objectives and goals) is the following quote: "The absence of any commitment to the hierarchy of controls is another manifestation of the priority of platforms over people."[13] Paraphrasing this statement, it could be said that the absence of a commitment to inherent safety is a manifestation of placing priority on production over safety. This can be expressed in a more positive, encouraging manner by saying that management promotion and adoption of the principles of inherent safety wherever possible are visible demonstrations of leadership and commitment to effective process-safety management. Some general concluding thoughts in this regard are given in Section 11.3.

11.2.2 PSM Element 2—Process Knowledge and Documentation

Information necessary for the safe design, operation and maintenance of any facility should be written, reliable, current and easily accessible by people who need to use it.[7]

There are several implications of inherent safety for this element that can be identified by use of the guide words in Table 11.3 and the checklist in Table 11.4. For example, the guide-word substitute can be applied with good effect to the documentation for component 2.1, *chemical and occupational health hazards*. (Throughout this chapter, the word *component*, when used in this context, refers to the items in the right-hand column of Table 11.1.) Similarly, component 2.2, *process definition/design criteria*, contains subcomponents for which the documentation could be reviewed in light of the inherent-safety guide words (e.g., *intensify*, as applied to *maximum intended inventory*). Component 2.5, *normal and upset conditions (operating procedures)*, contains the requirement that procedures be current, accurate, and reliable. The guide word *simplify* is especially appropriate here, as demonstrated by Figures 11.3 and 11.4.[4] (Recall the discussion in Section 9.7 on unfriendly instructions.)

An organization had decided to offer free antivirus software to its employees and posted the instructions shown in Figure 11.3 on its Web site for personnel to follow to install the new software. The download button was the color red (presumably to draw attention to it); several people in this organization clicked on the button to download the software without reading further. Unfortunately, they missed the fine print (Figure 11.3) advising them to uninstall any existing antivirus software on their computer. The results were numerous computer crashes and much "process time" spent to rectify the computer problems, both by the people directly affected and by computer support personnel.

The business interruption was significant enough that the organization reissued the instructions as shown in Figure 11.4. In this second set of instructions, the steps have been reordered and have been numbered, and the consequence of not first uninstalling existing antivirus software is better explained. The red download button is still present, but the "operator's" eye is not as likely to be immediately drawn to it as in Figure 11.3. The instructions given in Figure 11.4 are more accurate and reliable than those in Figure 11.3. The new instructions are simpler to follow. (Note that this example could also be discussed equally as well under PSM elements 4, 5, and 7.)

Also relevant here, from a different perspective, is component 2.7, *company memory (management of information)*. The intention of this component is to ensure that knowledge and information gained from plant experience likely to be important for the future safety of a facility is well documented so it is not forgotten or overlooked as personnel and organizational changes occur.[7] Hendershot[14] argues that this is especially critical when dealing with inherent safety and inherently safer design features. He gives several examples where such features have essentially

Antivirus Software

For Windows XP/2000/NT

DOWNLOAD

(Size 1.66 MB)

Windows 95/98 users click here

Macintosh users click here

Your **NetID** and **password** will be required when you click to download this software.

Note: Other manufacturers' antivirus products should be uninstalled prior to installing the new software. Failure to do so may result in the new software attempting to uninstall some antivirus products that currently exist on your machine.

Windows 95/98 users please install Windows 95/98 Upgrade at the bottom of this page...

What to Expect?...*more instructions given*

Figure 11.3 **First set of instructions concerning software installation. (From P. R. Amyotte, A. U. Goraya, D. C. Hendershot, and F. I. Khan. "Incorporation of Inherent Safety Principles in Process Safety Management,"** *Process Safety Progress* **26, no. 4 (2007): 333–346. With permission.)**

been put at risk because the reasons they were implemented were not clearly and adequately documented. This results in compromising of facility safety when future modifications are made by people who do not understand the intent of the original designer. (Thus, this point is also pertinent to PSM element 5 on management of change.) As noted earlier in Section 8.1.6, inherently safer design features are particularly susceptible to lapses in corporate memory because they are such integral parts of the design that their purpose may not be obvious, unlike an add-on device such as a high-pressure alarm.[14]

11.2.3 PSM Element 3—Capital Project Review and Design Procedures

Many industrial practitioners hold the opinion that careful attention to this element can have the greatest impact on the effectiveness of process-safety management.[15] Because of the importance of considering inherent safety early in the design sequence, when changes can most readily be made, inherent-safety considerations are particularly important in dealing with component 3.2, *hazard reviews* (e.g., a

Antivirus Software

Windows XP/2000/NT/ME Windows 95/98 Macintosh Dialup Connection

For Windows XP/2000/NT/ME

1. Uninstall ALL existing antivirus software.

- Failure to remove other antivirus products prior to installing the new software will cause problems with computer system.

- If you have any concerns, please seek advice from someone in your local IT support group, a workgroup manager or the help desk staff before proceeding.

2. Download and install the new antivirus software product

- Click the Download button. Your **NetID** and **password** will be required.

- Select "Open" if available or choose "Save" and select your desktop as the location.

- Locate the "Product" icon on your desktop and double click to install.

DOWNLOAD

(For Windows XP/2000/NT/ME only, Size: 1.66 MB)

3. What to Expect?...*more instructions given*

Figure 11.4 Second set of instructions concerning software installation. (From P. R. Amyotte, A. U. Goraya, D. C. Hendershot, and F. I. Khan. "Incorporation of Inherent Safety Principles in Process Safety Management," *Process Safety Progress* **26, no. 4 (2007): 333–346. With permission.)**

---

preliminary hazard analysis), and component 3.5, *process-design and review-procedures*. This is where many of the techniques discussed in Section 10.4.2 and referenced in Table 10.4 would be applicable. Perhaps the most useful feature of these tools is their potential to facilitate a relative comparison of the hazards and ensuing risk from different processing options.[16]

An important event of the past decade, relevant to all aspects of process safety and particularly inherent-safety and design procedures, occurred during the period December 1–3, 2004. These dates will be familiar to many people reading this book because they follow exactly 20 years after the MIC release at Bhopal. The event was the International Conference on the 20th Anniversary of the Bhopal Disaster: *Bhopal Gas Tragedy and Its Effects on Process Safety*, organized by Professor Jai P. Gupta of the Indian Institute of Technology, Kanpur, and held on the campus of IIT Kanpur. Many of the papers presented at the conference subsequently appeared in a special issue of the *Journal of Loss Prevention in the Process Industries*, volume 18 (numbers 4–6), published in 2005. Several of the quantitative indexing approaches referenced in Table 10.4 appear in this special issue.

As described in Section 10.2, qualitative as well as quantitative consideration of inherent safety in conducting *hazard reviews* (component 3.2) for selecting process options can be highly beneficial. The following example[17] further illustrates this point by making use of the inherent-safety guide words (Table 11.3) and checklist (Table 11.4). The case considered is the handling of dry additive at a polyethylene production facility. Key principles highlighted are intensification (minimization), substitution, and attenuation (moderation); additionally, the example illustrates the economic trade-offs and interrelationship among inherent, engineered, and procedural safety measures.

In the late 1970s, dry additive was received at the plant in heavy 50-kg containers. Operators were required to scoop additive into a feeder that supplied the additive to a preblender to ultimately mix additive with polyethylene resin to achieve certain resin properties. This activity caused concern over back strains as well as the need to wear respirators to control exposure to the additive dust. In the late 1980s and early 1990s, an effort was made to improve working conditions and efficiency. A capital project was proposed to pneumatically convey additive to the feeder. The additive is granular and has a minimum ignition energy of less than 10 mJ; it is therefore ignition-sensitive and must be treated carefully in a manner similar to flammable gases and liquids, which also have very low ignition energies. Chapter 13 discusses the hazards of explosible powders in detail.

A suggestion was made to consider the process with nitrogen as the conveying medium. A number of issues associated with this option soon materialized:

1. Higher operating costs using nitrogen (recycling nitrogen posed technical challenges; additionally, this would have been the first attempt at the site in conveying solids using nitrogen medium)
2. The need to monitor and control oxygen content so as not to exceed the limiting oxygen concentration
3. A requirement for greater operator attention if a manual monitoring approach was adopted
4. Prohibitively high project costs for automatic monitoring and alarms
5. Possible asphyxiation should nitrogen vent inside a building

Because of these concerns, the nitrogen-conveying option was not considered feasible. Operations returned to manual handling of the dry additive, but this time using smaller containers (25 kg) to deal with back-strain concerns. The site was also continuously working with the additive supplier to identify better approaches (e.g., to reduce manual handling). This gave rise to experimenting with a pellet-like version of the same additive in a pneumatic conveying system. This would, in theory, also remove the concern over the presence of a flammable dust cloud having a low ignition energy inside an air-conveying system. However, the mechanical energy of the pneumatic conveying system easily broke down the pellet-like additive into a very fine powder, causing extensive dust buildup and resulting in operating problems due to plugging. The dust-explosion concern was also reintroduced.

In the late 1990s, the site installed its current system, which involves the use of supersacks and tote tanks. Supersacks from the supplier are emptied by gravity into tote tanks at grade level. These totes are then taken by elevator to the appropriate floor, where they are placed on top of a feed-pipe system that connects to the additive feeder. The feed-pipe system is filled with additive by opening a slide valve located below the tote. The slide valve is opened slowly to avoid disturbing the sensitive operation of the feeder. This also helps with minimizing the formation of dust clouds inside the pipe system as it is filled. This current option has eliminated the following safety concerns:

1. Repetitive manual handling by operators and associated back strain
2. The need to wear respirators to avoid inhalation of dust
3. Excessive static charging and flammable dust cloud formation that would have been associated with pneumatic conveying, and the possible ignition of the ignition-sensitive dust inside an air-conveying system

Comments can also be made for PSM element 3 with respect to components 3.3, *siting*, and 3.4, *plot plan*. In siting a proposed expansion or new plant, the exposure hazard to and from adjacent plants or facilities is a critical consideration; similarly, the location of control rooms, offices, and other buildings should be carefully considered in conducting a plot-plan review.[7] This is in direct accordance with the discussion in Section 9.1 on avoiding knock-on or domino effects, and the use of unit segregation in the hierarchy of controls (Figure 1.1 in Section 1.3). Well-known examples where greater attention to facility siting and plot-plan review (temporary as well as capital) would have assisted with consequence mitigation include the administration and control buildings at Flixborough[18] and the contractor trailers at the BP Texas City refinery.[19]

11.2.4 PSM Element 4—Process-Risk Management

The PSM guide comments that component 4.1, *hazard identification*, is the most important step in process-risk management:

> If hazards are not identified, they cannot be considered in implementing a risk reduction program, nor addressed by emergency response plans.[7]

Several techniques are referenced in the guide for identifying and assessing hazards, including the Dow Fire and Explosion Index (F&EI) and Chemical Exposure Index (CEI), Hazop, What-If, and Checklist.[7] Section 10.4.2 of the current book discusses the inherent-safety aspects of the Dow indices, and Hazop is presented from an inherent-safety perspective throughout Section 10.2. Some of the quantitative methodologies referenced in Table 10.4 and applicable to PSM element 3 could also have relevance to PSM element 4; the key difference is that the review is no longer at the capital-project/early-design stage. For example, the Integrated Inherent Safety Index (I2SI) technique (Section 10.4.2) employs a structured guide-word approach in a Hazop-type manner.

PSM element 4 also offers a qualitative opportunity to bring inherent safety into components 4.1, *hazard identification*, and 4.3, *reduction of risk*, via the What-If/Checklist technique (which is simply a combination of the separate What-If and Checklist methods). As well described by the Center for Chemical Process Safety,[20] What-If analysis is a brainstorming approach in which a team asks questions and voices concerns about possible undesired events. The purposes are:

1. To identify hazards, hazardous situations, and specific events that could produce undesirable consequences
2. To examine the currently available safety measures to deal with the identified hazards and events
3. To suggest alternatives for risk reduction based on inherent as well as engineered and procedural safety

What-if analysis results are typically presented as a tabular listing of hazardous situations, consequences, existing safeguards, and options for risk reduction. An illustrative What-If example drawn from the everyday life activity of driving in a residential area is given in Table 11.5. There is a natural progression from the What-If scenario, to potential consequences of the event, to safety measures that might typically be employed, to recommendations for further safety measures (based in this case on inherent and passive safety).

In the current context, the key feature of the simple example in Table 11.5 is an attempt to explicitly incorporate inherent safety in the analysis. This can be done by the guide-word/checklist approach (Tables 11.3 and 11.4) in the following ways:

1. Once a number of What-If scenarios have been identified and analyzed by the team, checklists (inherent-safety-based and otherwise) can be consulted to determine if new hazards are identified. These new hazards can then be analyzed by completion of the What-If table.
2. In determining whether to make recommendations for consideration of further safeguards, the team can consult the checklists to see if new safety measures are identified.

Thus, checklists can be used at either the front or back end of the What-If methodology. In this manner, the Checklist analysis is combined with the What-If analysis to yield a What-If/Checklist method that combines the creative, brainstorming features of What-If with the systematic, rigorous features of Checklist.

Table 11.5 What-If Analysis Example with Inherent and Passive Safety Considerations

What If	Consequences/ Hazards	Existing Safeguards	Recommendations
• A pedestrian suddenly crosses the street in front of your car?	• You may hit the pedestrian, causing injury • You may hit another vehicle, causing damage • You may stop suddenly, causing injury in your vehicle	• Being attentive in residential areas • Always keeping a safe distance from other vehicles • Wearing seatbelt at all times	• Do not drive through residential areas except when necessary (Substitute) • Drive slowly in residential areas (Attenuate) • Consider providing fences between roads and areas with likely pedestrian exposure, for example, schools or playgrounds (Passive)

Source: Adapted from P. R. Amyotte et al., "Incorporation of Inherent Safety Principles in Process Safety Management," *Process Safety Progress* 26, no. 4 (2007): 333–346.

Further, the inherent-safety guide-word/checklist approach helps to ensure that inherent-safety considerations are explicitly considered in identifying both hazards and safety measures.

This is essentially the recommendation of the PSM guide: "Following risk evaluation, steps must be taken to reduce those risks which are deemed unacceptable. Such steps might include: *inventory reduction, alternative processes, alternative materials,* improved training and procedures, protective equipment, etc."[7] (with italics added to emphasize the inherent-safety suggestions related to intensification and substitution).

This type of simple, straightforward hazard analysis can be performed by virtually any plant worker engaged in any plant activity at any stage of a plant life cycle. It is ideally suited to what are commonly known in industry as field-level hazard (or risk) assessments, but this time with an inherent-safety emphasis.

Additional comments can be made for this PSM element with respect to components 4.6, *encouraging client and supplier companies to adopt similar risk-management practices,* and 4.7, *selection of businesses with acceptable risk.* The former component is particularly important because of the common practice of outsourcing or contracting of engineering services. This is often the case on large projects, where partnerships and joint ventures are formed, but may also apply to smaller projects through the awarding of subcontracts. Success on these projects is in large measure determined by the degree of commonality in risk-management practices among the different parties. Not the least of the concerns is whether there is a common set of expectations for safety performance and risk-awareness.[21] By extension, differences in approach to inherent safety and the corporate value placed on this concept can be problematic.

With respect to the latter component, *selection of businesses with acceptable risk,* helpful commentary is given in *Making EHS an Integral Part of Process Design*.[22] In explaining the business case for managing process safety, it is noted that effective PSM means "doing the right things right" (as per Peter Drucker's quote at the chapter start), thus leading to increases in revenue and productivity and reductions in product cost. It is further noted that for smaller companies, because of product stewardship requirements, demonstration of good process-safety management practices may be a prerequisite to doing business with larger companies more versed in PSM. Again, by extension, a mismatch with respect to the value placed on inherent safety may lead to missed business opportunities.

11.2.5 PSM Element 5—Management of Change

A system to manage change is critical to the operation of any facility. A written procedure should be required for all changes except replacement-in-kind. The system should address: a clear definition of change (scope of application); a description and technical basis for the proposed change; potential impact of the proposed change in health, safety and environment; authorization requirements to make the change; training requirements for employees or contractors following the change; updating of documentation including process safety information, operating procedures, maintenance procedures, alarm and interlock settings, fire protection systems, etc.; and contingencies for "emergency" changes.[7]

Simply put, inherent safety involves change, and change in the process industries must be managed. Potential hazards brought about by inherent-safety changes must therefore be identified and the ensuing risk reduced to an acceptable level. This is as important as the concept of looking for inherent-safety opportunities when making a process change.

A generic inherent-safety-based protocol for management of change (MOC) is shown in Figure 11.5.[4] This sequence of steps is based on the MOC process presented by Kelly.[23] It is important to note that MOC requires a clear statement as to the level in the supervisory/management hierarchy at which approval can be given. Best practice is that change must be approved by the professional level of management (or professional level of supervision in the United States, where

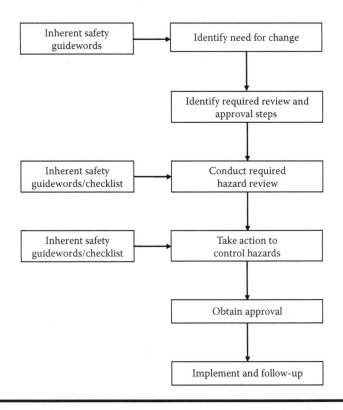

Figure 11.5   An inherent-safety-based management of change protocol. (From P. R. Amyotte, A. U. Goraya, D. C. Hendershot, and F. I. Khan. "Incorporation of Inherent Safety Principles in Process Safety Management," *Process Safety Progress* **26, no. 4 (2007): 333–346. With permission.)**

personnel who would be called managers in the United Kingdom are more likely called supervisors or superintendents). Equally as critical is the need for the authorizer to inspect the finished job to ensure that what has been done is what he or she authorized.

By recommending the use of inherent-safety guide words in the first step, *identify need for change*, the protocol recognizes that inherent safety is both a driving force for MOC and an opportunity during MOC. Use of the guide words and a checklist during the hazard-review and -control steps is again a recognition that the techniques used for these purposes are easily adaptable to explicit incorporation of inherent safety (such as the What-If/Checklist approach described in Section 11.2.4).

As an example, Figures 11.3 and 11.4 may again be considered, this time from an MOC perspective. The provision of systemwide antivirus software represented a change for the organization from the previous practice of computer users purchasing and installing such software individually and from a number of different vendors. The change was not adequately managed because the potential hazards arising from the change were not identified, and steps were not taken to reduce the risk of downloading without first uninstalling existing antivirus software. Use of the guide word *simplify* (Table 11.3) and the checklist question, "Are all manuals, guides and instructional material clear and easy to understand?" (Table 11.4), might have led to the following scenario: "What if someone clicks on the download button before uninstalling existing software?" Likely, risk-reduction measures such as those eventually taken after the fact would have been implemented as preventive measures. This rather simple example also illustrates the overlap between PSM elements and how inherent-safety considerations in a particular element can have multiple benefits throughout the suite of PSM tools.

Further consideration of the antivirus software upgrade process in a management-of-change review might lead to the following question: "Can the hazard of installing new antivirus software without removing previously installed antivirus software be eliminated?" It would seem that this might be possible in many cases; one could possibly design an installation routine that eliminates the hazard. Perhaps the installation program for the new antivirus software could canvass the computer hard disk for previous installations of commonly used antivirus software, and refuse to install if it finds one. Or, it could check for antivirus software already running; versions of security software are capable of monitoring the status of many antivirus programs, so presumably an installation program could check this same information. Some consumer software does this sort of thing now; for example, the installation routine for commercially available personal finance software checks for the presence of a previous version and uninstalls that previous version if found (after getting permission from the user).

In the process industries, Flixborough stands as a classic example of the need for both effective management of change and the use of inherently safer design. It is vital that these two "ways of thinking" remain linked to one another in the management of process safety. Change seems inevitable in life, and the advice of Hansen and Gammel[24] is worth heeding. These authors state that while MOC is critical to process safety, it is a concept that, if well implemented, could help prevent accidents in many other industries as well.[24]

11.2.6 PSM Element 6—Process and Equipment Integrity

Procedures for fabricating, inspecting and maintaining equipment are vital to process safety. Written procedures should be used to maintain ongoing integrity of process equipment such as: pressure vessels and storage tanks; piping, instrumentation and electrical systems; process control software; relief and vent systems and devices; emergency and fire protection systems; controls including monitoring devices and sensors, alarms and interlocks; and rotating equipment. A documented file should be maintained for each piece of equipment.[7]

As illustrated in Figure 11.2, this element is of obvious importance in preventing process-related incidents. Figure 11.1 shows the particular significance of components 6.5, *preventive maintenance*, and 6.7, *maintenance procedures*. With respect to the former component, risk-based inspection, integrity modeling, and integrity management represent current industry initiatives. There are clear inherent-safety implications here, in the sense that these approaches seek to optimize inspection and maintenance schedules for process equipment. They also afford an examination of *materials of construction* (component 6.2), thereby highlighting possible substitution in terms of parts replacement.

Alarm and instrument management (component 6.8) becomes arguably easier when there are fewer alarms and instruments to manage. This is a key accomplishment of inherently safer design.

Concerning *maintenance procedures* (component 6.7), a 1997 safety alert from the U.S. Environmental Protection Agency[25] offered the advice that facilities with storage tanks containing flammable vapors should review their equipment and operations in the following areas:

1. Design of atmospheric storage tanks
2. Inspection and maintenance of storage tanks
3. Hot-work safety
4. Ignition source reduction

While the middle two items in this list are largely procedural in nature, the first and last items have inherent-safety overtones (simplification and intensification, respectively). Additional inherent-safety benefits may include a reduced frequency of *opening of process equipment* (a subcomponent of *maintenance procedures*).

11.2.7 PSM Element 7—Human Factors

> Human factors are a significant contributor to many process accidents. Three key areas are operator–process/equipment interface, administrative controls and human error assessment.[7]

This PSM element on human factors has a strong relationship with the principles of inherent safety, particularly simplification (Section 7.4). Component 7.1, *operator–process/equipment interface*, refers to issues such as:[7]

1. The design of equipment increasing the potential for error (e.g., confusing equipment, positioning of dials, color coding, different directions for on/off, etc.)
2. The need for a task analysis (a step-by-step approach to examine how a job will be done) to determine what can go wrong during the task and how potential problem areas can be controlled

This component therefore affords ample opportunities for use of the What-If/Checklist technique described in Section 11.2.4. This relatively straightforward tool would likely have identified the need for consideration of human factors in the example given by Figures 11.3 and 11.4, as well as in a case reported in the media of a $225-million typo. This latter example involved the loss to a securities company of 27 billion yen (US$225 million) on a stock trade because of a "typing error."[26]

From a technical perspective, an inherent-safety-based What-If/Checklist task analysis may have identified the hazard posed by the design of a particular valve actuator at the Giant refinery in New Mexico.[27] The plug valve in question was originally designed to be opened and closed by a gear-operated actuator. This actuator was replaced by a valve wrench (2-ft-long bar) that was inserted into a square collar. (Thus, this is also an MOC incident.) Although there was a position indicator on the valve itself, operators had become accustomed to determining whether the valve was open or closed by the orientation of the valve wrench. If the wrench was perpendicular to the direction of flow through the valve, then the valve was thought to be closed. Reliance on this "on/off" determination ultimately proved to be flawed because of the fact that the wrench collar was removable and could be repositioned on the valve stem in different directions. On the day of the incident, the valve was thought to be closed when it was in fact open, and a flammable liquid release occurred, followed by a fire and explosion.

The description of component 7.2 of this PSM element, *administrative control versus hardware control*, contains these statements:

> Hazards may be controlled by the use of procedures or by the addition of protective equipment. This balance is often a matter of company culture and economics. If procedures are well understood, kept current and are used, then they are likely to be effective. Similarly protective systems need regular testing and maintenance to be effective. The problem of administrative versus hardware controls should be considered and a balance selected by conscious choice rather than allowing it to happen by default.[7]

This description may leave some readers with the unfortunate impression that only procedural (administrative) and engineered (add-on) measures are available, or are effective, for hazard control. Although well-written and understood procedures will be inherently safer, there is much more to inherent safety than facilitating effective procedural safety. This component is one of the most important places in the explanation of PSM where explicit use of the term *inherent safety* would be highly beneficial.

The third and final component of human factors, 7.3 *human error assessment*, is perhaps one of the more challenging areas of process-safety management. Human error assessment alone can seem daunting, let alone trying to incorporate the

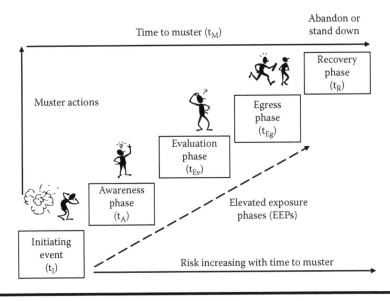

Figure 11.6 **Graphical representation of the phases comprising a muster sequence offshore. (From P. R. Amyotte, A. U. Goraya, D. C. Hendershot, and F. I. Khan, "Incorporation of Inherent Safety Principles in Process Safety Management," in** *Proceedings of 21st Annual International Conference, Center for Chemical Process Safety, American Institute of Chemical Engineers* **(Orlando, FL, 2006), 175–207. With permission.)**

principles of inherent safety. Nevertheless, human error assessment is becoming increasingly important in industry and is a growing area of concern for the public and for regulators.

A recent example from the offshore industry illustrates the quantitative assessment of human error probability and the use of these results to help achieve an inherently safer design. DiMattia et al.[28] and Khan et al.[29] describe a method for determining human error probabilities during the process of emergency musters on offshore oil and gas production platforms. This process consists of 18 separate actions or tasks (beginning with alarm detection) grouped into four distinct phases, as shown in Figure 11.6.[3] The final phase involves the tasks performed in the temporary safe refuge (TSR) before standing down or moving on to the platform abandonment stage. Each phase of the muster has an associated elapsed time (e.g., t_{Eg} for the egress phase), which collectively make up the total time of muster (t_M).

An expert judgment technique known as SLIM (Success Likelihood Index Methodology), developed in the nuclear industry, was used by DiMattia et al.[28] Factors affecting human performance (performance-influencing factors such as stress, training, and experience) were quantitatively analyzed by means of a team of expert judges and the SLIM technique to yield a human error probability for each of the 18 muster tasks and 3 different muster initiators (man overboard, gas release, and fire and explosion). These data were then generalized to any muster initiator and scenario by a questionnaire and reference graph methodology developed by Khan et al.[29]

The usefulness of human error probability data was illustrated by Khan et al.[29] in the following manner. First, a consequence table was developed and used to qualitatively assess the impact of not completing a given muster task. This severity ranking was then paired with the quantitative probability value by means of a risk matrix to identify those tasks most in need of risk-reduction measures (inherent, engineered, and procedural). For example, failure to detect the muster alarm (first action) in the fire and explosion scenario has a probability of 0.4 and consequences ranging from delay of time to muster through to loss of critical time needed to respond to the problems that initiated the alarm, and possible loss of life. Risk-reduction measures from an inherent-safety perspective were identified as:

1. Eliminate obstructions near the alarms.
2. Minimize the amount of noise-producing machinery when possible.
3. Substitute alarms with ones that give an easily recognizable tone.
4. Reduce the electrical dependence of the alarms that are used on the platform.
5. Simplify the alarms, making maintenance easier, as well as lowering the possible risk of an alarm malfunctioning.

Quantitative human error assessment is therefore not only possible, but can be valuable in attempting to reduce risk by employing the principles of inherent safety. What is required in this application is a scientifically rigorous method of determining probability data for human error, such that objectivity is brought to an otherwise potentially subjective process. The topic of human/operator reliability is briefly touched upon further in Section 15.3.

11.2.8 PSM Element 8—Training and Performance

> People need to be trained in the right skills and to have ongoing retraining to maintain these skills.[7]

Embodied within the components of this PSM element (Table 11.1) is the management cycle comprised of *plan, do, check, act*. The activities in the training protocol given by DiBerardinis[30] are particularly appropriate in this regard:

1. Conducting a needs analysis
2. Setting learning objectives
3. Deciding on the method of presentation and delivery of training
    a. Styles: lectures, group discussion, hands-on exercises, self-learning, etc.
    b. Materials: overheads, slides, videos, workbooks, etc.

4. Evaluating the instruction
5. Providing feedback on the instruction

An important area to examine when assessing training needs is accident investigation reports (Section 10.5). These will often highlight the requirement for more attention to inherent safety, possibly with respect to training in the use of inherent-safety principles.

Excellent advice is given by Felder and Brent[31] concerning the setting of instructional (learning) objectives. They comment that in setting such objectives, there are four leading verbs that should be avoided: *know, learn, appreciate,* and *understand.* Thus, while it would be desirable for plant employees to know, learn, appreciate, and understand the principles of inherent safety, these are not valid instructional objectives because it is not possible to directly see whether they have been done. It is necessary to consider what trainees should be asked to *do* to demonstrate their knowledge, learning, appreciation, and understanding of inherent safety, and then make those activities the instructional objectives.[31]

Felder and Brent[31] further describe the concept of using action verbs to set instructional objectives. Using their breakdown according to *Bloom's Taxonomy of Educational Objectives,* appropriate instructional objectives for inherent safety would be:

1. **Knowledge:** *list* the key principles of inherent safety; *state* the preferred order of safety measures in the hierarchy of controls.
2. **Comprehension:** *explain* in your own words the concept of attenuation as applied to inherent safety.
3. **Application:** *calculate* the Dow Fire and Explosion Index for several material inventories.
4. **Analysis:** *identify* the inherent-safety features in a given design; *explain* the inherent-safety causation factors for a given incident description.
5. **Synthesis:** *formulate* an inherently safer alternative to a given design or processing route; *make up* a case study (technical or from everyday life) involving some aspect of inherently safer design.
6. **Evaluation:** *determine* which of several chemical reactor designs is inherently safer and explain your reasons; *select* from available techniques for measuring the inherent-safety features of a design and justify your choice.

With respect to training resources on inherent safety, there are now many journal papers, presentations, and texts available, as witnessed by the citations at the end of each chapter in this book. Gupta has also published details of his recommended content for a course on inherently safer design.[32] Additional resources include the Loss Prevention Bulletins published by the Institution of Chemical Engineers (IChemE, U.K.) and the incident investigation reports regularly posted by the U.S. Chemical Safety Board (CSB) on its Web site (www.csb.gov). Many of these are full of inherent-safety lessons, as discussed in Chapter 14. Finally, there are excellent training packages on inherent safety and inherently safer process design available from technical societies such as the American Institute of Chemical Engineers[33] and the Institution of Chemical Engineers.[34]

A final comment is made for this element with respect to the continuous improvement loop necessary in process-safety management. In conducting training exercises, especially those that test the performance of the human body, it is important to note when human limitations occur and to feed these observations back into the design of the workplace and its systems. In this manner, an inherently safer design can be attempted so that human-error acceptance is replaced by human-factor consideration.

11.2.9 PSM Element 9—Incident Investigation

The interaction between this PSM element and inherently safer design was examined by Goraya et al.[10] in their development of an inherent-safety-based incident investigation methodology. In Figure 11.7,[10] one may see the (hopefully by now) familiar approach of guide words and checklists being brought into a generic protocol at appropriate places to again explicitly incorporate the principles of inherent safety.

Other features of the methodology shown in Figure 11.7 include the use of:

1. A basic framework adapted from that presented by Rosser[35] with additional input from the Institute of American Chemical Engineers[36]
2. An integrated approach in considering all potential categories of loss[37]
3. Evidence classification into data categories[38]
4. The domino loss causation model for identification of causal factors[39]
5. A layered investigation approach for making recommendations[40]

The final item in this list is critical to the integration of inherent-safety considerations within an investigation protocol. It is well understood that the root causes of process incidents are typically management system deficiencies. And it is of course necessary to take immediate action to remove existing hazards following an incident. But inherently safer design, as emphasized in Section 10.5, requires an attempt to avoid hazards and to permanently remove them whenever possible.

11.2.10 PSM Element 10—Company Standards, Codes, and Regulations

> A *management system* is needed to ensure that the various internal and external published guidelines, standards and regulations are current, disseminated to appropriate people and departments, and applied throughout the plant.[7]

The PSM guide[7] breaks this element into two components: external to the company and internal to its operations. External codes and regulations include legislated

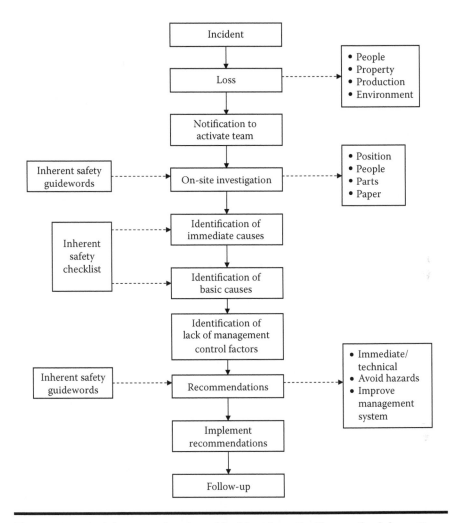

Figure 11.7 **An inherent-safety-based incident investigation methodology. (From A. Goraya, P. R. Amyotte, and F. I. Khan, "An Inherent Safety-Based Incident Investigation Methodology,"** *Process Safety Progress* **23, no. 3 (2004): 197–205. With permission.)**

items; these have been discussed earlier in Section 10.3. A brief comment here is the common theme in regulations globally to refer to specified threshold quantities of hazardous materials. This is an obvious recognition of the benefits brought about by intensification of inventories.

With respect to internal standards, examples do exist where companies have incorporated inherent safety into process-design principles[41] as well as specific elements of process-safety management.[42] Renshaw[43,44] describes the Rohm and Haas major accident prevention program, which uses checklists and consequence analysis to encourage inherent-safety considerations and understand the potential reduction in consequences from alternatives based on inherent safety. One could comfortably surmise that the use of inherently safer design principles in major chemical companies[45,46] occurs, at least in part, because of internal standards promoting and mandating their consideration.

11.2.11 PSM Element 11—Audits and Corrective Actions

> The purpose of safety audits is to determine the status and effectiveness of safety management efforts versus goals and also the progress toward those goals.[7]

It stands to reason that if the principles of inherent safety are employed as a key means of ensuring effective process-safety management, then the PSM system should be audited for visible signs of inherent-safety naming and usage. Expressed differently, and linking back to PSM element 1, given that one of the goals of a PSM system is to demonstrate leadership on process-safety matters, then a system audit should seek evidence of a commitment to inherent-safety understanding and usage.

11.2.12 PSM Element 12—Enhancement of Process-Safety Knowledge

> A management system for process safety should be designed for continuous improvement. Safety requirements are becoming more stringent, while knowledge of systems and technology is growing, e.g., consequence modelling techniques. Safe operation of a process plant calls for personnel to stay abreast of current developments, and for safety information to be readily accessible.[7]

In the last sentence above, this element refers to staying abreast of current developments; this surely must include inherent safety and the increased level of research and practice activity over the past decade or so. As noted in Section 11.2.8, a significant number of resources on inherent-safety principles and usage are now available. Any discussion of the enhancement of process-safety knowledge in the twenty-first century must include inherent safety and inherently safer design.

11.3 Safety Culture

To conclude this chapter, some thoughts are given on the role of safety culture in today's process-safety world. There can be no doubt it is an important topic, especially since the BP Texas City incident in 2005 (Section 9.1).[47] Recent emphasis has also been placed on safety culture in other fields (e.g., occupational health and safety[48]) and in other applications (e.g., offshore safety[49]). Section 11.2.1 has addressed safety culture from the perspective of management leadership and accountability within a process-safety management system. Reference 9 on risk-based process safety (RBPS) management lists the first of its four RBPS pillars as committing to process safety, and emphasizes the need to develop and sustain a culture that embraces process safety.

It is beyond the scope of the current book to provide an extensive review of the literature on process-safety culture. Here it is sufficient to say that two key points have emerged:

1. Typical occupational safety indicators such as lost-time injuries (LTIs) are inappropriate as primary indicators with respect to process safety.
2. Leading indicators are generally viewed as more useful than lagging indicators.

Similar to the development of methods to measure the inherent "safeness" of a process (Section 10.4.2), safety-culture metrics is currently an area of significant interest in both academia and industry. Hopkins[50] has recently commented that perhaps the most important consideration for process-safety indicators is that they measure the effectiveness of the various controls comprising the risk-control system. This affords an opportunity to link back to the elements making up the process-safety management system and assess the level of commitment to the principles of inherent safety within that system. Little work appears to have been done in this area, which has been identified by Glendon[51] as a major challenge for industry (i.e., the linking of safety-culture methodologies with process-safety approaches and broader systems safety and risk-management concepts).

The coming years will undoubtedly see advances in the area of metrics for process-safety culture; this will be a welcome development, especially where the measurement tools address inherently safer design. Perhaps, however, there is something to be gained by looking to the past as well as to the future. Many examples have been given in this book of inherent-safety applications dating centuries back. It would seem that while it has not necessarily been named as such, safety culture has also benefited from consideration over time.

One of the first recorded accounts of a dust explosion was written by Count Morozzo, who gave a detailed account of an explosion in a flour warehouse in Turin, Italy.[52] In the final paragraph of his report, the count writes:

> Ignorance of the fore-mentioned circumstances, and a culpable negligence of those precautions which ought to be taken, have often caused more misfortunes and loss than the most contriving malice. It is therefore of great importance that these facts should be universally known, that public utility may reap from them every possible advantage.[53]

The above passage makes an eloquent case for the importance of incident investigation and the sharing of lessons learned—and for a strong safety culture. It is instructive to also note that it was written over 200 years ago.

And with a slight adaptation to the subject of safety (given in brackets below), the words of Benjamin Franklin[54] from 1758 give further credence to the notion that cultural actions, not just system and process factors, must be given their due consideration:

> Friends and Neighbors, the [impositions of the safety regulator] are indeed very heavy, and if those laid on by the Government were the only ones we had to pay, we might more easily discharge them; but we have many others, and much more grievous to some of us. We are taxed twice as much by our Idleness, three times as much by our Pride, and four times as much by our Folly; and from these taxes the Commissioners cannot ease or deliver us.

References

1. G. I. J. M. Zwetsloot and N. A. Ashford, "The Feasibility of Encouraging Inherently Safer Production in Industrial Firms," *Safety Science* 41, no. 2-3 (2003): 219–240.
2. N. A. Ashford and G. Zwetsloot, "Encouraging Inherently Safer Production in European Firms: A Report from the Field," *Journal of Hazardous Materials* 78, no. 1-3 (2000): 123–144.
3. P. R. Amyotte, A. U. Goraya, D. C. Hendershot, and F. I. Khan, "Incorporation of Inherent Safety Principles in Process Safety Management," in *Proceedings of 21st Annual International Conference, Center for Chemical Process Safety, American Institute of Chemical Engineers* (Orlando, FL, 2006), 175–207.
4. P. R. Amyotte, A. U. Goraya, D. C. Hendershot, and F. I. Khan, "Incorporation of Inherent Safety Principles in Process Safety Management," *Process Safety Progress* 26, no. 4 (2007): 333–346.
5. "New Integrated Management System Attempts to Link Environment, Health, Safety and Process Management," *Workplace Environment Health & Safety Reporter* 7, no. 1 (2001): 1166.
6. P. R. Amyotte and D. J. McCutcheon, "Risk Management: An Area of Knowledge for All Engineers" (discussion paper prepared for Canadian Council of Professional Engineers, 2006).
7. CSChE, *Process Safety Management*, 3rd ed. (Ottawa, ON: Canadian Society for Chemical Engineering, 2002).
8. Center for Chemical Process Safety, *Guidelines for Technical Management of Chemical Process Safety* (New York: American Institute of Chemical Engineers, 1989).
9. Center for Chemical Process Safety, *Guidelines for Risk Based Process Safety* (Hoboken, NJ: John Wiley & Sons, 2007).
10. A. Goraya, P. R. Amyotte, and F. I. Khan, "An Inherent Safety-Based Incident Investigation Methodology," *Process Safety Progress* 23, no. 3 (2004): 197–205.

11. Center for Chemical Process Safety, *Inherently Safer Chemical Processes: A Life Cycle Approach*, 2nd ed. (Hoboken, NJ: John Wiley & Sons, 2009).

12. S. Griffiths, "Leadership, Commitment & Accountability: The Driver of Safety Performance," in *Process Safety and Loss Management Symposium, 55th Canadian Chemical Engineering Conference, Canadian Society for Chemical Engineering* (Toronto, 2005).

13. A. Hopkins, *Safety, Culture and Risk: The Organizational Causes of Disasters* (Sydney: CCH Australia Limited, 2005).

14. D. C. Hendershot, "Tell Me Why," *Journal of Hazardous Materials* 142, no. 3 (2007): 582–588.

15. G. Creedy, private communication, 2005.

16. G. Phillips, private communication, 2005.

17. M. Marta, private communication, 2005.

18. R. E. Sanders, "Designs That Lacked Inherent Safety: Case Studies," *Journal of Hazardous Materials* 104, no. 1-3 (2003): 149–161.

19. U.S. Chemical Safety Board, Washington, DC, 2006, http://powerlink.powerstream.net/002/00174/051222bp/BPAnimations.asx.

20. Center for Chemical Process Safety, *Guidelines for Hazard Evaluation Procedures*, 2nd ed. (New York: American Institute of Chemical Engineers, 1992).

21. J. C. Horwood and P. R. Amyotte, "Impact of Expectations and Culture on Project Safety Performance," IChemE Symposium Series No. 153, in *12th International Symposium on Loss Prevention and Safety Promotion in the Process Industries* (Edinburgh, U.K., 2007).

22. Center for Chemical Process Safety, *Making EHS an Integral Part of Process Design* (New York: American Institute of Chemical Engineers, 2001).

23. B. D. Kelly, in *Management of Change in Process Plants: A Participative Workshop* (Calgary, AB, Canada, November 2000).

24. M. D. Hansen and G. W. Gammel, "Management of Change: A Key to Safety—Not Just Process Safety," *Professional Safety* 53, no. 10 (2008): 41–50.

25. EPA, "Catastrophic Failure of Storage Tanks" (Chemical Safety Alert, U.S. Environmental Protection Agency, Washington, DC, 1997).

26. "Tokyo Market Suffers after $225-Million Typo," *Chronicle Herald* (Halifax, Canada), 10 December 2005.

27. CSB, "Oil Refinery and Explosion (Giant Industries' Ciniza Oil Refinery)" (Case Study, U.S. Chemical Safety Board, Washington, DC, 2005).

28. D. G. DiMattia, F. I. Khan, and P. R. Amyotte, "Determination of Human Error Probabilities for Offshore Platform Musters," *Journal of Loss Prevention in the Process Industries* 18, no. 4-6 (2005): 488–501.

29. F. I. Khan, P. R. Amyotte, and D. G. DiMattia, "HEPI: A New Tool for Human Error Probability Calculation for Offshore Operation," *Safety Science* 44, no. 4 (2006): 313–334.

30. L. J. DiBerardinis, ed., *Handbook of Occupational Safety and Health*, 2nd ed. (New York: John Wiley & Sons, 1999).

31. R. M. Felder and R. Brent, "Objectively Speaking," *Chemical Engineering Education* 31, no. 3 (1997): 178–179.

32. J. P. Gupta, "A Course on Inherently Safer Design," *Journal of Loss Prevention in the Process Industries* 13, no. 1 (2000): 63–66.

33. SACHE, *Inherently Safer Plants* (New York: Center for Chemical Process Safety, American Institute of Chemical Engineers, 1996), Slide Package, Safety and Chemical Engineering Education.

34. IChemE, *Inherently Safer Process Design*, Training Resource STP001 (Rugby, U.K.: Institution of Chemical Engineers, 2005).

35. B. W. J. Rosser, "Incident Investigation: The Most Important Element in Any H.S.E. Management System," in *Proceedings of the 4th Biennial Process Safety and Loss Management Conference* (Calgary, AB, Canada, 2000).

36. Center for Chemical Process Safety, *Guidelines for Investigation of Chemical Process Incidents*, 2nd ed. (New York: American Institute of Chemical Engineers, 2003).

37. L. Wilson and D. McCutcheon, *Industrial Safety and Risk Management* (Edmonton: University of Alberta Press, 2003).

38. L. Wilson, *Basic Learnings in Industrial Safety and Loss Management* (Edmonton: Association of Professional Engineers, Geologists and Geophysicists of Alberta, 1998).

39. F. E. Bird and G. L. Germain, *Practical Loss Control Leadership* (Loganville, GA: Det Norske Veritas, 1996).

40. T. A. Kletz, "Layered Accident Investigations," *Hydrocarbon Processing* November (1979): 373–382.

41. NOVA, "LPS 2.2: Inherently Safer Designs," Group of Loss Prevention Standards 2 (Plant Design) (Red Deer, AB, Canada: NOVA Chemicals Corp., 2001).

42. W. Bain, private communication, 2005.

43. F. M. Renshaw, "A Major Accident Prevention Program," *Plant/Operations Progress* 9 (1990): 194–197.

44. F. M. Renshaw, "A Major Accident Prevention Program: Ten Years of Experience," *Process Safety Progress* 23, no. 2 (2004): 155–162.

45. C. C. Clements, "Application of Inherently Safer Process Concepts in DuPont," in *American Institute of Chemical Engineers 2005 Spring National Meeting, Process Plant Safety Symposium* (Atlanta, 2005).

46. T. Overton and G. M. King, "Inherently Safer Technology: An Evolutionary Approach," *Process Safety Progress* 25, no. 2 (2006): 116–119.

47. D. C. Hendershot, "Process Safety Culture," *Journal of Chemical Health and Safety* 14, no. 3 (2007): 39–40.

48. J. A. Erickson, "Corporate Culture: Examining Its Effects on Safety Performance," *Professional Safety* 53, no. 11 (2008): 35–38.

49. S. Antonsen, "The Relationship between Safety Culture and Safety on Offshore Supply Vessels," *Safety Science* 47, no. 8 (2009): 1118–1128.

50. A. Hopkins, "Thinking about Process Safety Indicators," Working Paper 53, National Research Centre for OHS Regulation, Australian National University (paper prepared for presentation at the Oil and Gas Industry Conference, Manchester, U.K., November 2007).

51. I. Glendon, "Safety Culture and Safety Climate: How Far Have We Come and Where Should We Be Heading?" *Journal of Occupational Health and Safety: Australia and New Zealand* 24, no. 3 (2008): 249–271.

52. P. R. Amyotte and R. K. Eckhoff, "Dust Explosion Causation, Prevention and Mitigation: An Overview," *Journal of Chemical Health and Safety*, 17, no. 1, January/February 2010: 15–28.

53. C. Morozzo, "Account of a Violent Explosion Which Happened in the Flour-Warehouse, at Turin, December the 14th, 1785; To Which Are Added Some Observations on Spontaneous Inflammations," from the Memoirs of the Academy of Sciences of Turin (London: The Repertory of Arts and Manufactures, 1795).
54. Benjamin Franklin, quoted in *Economic Briefing*, No. 4, HM Treasury, London, September 1992, p. 7.

Chapter 12

Friendlier Plants and the Nuclear Industry

Can a new type of fission reactor solve the nuclear power
problems of U.S. electric utilities? Is it realistic to believe that
manufacturers will build, federal regulators will allow, and
utilities will buy such radically different reactors? A growing
coalition of manufacturers, utility planners, and engineers thinks
the answer may very well be yes.

—L. M. Lidsky[1]

12.1 New Types of Reactors

There is one sense in which a nuclear reactor is already inherently safe. It cannot
explode like an atomic bomb (the nuclear equivalent of the general physics axiom:
You can't push on a rope). No protective equipment prevents such an explosion
because the reactors are inherently incapable of exploding in this way. However,
virtually all commercial reactors could suffer a runaway or meltdown if their layers
of protective equipment all failed at the same time—a very unlikely coincidence—
or were neglected, or switched off, as at Chernobyl.[2] What, then, will be the result
of applying the philosophy of this book to the nuclear industry? Will we have to
stop building nuclear reactors and make our electricity in other ways?

The answer is no, for there are designs of nuclear reactor that are inherently
incapable of overheating to the same extent as water-cooled reactors. They are not

dependent, or are less dependent, on engineered protective systems that might fail or be neglected.

Gas-cooled reactors are inherently safer than water-cooled ones because, if coolant pressure is lost, convection cooling assisted by a large mass of graphite will prevent them from overheating to the same extent.[3] Fast breeder reactors are inherently safer than thermal reactors. According to Marsham,[4] "The unique capability of fast reactors to remove heat from the fuel by natural circulation without external power or water supplies has been fully demonstrated … as have the characteristics of the system which inherently cause large power reductions if for some reason temperatures start to increase. In addition we are finding that the radiation levels experienced by the operators are barely discernible above the natural radiation levels." (Note: The terms *fast* and *thermal* refer to the energy of the neutrons. In nearly all existing commercial reactors, gas- and water-cooled, the neutrons have low or thermal energies.)

The fast reactor uses liquid metal, usually liquid sodium, as the coolant. On the other hand, the coolant pressure is low, which means that any leaks are likely to be small. The coolant is contained in a double-walled vessel, and thus if there is a leak, the core will not be uncovered. Yokoyama et al.[5] describe a fast reactor technology utilizing zirconium alloy metal fuel with liquid sodium coolant in which sodium boiling during an unprotected loss of coolant flow (ULOF) event is eliminated by an inherent-safety feature termed *self-controllability*.

Weinberg and Spiewak[6] and Lidsky[1] have described other types of inherently safer nuclear reactors under development in the latter part of the twentieth century, in which a core meltdown is a physical impossibility.

In the high-temperature gas (HTG) reactor, a small core containing graphite to moderate (that is, slow down) the neutrons is cooled by high-pressure helium. In conventional reactors, the heat developed by radioactive decay of fission products can, if the cooling has failed, melt the core after the chain reaction has been stopped. In the HTG reactor, the high-temperature resistance of the fuel and the high surface-to-volume ratio ensure that the afterheat is lost by radiation and conduction to the environment and that the core temperature does not exceed a safe level. The reactors are small, and a typical power station might contain ten.

One design of HTG is called a pebble-bed reactor because the core consists of hundreds of thousands of graphite spheres, about the size of tennis balls, containing particles of uranium oxide coated with silicon carbide. The fission products are retained within the spheres cooled by the helium, which is used to raise steam. Fresh spheres are added to the top of the reactor, move slowly down, and are discharged from the bottom. Each sphere is like a miniature pressure vessel. The design is relatively expensive, but it should be possible to mass produce the units in factories rather than to have to construct them on-site.

A recent paper describes the use of a pebble-bed nuclear reactor in conjunction with an open-cycle gas turbine under the Netherland's NEREUS project.[7] The NEREUS acronym represents a set of admirable inherent-safety goals: *N*atural safe, *E*fficient *R*eactor, *E*asy to operate, *U*ltimately simple and *S*mall.

Another inherently safer design is the Swedish Process Inherent Ultimate Safety (PIUS) reactor, which is a water-cooled reactor immersed in boric acid solution. If the coolant pumps fail, the boric acid solution enters the core by natural convection. The boron absorbs neutrons and stops the chain reaction, and the water removes the residual heat. No makeup water is needed for a week.

These alternatives to water-cooled reactors have lower power densities and thus tend to be more expensive but, on the other hand, contain less additional protective equipment and give the operator longer time in which to react. Franklin writes,[8] "When operators are subject to conditions of extreme urgency ... they will react in ways that lead to a high risk of promoting accidents rather than diminishing them. This is materially increased if operators are aware of the very small time margins that are available to them.... It is much better to have reactors which, even if they do not secure the last few percent of capital cost effectiveness, provide the operator with half an hour to reflect on the consequences of the action before he needs to intervene."

In 1989, a group of U.S. and U.K. organizations put forward a proposal for the joint development of a small (300 MW) water-cooled reactor, the Safe Integral Reactor (SIR).[9] This size was chosen so that the steam generator and the water pumps, and not just the core, could be installed in the containment vessel. According to Hayns,[9] the power density would be "about half that of current large plant and makes an important contribution to increased safety margins and operating flexibility ... for all transient events the reactor is essentially self-regulating." Convection cooling will be sufficient to remove the afterheat, and "there are no transient events which threaten the core and hence no diverse emergency shutdown system should be necessary." However, as of the late 1990s, development had been abandoned. A similar Italian design has been described.[10,11]

It is interesting to note that work on the Italian design described in Reference 10 has continued into the twenty-first century with further development of the 600 MWth MARS (Multipurpose Advanced "inherently" Safe Reactor).[12] An area of current interest is the search for a simplified solution in terms of fuel management. A candidate approach is a once-through fuel cycle in which the fuel assemblies are loaded in the core and remain in the same position until eventual unloading.[12]

As mentioned in Section 9.5, the Chernobyl boiling-water reactor was particularly unfriendly because, at low power outputs, below 20% of maximum load, it had a positive power coefficient; that is, if it got too hot, the rate of heat production increased, and the reactor got hotter still. The operators were told not to go below 20% output, but there was no added protective equipment to prevent them from doing so, and when they did so, for what they thought were good reasons (to complete an urgent experiment), a runaway occurred.[2]

No other commercial design of reactor has a positive power coefficient (though some U.S. military reactors do), and the Chernobyl design has now been modified. The fast reactor has a particularly large negative coefficient.

The Chernobyl reactor may be likened to a marble on an inverted (convex-up) saucer. If the marble moves from the equilibrium position, the forces causing

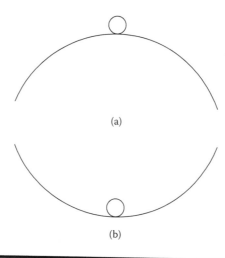

(a)

(b)

Figure 12.1   Comparison of reactors: (a) The Chernobyl nuclear reactor, at low outputs, resembled a marble on a convex surface. If the temperature rose, the heat output increased, and the temperature rose uncontrollably. (b) Other reactors resemble a marble on a concave surface. As the temperature rises, the heat output falls.

it to move increase. Other reactors resemble a marble on a concave-up saucer (Figure 12.1).

Weinberg and Spiewak[6] argue that in a typical pressurized water reactor (PWR), the commonest type, there will be a core melt accident once in 10,000 years (though less often in the latest U.K. design). This does not seem very often, but if in the future there are 5000 reactors, about 10 times the present number, there will be a meltdown every two years. Only one in 10 or 100 of these will result in a discharge of radioactive materials of Chernobyl size, but even so, it is doubtful that the public will accept it. More important, the calculations assume that the reactors are well designed and are operated as intended by the designers. In the light of Chernobyl and Three Mile Island, and incidents in other industries such as Flixborough (Section 3.2.4) and Bhopal (Section 3.6), is this a realistic assumption?

In the short term, the PWR, with its complex, replicated additional safety systems, may be the right answer for the West. It can be argued that the operating companies have the technical competence and commitment to see that the added safety systems are properly used and maintained. But the advantages of the inherently safer designs are so great that perhaps we will be building conventional PWRs for only a few more decades. For countries that lack the resources, culture, or commitment necessary to maintain complex added safety systems, conventional PWRs may not be the answer even today, and those countries should perhaps wait until inherently safer designs are available. Inherently safer designs may overcome public resistance to nuclear power, but this has been questioned (Section 10.3).

Although I feel confident that a major nuclear accident in the West is extremely improbable, I have to admit that, before the explosion at Flixborough and the toxic release at Bhopal, I would have made similar remarks about companies under the control of Dutch State Mines and Union Carbide. However, both these accidents occurred in overseas plants, which tend to get less supervision than domestic ones. One is reminded again of the admonition of Edwards (Section 4.2.4): "We should export inherent safety, not risk."[13]

12.2 Features of Nuclear Plants

Features of nuclear plants already discussed are the use, in some plants, of convection cooling (Section 9.11) and of highly reliable equipment requiring little or no maintenance (Section 6.3.2). As a result of robust construction, the gas-cooled reactor at Calder Hall, United Kingdom, the world's first commercial reactor, was approved for operation into the twenty-first century[14] and actually ceased power generation only in 2003.

Forsberg and Weinberg[15] list five PRIME features desirable in a nuclear reactor:

1. *P*assive safety (Section 9.11)
2. *R*esilient operation (that is, the safety systems do not interfere with the operation and maintenance of the plant, and thus there is no incentive to bypass them. It is better to remove the reasons for bypassing them than to make bypassing difficult. If operators are determined to bypass a device, they will usually find a way to do so in the end. Recall the nitrogen asphyxiation incident described in Section 9.2.)
3. *I*nherent safety
4. *M*alevolence resistance
5. *E*xtended safety (that is, the ability to walk away from the plant in the knowledge that it is safe to do so because, however many active protective devices fail, there will be no disaster)

In a lengthy report, Forsberg et al.[16] have described many examples of these features. Some had not yet been tried out as of the late 1990s and may have fallen by the wayside. Several are used in the joint U.K.–U.S. design for a Safe Integral Reactor described in Section 12.1. Some of the suggestions may have possible applications elsewhere and are described below in this section in the hope that they may stimulate a few thoughts. Sweden now seems to be a primary source of new ideas in nuclear technology, especially in light of recent government decisions to permit replacement of existing nuclear reactors (thus effectively reversing the previous nuclear phase-out policy). Japan is also important in this regard (e.g., Reference

5), but the United States appears to have lost its lead. As described by Ingersoll,[17] the United States is experiencing the formative stages of a "nuclear renaissance" or second nuclear era (the latter term having been coined by Weinberg in the early 1980s[6]). Ingersoll contends that reactors deliberately built small have a significant role to play in this renaissance by enhancing the safety, security, operational flexibility, and economic attractiveness of nuclear power generation.[17]

The advantages of convective cooling have already been mentioned. Extending this idea, by half burying a reactor next to the ultimate heat sink, the cooling pond, emergency convective cooling can take place directly to the pond (p. 6-75, with page numbers referring to the report of Forsberg et al.[16]).

Other ideas for improving cooling are the following:

1. Diesel generators for supplying power for emergency cooling are expensive, unreliable, and easily neglected; batteries are better; gravity is better still.
2. In the PIUS design, if pumped water circulation stops, the reactor can be flooded by gravity with borated water, which both cools and absorbs neutrons. The borated water can be kept out by a fluidic diode, a check valve with no moving parts, that "opens" when the pressure falls and allows circulation by convection (Section 7.9 and pp. 6-9 and 6-50[16]).
3. If water circulation is lost, steam will be formed and makeup water needed. The steam flow could pass through an injector that sucks in the water (p. 6-60).

The report includes many ideas for preventing overheating in reactors, such as the following:

1. Metal hydrides could be placed in tubes in a reactor core. The hydrogen atoms are a moderator; that is, they slow down neutrons so that they react. If local overheating occurs, the hydrides decompose, the hydrogen migrates, and the reaction rate falls (p. 5-21).
2. Reaction stoppers (neutron poisons) could be placed inside fuel elements. Normally they would be shielded by the fuel but would be released if the fuel overheats (p. 5-68).
3. Tin can provide passive negative feedback; as it gets hotter it absorbs more neutrons and slows down the nuclear reaction (p. 5-57).
4. The control rods controlling the core temperature are usually held up by electric power; an active protective system has to send a signal for them to be lowered. If they are held up by water pressure, they will fall if pressure is lost (p. 5-29). If they are held up by thermal fuses, they will fall if the temperature rises (pp. 5-44 and 5-49).
5. Better fuel claddings: the materials used today can withstand about 600°C, although the fuel can withstand more. If the cladding could withstand higher temperatures, heat output, which falls as the temperature rises, would become

zero, and convection and radiation cooling would increase. Ceramic materials, which can withstand 1600°C, might be the answer. There may be little hope that they will ever be strong enough to be made into long, thin tubes, but the fuel could be made into small balls, as described in Section 12.1 (and in pp. 5-61 and 7-3 of the report). If the fuel pellets could be made even smaller, a fluidized bed reactor, with water as the fluidizing medium, might be possible (p. 6-29).

6. Better temperature distribution means that more heat can be absorbed before the fuel and its container disintegrate. One suggestion is to make the fuel in the form of a disk with a hole in the middle (a bagel reactor?) interleaved with disks of the graphite moderator. There would be holes in the fuel disks but not the graphite ones because fuel is a poor conductor of heat, and the insides of the disks would be the hottest part (p. 6-65).

The report also describes some ideas for mitigating the consequences of a melt-down, should one occur.

One can see some of these features (e.g., water acting as a thermal sink) in the design of the CANDU (CANada Deuterium Uranium) reactor, a type of pressurized heavy-water reactor (PHWR). The basic CANDU design incorporates natural uranium as the fuel and heavy water as both moderator and coolant; the design for the Advanced CANDU Reactor (ACR) under development calls for slightly enriched uranium fuel in conjunction with heavy water as moderator and light water as coolant.[18] Nguyen et al.[18] give the following inherent-safety features of the CANDU design:

1. The reactor core consists of several hundred pressure tubes containing the fuel and heavy-water coolant; each pressure tube is contained within a calandria tube, the space between the tubes being filled with an insulating gas that reduces normal heat loss to the heavy-water moderator surrounding the calandria tubes. The channelized fuel design precludes a core melt when the coolant is at high pressure when engineered depressurization is unavailable. Should a pressure tube fail in this scenario, rapid depressurization of the primary coolant would occur, eliminating the possibility of direct containment heating.

2. The multiple and separate volumes of water surrounding the fuel mean that this large water inventory can remove decay heat from the fuel for many hours and thus provide the opportunity for timely operator intervention. The moderator (heavy water) surrounding the fuel channels is at low temperature and low pressure, and can act as a heat sink to help maintain integrity of the fuel channels when internal cooling has failed.

3. Containment systems, including the externally cooled calandria vessel itself, are robust and afford protection from the consequences of severe accidents.

12.3 Criticality

Another aspect of inherent safety in the nuclear industry is criticality. The amount of radioactive material in one place must not exceed the critical level. Keeping the total stock below the critical level is the ideal, but is not always possible. Exclusion of moderators, which slow down the neutrons so that they are more effective in causing fission, is sometimes possible. An effective method of control is to handle the radioactive materials in cylinders with a diameter below the minimum at which criticality can occur, or in slabs somewhat thinner than the minimum critical thickness. Safety is then inherent and is not dependent on control of the stocks of radioactive materials and moderators or the presence of shields.[19] (Note: If the critical level is exceeded, there will not be an explosion but, rather, an unwanted heat release, typically several kilojoules, and a burst of neutrons, which could be fatal to anyone nearby.)

12.4 Conclusion

If chemical engineers have been slow to realize the advantages of inherently safer designs and put them into practice (Sections 10.1.2 and 10.7), they have been speedy in comparison with nuclear engineers. Despite Chernobyl, nuclear engineers are still living in the nuclear equivalent of a pre-Flixborough and pre-Bhopal age, confident of their ability to keep nuclear hazards under control. There is no doubt that traditional designs of nuclear power plant *can* be managed safely. In the light of Flixborough and Bhopal, are we convinced that they *will* be managed safely, everywhere, all the time?

The nuclear industry, not unlike the chemical process industries, faces continued challenges related to on-site security, off-site safety, and public acceptance. Add in the additional issues of waste management and disposal, and the opportunities for abatement of greenhouse gas emissions, and it is clear—or it should be—that the path forward lies in inherently safer designs.

References

1. L. M. Lidsky, "The Reactor of the Future," *Technol. Rev.* 87 (1984): 52–56.
2. T. A. Kletz, "Chernobyl," in *Learning from Accidents*, 3rd ed. (Oxford, U.K.: Butterworth–Heinemann, 2001), chap. 12.
3. A. M. Weinberg, "Three Mile Island in Perspective: Keynote Address," in *The Three Mile Island Nuclear Accident*, ed. T. H. Moss and D. L. Sills (New York: New York Academy of Science, 1981), 2.
4. T. Marsham, "The Future of the Breeder Reactor and Other Advanced Reactor Concepts in OECD," *Atom* 321 (1983): 146–147.

5. T. Yokoyama, H. Ninokata, and H. Endo, "Metal Fueled Long Life Fast Reactor Cores with Inherent Safety Features," in *Global 2007: Advanced Nuclear Fuel Cycles and Systems* (Boise, ID, 2007), 81–86.

6. A. M. Weinberg and I. Spiewak, "Inherently Safe Reactors and Second Nuclear Era," *Science* 224, no. 4656 (1984): 1398–1402.

7. G. A. K. Crommelin and W. F. Crommelin, "Inherently Safe Nuclear Power Generation with Electrically Coupled Compressor and Turbo Expander(s)," in *Proceedings of ASME Turbo Expo 2004: Power for Land, Sea, and Air, Volume 7: Biomass and Alternative Fuels* (Vienna, Austria, 2004), 281–287.

8. N. Franklin, "The Accident at Chernobyl," *Chem. Eng. (U.K.)* 430 (1986): 17–22.

9. M. Hayns, "The SIR Project," *Atom* 392 (1989): 2–8.

10. M. Cumo, *600 MWth MARS Nuclear Plant: Design Progress Report* (Rome, Italy: University Department of Nuclear Engineering and Energy Conversion, 1996).

11. A. Naviglio, "Reactors with Passive and Inherent Safety Features," in *Nuclear Power Plant: Distant Learning Package*, ed. M. Cumo and N. Afgan (Rome, Italy: Distant Learning Centre of University "La Sapienza," 1995).

12. H. Golfier, S. Caterino, C. Poinot, M. Delpech, G. Mignot, A. Naviglio, and A. Gandini, "Multipurpose Advanced 'inherently' Safe Reactor (MARS): Core Design Studies," in *PHYSOR-2006, ANS Topical Meeting on Reactor Physics* (Vancouver, BC, Canada, 2006).

13. D. W. Edwards, "Export Inherent Safety NOT Risk," *Journal of Loss Prevention in the Process Industries* 18, no. 4-6 (2005): 254–260.

14. Health and Safety Executive (U.K.), "Calder/Chapelcross: Go Ahead for 50 Years," *Nucl. Saf. Newsletter* no. 11 (1996): 2.

15. C. W. Forsberg and A. M. Weinberg, "Advanced Reactors, Passive Safety, and Acceptance of Nuclear Energy," *Annual Review of Energy* 15 (1990): 133–152.

16. C. W. Forsberg, D. L. Moses, E. B. Lewis, R. Gibson, R. Pearson, W. J. Reich, G. A. Murphy, R. H. Stauton, and W. E. Kohn, *Proposed and Existing Passive and Safety-Related Structures, Systems and Components (Building Blocks) for Advanced Light-Water Reactors*, Report No. ORNL-6554 (Springfield, VA: National Technical Information Service, 1989).

17. D. T. Ingersoll, "Deliberately Small Reactors and the Second Nuclear Era," *Progress in Nuclear Energy* 51, no. 4-5 (2009): 589–603.

18. T. Nguyen, R. Jaitly, K. Dinnie, R. Henry, D. Sinclair, D. J. Wilson, and M. O'Neill, "Development of Severe Accident Management Guidance (SAMG) for the Canadian CANDU 6 Nuclear Power Plants," *Nuclear Engineering and Design* 238, no. 4 (2008): 1093–1099.

19. G. Walker, "Critical Safety," *Atom* 323 (1983): 196–199.

Chapter 13

The Role of Inherently Safer Design in Dust Explosion Prevention and Mitigation

It was prettily devised of Aesop, "The fly sat upon the axletree of the chariot-wheel and said, what a dust do I raise."

—Sir Francis Bacon

Write your injuries in dust, your benefits in marble.

—Benjamin Franklin

Chapter 12 dealt with examples of inherent safety and friendlier plant design in the nuclear industry. Here we cover the usage of inherently safer design not in a specific industry per se, but rather in a particular application area within industry (the process industries and otherwise). Explosion prevention and mitigation, with an emphasis on dust explosions, is the subject of Chapter 13.

An overview of the occurrence and causation factors of dust explosions is first given. The hierarchy of controls (Section 1.3) is applied to the problem of dust explosion control, and examples are given of inherently safer design by principle (intensification, substitution, etc.). As with Chapter 11 on process-safety management, these additional examples are intended to supplement those provided in earlier chapters. Typical prevention and mitigation measures used in industry are then categorized within the hierarchy of controls. Brief concluding remarks are made to indicate possible future work on other process-safety concerns.

13.1 Dust Explosion Overview

One of the most comprehensive sources of information on all aspects of dust explosions is the text by Eckhoff.[1] Several new reviews of the field have also been published from a number of different perspectives: case histories, causes, consequences, and control of dust explosions;[2] the role of powder science and technology in understanding dust explosion phenomena;[3] and the status of developments in basic knowledge and practical application with respect to dust explosion prevention and mitigation.[4] A recent overview article by Amyotte and Eckhoff[5] forms the basis for much of the discussion in this section.

A dust explosion can occur when particulate solid material is suspended in air and a sufficiently energetic ignition source is present. The consequences are often similar to those arising from a gas explosion in terms of impact on people, physical assets, and business production. While most industrial practitioners are familiar with at least the basic concepts of gas explosions (e.g., the need for a fuel, oxidant, and ignition source), the same cannot be said for dust explosions.

The primary distinguishing factor between dust and gas explosions is the phase of the fuel itself—solid versus gaseous. Particle size is therefore a dominant issue in efforts aimed at preventing dust explosions and mitigating their consequences. (Section 13.5.2.) The National Fire Protection Association (NFPA) in the United States defines a dust as any finely divided solid, 420 μm or 0.017 in. or less in diameter (i.e., material capable of passing through a U.S. no. 40 standard sieve).[6] Since the range of explosible particle sizes for a given material can be quite large, this definition highlights the importance of considering the particle size distribution in addition to a mean or median particle diameter. Further, the shape for which a given material poses a dust explosion hazard may not be limited to spherical or near-spherical particles, but could include flakes, fibers, and flocculent forms.

13.1.1 Dust Explosion Occurrence

Dust explosions have a long history of recorded occurrence, as evidenced by Count Morozzo's account of the 1785 flour warehouse explosion in Turin, Italy (Section 11.3).[7] In the mid-1800s, Michael Faraday and colleagues elucidated the key role of coal dust in the devastating explosion in the Haswell, United Kingdom, coal mine.[5] Fast-forwarding to the twenty-first century, dust explosions remain a persistent and damaging industrial occurrence. The U.S. Chemical Safety Board (CSB) has recently completed a series of reports dealing with investigations into the causes of serious dust explosion incidents that occurred in the United States during 2003.[8–10]

A fourth CSB report[11] gives the findings of a study of dust explosions in general industry that was initiated following the three catastrophic incidents in 2003. At the time of writing, the CSB was engaged in investigating the 2008 explosion and fire at the Imperial Sugar refinery near Savannah, GA.

Dust explosion incidents are not, however, restricted to coal mines and food-processing facilities; nor are they restricted to the scenario of an industrial disaster. Frank[12] gives incident data reported by the U.S. CSB and FM Global to illustrate that dust explosions have occurred, for example, in the following industries with the indicated typical commodities:

1. Wood and paper products (dusts from sawing, cutting, grinding, etc.)
2. Grain and foodstuffs (grain dust, flour)
3. Metal and metal products (metal powders and dusts)
4. Power generation (pulverized coal, peat, and wood)
5. Rubber
6. Chemical process industries (acetate flake, pharmaceuticals, dyes, pesticides)
7. Plastic/polymer production and processing
8. Mining (coal, sulfide ores, sulfur)
9. Textile manufacturing (linen flax, cotton, wool).

13.1.2 *Dust Explosion Causation*

Because risks are determined by assessment of the probable consequences of identified hazards, thorough hazard identification is key to the effective management of risk; one cannot manage the risk arising from a hazard that has not been identified. With respect to dust explosions, it is essential to determine whether a given material actually constitutes an explosion hazard and, if so, the degree of hazard as represented by various explosibility parameters. Only then can appropriate risk-reduction measures aimed at prevention and mitigation be devised and implemented.

Thus, when identifying dust explosion hazards, one is inevitably drawn to an examination of the material itself in an attempt to answer questions such as:

Can the dust yield an explosion when dispersed as a cloud in air?
How high is the resulting overpressure if the explosion occurs at constant volume?
How quickly does the pressure rise if the explosion occurs at constant volume?
What concentration of airborne dust is needed for an explosion?
How much energy, or how high a temperature, is needed for ignition?
What minimum percentage of oxygen in the atmosphere is required to sustain flame propagation in the dust cloud?

These questions are addressed by determining the basic explosibility parameters of the dust, as shown in Table 13.1. It is important to recognize that these parameters are not fundamental properties of a given material. They are strongly

Table 13.1 Important Dust Explosibility Parameters and Their Application

Parameter	Typical Units	Description	Risk Component Addressed	Example Industrial Applications [as per Eckhoff[1]]
P_{max}	bar(g)	Maximum explosion pressure in constant-volume explosion	Consequence severity	Containment Venting Suppression Isolation Partial inerting
$(dP/dt)_{max}$	bar/s	Maximum rate of pressure rise in constant-volume explosion	Consequence severity	As per P_{max}
K_{St}	bar·m/s	Volume-normalized (or standardized) maximum rate of pressure rise in constant-volume explosion	Consequence severity	As per P_{max}
MEC	g/m³	Minimum explosible (or explosive) dust concentration	Likelihood of occurrence	Control of dust concentrations
MIE	mJ	Minimum ignition energy of dust cloud (electric spark)	Likelihood of occurrence	Removal of ignition sources Grounding and bonding
MIT	°C	Minimum ignition temperature of dust cloud	Likelihood of occurrence	Control of process and surface temperatures (dust clouds)
LIT	°C	Minimum ignition temperature of dust layer or dust deposit	Likelihood of occurrence	Control of process and surface temperatures (dust layers)

(continued)

Table 13.1   Important Dust Explosibility Parameters and Their Application (Continued)

Parameter	Typical Units	Description	Risk Component Addressed	Example Industrial Applications [as per Eckhoff[1]]
MOC (LOC)	vol%	Minimum (or limiting) oxygen concentration in the atmosphere for flame propagation in dust cloud	Likelihood of occurrence	Inerting (with inert gas)

Source: P. R. Amyotte and R. K. Eckhoff, "Dust Explosion Causation, Prevention and Mitigation: An Overview," *Journal of Chemical Health and Safety* 17, no. 1, January/February 2010: 15–28.

dependent on both material characteristics (e.g., moisture content and particle size, shape, and porosity) and experimental conditions (e.g., vessel volume, turbulence of the dust cloud, and applied ignition energy).

Most of the parameters listed in Table 13.1 are self-explanatory, with the exception of K_{St}, the volume-normalized maximum rate of pressure rise. The parameter $(dP/dt)_{max}$ is of course dependent on the volume of the explosion chamber, and is therefore of limited use on its own. For scaling to larger volumes, maximum rates of pressure rise are normalized by multiplying by the cube-root of the explosion chamber volume, V (i.e., $K_{St} = (dP/dt)_{max} \cdot V^{1/3}$).

This equation for K_{St} is sometimes referred to as the cubic or cube-root *law* and K_{St} itself as the *dust constant*. (The subscript *St* derives from the German word for dust, *staub*.) It is preferable, however, to refer to the equation as the cubic *relationship* and K_{St} as the *volume-normalized (or standardized) maximum rate of pressure rise*, or simply as K_{St}. There is nothing fundamental (in the sense of an inviolable law) or constant about either the K_{St} equation or the parameter itself.

Values of explosibility parameters for many materials can be found in the literature (e.g., Eckhoff and NFPA 68)[1,6] or online databases (e.g., BGIA).[13] Such values should, however, only be used as indications, and not as the ultimate basis for design of actual safety measures in industry (which should be based on test data for the actual dust in question).[1] Further, it is absolutely essential to use only data obtained from the authorized standard test methods for determining the various parameters. (An example would be the use of K_{St} values for sizing of areas of explosion vents and for design of explosion-isolation and explosion-suppression systems according to current standards.) Applicable test methods in this regard include those falling under the auspices of the American Society for Testing and Materials

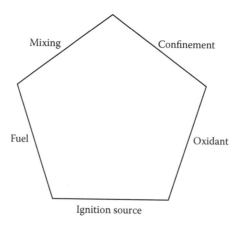

Figure 13.1   The explosion pentagon.

(ASTM), International Organization for Standardization (ISO), and European Committee for Standardization (CEN).

Dust explosion risk arises when the requirements of the explosion pentagon are satisfied. As described by Kauffman[14] and illustrated in Figure 13.1, the explosion pentagon expands the basic fire triangle to include mixing of the fuel and oxidant and confinement of the resultant mixture. The first of these additional components illustrates the previously mentioned key difference between dust and gas explosions—a solid rather than a gaseous fuel. A gas explosion therefore involves a homogeneous system in which the smallest entities of fuel and air are separated only by molecular distances. Thorough mixing of fuel and oxidant is readily achieved, and gravitational effects are negligible. However, in a dust/air mixture, the dust particles are strongly influenced by gravity; an essential prerequisite for a dust explosion is the formation of a dust/oxidant suspension. Once combustion of this mixture occurs, confinement (partial or complete) permits an overpressure to develop, thus enabling a fast-burning dust flame to transition to a dust explosion.

Dust explosions usually occur in industry inside process vessels and units such as mills, grinders, and dryers, i.e., inside equipment where the conditions of the explosion pentagon are satisfied. Such occurrences are often called *primary* explosions, especially if they result in *secondary* explosions external to the process unit. The reason for the majority of dust explosions being initiated in this manner is that the range of explosible concentrations is orders of magnitude greater than the concentrations permitted in areas inhabited by workers. In other words, airborne concentration thresholds set by occupational hygiene considerations are much lower than the minimum explosible concentration, or MEC (Table 13.1).

Notwithstanding the discussion in the previous paragraph, dust explosions do occur in process areas, not just inside process units. A secondary explosion can be initiated due to entrainment of dust layers by the blast waves arising from a primary explosion. The primary event might be a dust explosion originating in a process unit, or could be any disturbance energetic enough to disperse explosible dust layered on the floor and various work surfaces. An example of such an energetic disturbance (other than a primary dust explosion) would be a gas explosion leading to a dust explosion. This is a well-documented phenomenon in the underground coal mining industry, where devastating effects can result from the overpressures and rates of pressure rise generated in a coal dust explosion that has been triggered by a methane explosion. Housekeeping as an explosion prevention measure is discussed in Section 13.3 under the inherent-safety principle of intensification.

13.2 Use of the Hierarchy of Controls

It is both helpful and advisable to employ a framework for making appropriate choices when selecting dust explosion prevention and mitigation measures (e.g., those listed in the right-hand column of Table 13.1). Various frameworks, or heuristics, such as the familiar fire triangle and the previously described explosion pentagon (Section 13.1.2) are available for this purpose. Removal of a component from either polygon can act in a preventive or protective manner, depending on the particular component targeted. Elimination of electrostatic ignition sources by grounding and bonding is clearly preventive; explosion relief venting serves to mitigate the consequences of confinement.

What is missing in the above discussion, however, is guidance on which risk-reduction techniques are most effective and in what order the various techniques should be considered. Prevention is obviously preferred to mitigation, but both are likely to be required in a given application. The question, then, is which dust explosion control measures should receive priority? The answer will by now be obvious—inherent, passive engineered, active engineered and procedural—in that order. This is, of course, the hierarchy of controls (Section 1.3).

In a series of papers over the past several years, Amyotte and coworkers have attempted to establish the hierarchy shown in Figure 1.1 (Section 1.3) as an explicit framework for dust explosion prevention and mitigation.[15–18] Others have also made a case for an inherently safer approach to dust explosion control. Kao and Duh[19] offer several suggestions for inherent-safety checks on existing ABS (acrylonitrile/butadiene/styrene) plants, including the attenuation measures of using a lower mass flow rate and control of particle size (Section 13.5.2).

In their comprehensive review of explosion prevention and protection systems, Pekalski et al.[20] comment that, when designing or modifying a process, the concept of safety should already be present at the earliest stages of the design. In other words, one should not simply jump in and begin applying commercially available preventive and protective systems. Pekalski et al.[20] recommend the adoption of various modern design and safety management approaches, the first of which they list as inherently safer design.

Kaelin[21] has recently given the following practical suggestions to control dust explosion hazards:

1. Control of the spread of combustible dust atmospheres by use of local exhaust ventilation, management of dust deposits, and good housekeeping to minimize electrically classified areas requiring specialized equipment.
2. Elimination or control of ignition sources, including electrical and frictional sparks, hot surfaces, and electrostatic discharges.
3. Application of explosion safeguards such as explosion protection (containment, venting, and suppression), use of inert gas, and explosion isolation.

Although the term *inherent safety* is not used in the above listing, the principles are evident and there are at least overtones of the hierarchy of controls present.

As discussed throughout this book, early application of the hierarchy of controls in the design cycle affords many opportunities for systematic application of the inherent-safety principles of intensification, substitution, attenuation, simplification, and others. This is the thesis of the remainder of this chapter, in which several examples are given to demonstrate how inherently safer design can help reduce the dust explosion problem at its root source: the material itself, the processing route, the process equipment, and the work procedures employed.

It is important to note the use of the word *help* in the previous sentence; inherently safer design might not reduce the explosion risk to an acceptable level on its own. Engineered devices such as pressure relief valves, rupture discs, isolation valves, and suppression canisters, as well as procedural safeguards such as grounding and bonding, all have an important part to play in the hierarchy of controls. It is likely, though, that by applying the principles of inherent safety, one can either eliminate the need for some engineered and procedural measures or achieve a state of lessened reliance on engineered devices and procedures (which may fail or not be followed, respectively). If what you don't have can't leak, it stands to reason that what you don't have can't explode.

13.3 Intensification

Combustible dusts exist in industry as either the desired product or an unwanted by-product of the process undertaken. In either case, it is critical to minimize, whenever possible, the amount of dust available to participate in an explosion arising from normal operating conditions or an upset event. This means that consideration must be given to the formation of both dust clouds and dust layers. With respect to particulate dispersions, the defining parameter is the minimum explosible concentration (MEC), as described in Table 13.1. If airborne dust concentrations can be kept below the MEC of the material involved, then the fuel component of the explosion pentagon is removed, and dust explosions can, in theory, be prevented. As noted by Eckhoff,[1] this approach is not generally applicable in practice because of the large quantities of particulate material present in powder handling/processing equipment (in which most primary or initial dust explosions occur; Section 13.1.2). Exceptions do exist, and the hierarchy of controls embodies the idea of looking for fuel-intensification (minimization) opportunities as a regular, early feature of facility design. Eckhoff[1] identifies electrostatic powder painting and industry-specific dust-extraction systems as examples in this regard. In the latter case, it may be possible to minimize dust concentrations by controlling airflow, thus also invoking the principle of attenuation.

It is, however, entirely possible—and absolutely essential—to minimize fuel loadings in the case of dust layers. The occurrence of secondary dust explosions (Section 13.1.2) can be avoided by the removal of dust deposits from the workplace in a manner that limits the formation of dust suspensions (e.g., vacuuming with an explosion-proof device instead of sweeping). This point is well illustrated by Frank,[12] who makes a convincing argument for the critical importance of housekeeping in preventing dust explosions. The key role of housekeeping (i.e., hazardous material reduction) can also be seen in the following list of items identified by the U.S. CSB in its three recent investigations into dust explosion incidents:

1. Recommendation arising from dust explosion at West Pharmaceutical Services: "Ensure that spaces inaccessible to housekeeping are sealed to prevent dust accumulation."[8]
2. Root causes of CTA Acoustics dust explosion: "The use of metal tools, brooms, compressed air, and fans during line cleaning dispersed combustible dust in potentially explosive concentrations and also caused it to settle on elevated flat surfaces throughout the facility. The housekeeping program did not effectively remove combustible dust that accumulated in areas above production lines. CTA did not use [the principle of] minimizing flat surfaces to prevent accumulation of combustible dusts."[9]
3. Contributing cause of Hayes Lemmerz International-Huntington dust explosion: "Maintenance and housekeeping in the chip-processing area were inadequate, leading to significant flammable dust accumulations that contributed to secondary deflagrations at furnace no. 5."[10]

The required amount of layered dust that, once airborne, could sustain a secondary dust explosion is often grossly overestimated. For example, in the Westray mine explosion[22] described in Chapter 14, one of the contributing factors was the presence of coal dust layers several centimeters thick throughout the mine workings. In fact, the amount of coal dust that could be dispersed by an aerodynamic disturbance and then combusted with the available oxygen would have been significantly less than several centimeters.

This point is illustrated in a general manner by the expression: $C = (\rho_{bulk})$ (h/H), where ρ_{bulk} is the bulk density of a dust layer, h is the layer thickness, H is the height of the dust cloud produced from the layer, and C is the resulting dust concentration.[1,5] Thus, a 1-mm-thick layer of a dust of bulk density 500 kg/m³ on the floor of a 5-m high room will generate a cloud of average concentration 100 g/m³ if dispersed evenly all over the room. Such a concentration is on the order of the minimum explosible concentration for many explosible dusts. Partial dispersion up to 1 m above the floor yields a dust concentration of 500 g/m³; this is a concentration that is on the order of the optimum concentration (i.e., the concentration producing the most devastating overpressures and rates of pressure rise) for many explosible dusts.

Clearly, even seemingly harmless dust layers have the potential to rapidly escalate the risk of a dust explosion. This observation helps to explain the advice given by experienced industrial practitioners, such as: *There's too much layered dust if you can see your footprints in the dust or if you can write your initials in the dust.* These comments, although anecdotal, have a firm foundation in the chemistry and physics of dust explosions. They are therefore helpful reminders of inherent safety and intensification, which are similarly deeply rooted in the underlying chemistry and physics of a given process.

13.4 Substitution

The substitution principle can be applied to the dust explosion problem in several ways, e.g., as simply as substituting one work procedure for another. The previous example of using an explosion-proof vacuum in place of a broom for removing dust accumulations would apply here. Process equipment can also be made inherently safer by the replacement of bucket elevators and other mechanical conveying systems with dense-phase pneumatic transport; this could also be viewed as a form of attenuation, in that dust handling occurs under less hazardous conditions. In the latter option, the dust concentration is high enough that powder transport is assumed to occur at concentrations above the material's upper explosible limit (even though this limit may not be known precisely because of the experimental difficulties involved in its determination).

Kong[23] further illustrates the importance of the substitution principle in his analysis of a dust explosion initiated by a propagating brush discharge that occurred on a glass-lined pipe at the bottom of a hopper. He comments that the unnecessary use of insulating materials (i.e., the glass lining) should have been avoided; this implies the need for substitution of process hardware with less hazardous materials of construction.

Substitution of a process route can also be beneficial in preventing and mitigating dust explosions. The work of Mintz et al.[24] illustrates the development of an inherently safer process for the manufacture of refractory brick materials. The preparation of powdered metals (explosible aluminum/magnesium alloys) for refractory use is an application where a chemically inert dust (burnt magnesite, MgO) is present in the final product. It was determined that the inert dust could be introduced at an earlier stage of the process in amounts sufficient to render the aluminum/magnesium alloys nonexplosible throughout virtually the entire manufacturing sequence. Substitution of an alternative process methodology, coupled with attenuation of the material hazard through admixture of solid inertants (Section 13.5.1), helped to achieve the goal of safer plant operation without the addition of engineered or procedural controls.

Substitution of the hazardous material itself (i.e., the explosible dust) will be difficult to achieve in most cases, given that the dust is often the actual product or a component of the feed stream to the process. Such opportunities do arise, however, as demonstrated by Amyotte et al.[25] in their investigation of petroleum coke as a partial replacement for pulverized coal in the feed to utility boilers. The motivation for fuel blending in this case is driven largely by economics, but there are also accompanying inherent-safety benefits. The lowering of the explosion overpressure and rate of pressure rise brought about by petroleum coke substitution could potentially lead to a reduction in venting and suppression requirements.

This effect on dust explosibility is illustrated in Figure 13.2[25] for the case of pure fuels and in Figure 13.3[25] for the case of mixtures of petroleum coke with two different coals and a mixture of the two coals (50 wt% of each). As the rate of pressure rise is a measure of consequence severity (Table 13.1), the safety benefits of substituting petroleum coke to the largest extent economically and technically feasible are thus demonstrated.

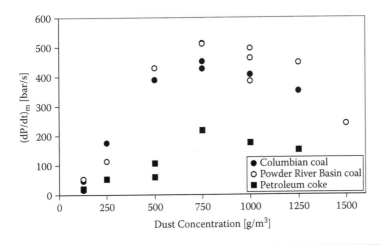

Figure 13.2 **Influence of fuel type on rate of pressure rise, $(dP/dt)_m$, of individual fuels (coal and petroleum coke). Tests were done using a Siwek 20-L chamber with 5-kJ ignition energy. (From P. R. Amyotte, A. Basu, and F. I. Khan, "Reduction of Dust Explosion Hazard by Fuel Substitution in Power Plants,"** *Process Safety and Environmental Protection* **81, no. 6 (2003): 457–462. With permission.)**

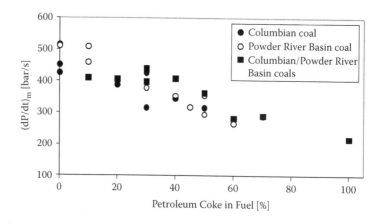

Figure 13.3 Reduction in rate of pressure rise, $(dP/dt)_m$, achieved by using blended fuels of coal and petroleum coke. Tests were done using a Siwek 20-L chamber with 5-kJ ignition energy and dust concentration of 750 g/m^3.

13.5 Attenuation

The risk of a dust explosion can be reduced, or in some cases eliminated, by processing the material in an alternative form (slurry, pellet, etc.) or by processing in solution if possible. These are all examples of attenuation (moderation), in which a bulk material is processed in a form that would not be considered a dust by any definition.[26] Caution must still be exercised, however, if the dust explosion hazard is reintroduced by particle size reduction in the case of pelletized materials, or by drying in the case of slurries and solutions.

In addition to being an excellent resource on intensification via housekeeping, Reference 12 provides an analysis of the serious dust explosion that occurred at the West Pharmaceuticals facility in Kinston, North Carolina, on January 29, 2003.[8] This plant manufactured rubber components for the health-care industry; in part of the process, rubber was coated with polyethylene powder to prevent sticking. The coating process employed the principle of attenuation, in that rubber strips were run through a tank containing a slurry of polyethylene powder in water. After this stage, the water was evaporated by fans to leave only the required powder coating on the rubber strips. This drying step reintroduced the dust explosion hazard by the generation of airborne polyethylene dust.[12]

13.5.1 Admixture of Solid Inertants

Another way to reduce the risk of a dust explosion is to mix a solid inertant with an explosible dust in sufficient quantity that the resulting mixture is rendered nonexplosible. Two approaches are possible.[27] The term *inerting* arises when focusing on *preventing* the occurrence of a dust explosion; use of an inert dust—premixed in sufficient amount—can lead to removal of the heat necessary for combustion. On the other hand, the term *suppression* arises when focusing on *mitigating* the consequences of a dust explosion. The intent is the same as with inerting: to remove the heat necessary for sustained combustion and thus to limit the generation of destructive overpressures in an enclosed volume. In the case of suppression, however, the inert dust is injected into the just-ignited explosible dust/air mixture rather than being intimately premixed with the explosible dust prior to ignition (as in the case of inerting). Inerting may therefore be defined as an inherent-safety measure (attenuation), while suppression is an active engineered measure.

As noted by Eckhoff,[1] inerting by solids admixture is not widely used in the process industries. This is due to the previously described need to intimately premix the explosible and inert dusts when the application is explosion inerting; this mixing can lead to unacceptable product contamination. Inerting is perhaps most prevalent in the underground coal mining industry, where inert rock or stone dust (e.g., limestone or dolomite) is mixed with coal dust generated during the mining process. (See Chapter 14.) Some industrial examples from applications other than coal mining do exist, however.[26]

The concept of an inerting level was applied to an explosible-dust/admixed-inertant formulation used in the food processing industry. In practice, the inertant was introduced into the manufacturing process initially for product quality purposes, and subsequently for explosion prevention purposes once its usefulness in this regard was identified. Laboratory-scale tests showed that the inerting level was only slightly higher than the usual percentage of inertant used for quality control. Raising the inertant percentage to the inerting level was determined by the company to have no adverse effects on the quality of the final product and also to be economically feasible. Thus, the manufacturing process was made inherently safer by means of the attenuation principle.

The above example is somewhat analogous to the case described earlier in Section 13.4 for the manufacture of refractory brick materials. Similarly, the partial substitution of petroleum coke for coal as a utility boiler fuel (Section 13.4) also employs the idea of attenuating the dust explosion hazard. Petroleum coke, although obviously not totally inert, does have a significantly lower volatile content than bituminous coal and therefore exerts an "inerting-like" influence when mixed as a fuel with coal dust.

Staying with the electrical utility industry, incomplete combustion and subsequent contamination of a waste stream can pose a serious explosion hazard. This is the problem of fuel carryover, as might arise in the case of contamination of fly ash with pulverized coal. The coal, if present in sufficient quantities in the mixture, can act as a fuel source for a potential explosion. Solutions to the risk posed by fuel carryover can, however, be identified by recognizing that contamination of a supposedly inert dust (fly ash) with an explosible dust (coal dust) is simply explosion inerting "turned around."[27] The underlying phenomena are the same, whether one is attempting to thermally inert coal dust with fly ash or to determine the percentage of coal dust that would render explosible a mixture of said coal dust and fly ash.[28] Hence, the concept of attenuation can be used to establish explosibility boundaries for fly ash/pulverized fuel mixtures.

A comprehensive series of tests has recently been conducted to determine the explosibility characteristics of aluminum and magnesium dusts generated during shredding processes involved in industrial recycling.[29] A portion of this work was aimed at determining the inerting effectiveness of calcium oxide (CaO) and calcium carbonate ($CaCO_3$) with respect to magnesium, using a device known as a modified Hartmann tube to determine minimum explosible concentration values. For a 50:50 inertant/Mg mixture, the MEC was just over 200 g/m^3 with CaO as the inertant and just under 300 g/m^3 with $CaCO_3$ as the inertant. The MEC for magnesium alone was measured as 100 g/m^3.

It was concluded that attenuation of the dust explosion hazard in this case was not significant enough to warrant full-scale application of inertants to the dust generated by industrial shredding as part of the recycling process. Inertant concentrations greater than 50 wt% would be required, and this was not deemed to be feasible. This example demonstrates that inherent-safety measures are exactly the same as engineered and procedural measures in at least one regard: They may not always be practical from a technical perspective. Further, even if found to be practical, a cost/benefit analysis may reveal that implementation is not possible. Clearly, trade-offs are necessary when attempting to reduce explosion risk by means of the hierarchy of controls.

The description given by Frank[12] of the Jahn Foundry explosion in Springfield, Massachusetts on February 25, 1999, contains a further illustration of attenuation. The shell-mold fabrication process involved mixing a phenol/formaldehyde resin with sand in the proportions of 5% resin to 95% sand.[12] Assuming the resin and sand particle size distributions were not too dissimilar, this level of inertant (sand) would surely be sufficient to prevent an explosion of the resin once the mold components were mixed so that the molds could be shaped and then cured in an oven. This appears to be a reasonable conclusion; Frank[12] cites two plausible explosion initiation scenarios, neither of which involves the resin/sand mixture: (a) accumulation of a natural-gas/air mixture in a curing oven, and (b) ignition of a resin cloud external to an oven by the hot oven surface.

13.5.2 Attention to Particle Size

The effects of an increase in explosible dust particle size are numerous and are well established; these include, for example, a decrease in the maximum explosion pressure, P_{max}, a potentially significant decrease in the maximum rate of pressure rise, $(dP/dt)_{max}$, and an increase in the minimum explosible concentration, MEC.[30] Increasing the mass mean diameter of an explosible dust by increasing the particle size throughout the entire size distribution can also significantly decrease the inerting level. Thus, from the perspective of attenuation, larger-size particles of an explosible dust are inherently safer than finer sizes of the same dust.

Increasing the particle size as an explosion prevention strategy is a classic application of inherent safety, in that it makes use of the underlying chemistry and physics of the materials, chemical transformations, and molecular separations involved in chemical processing.[26] Here, the chemistry and physics of a larger particle size lead to risk reduction. An increase in particle size brings about a decrease in particle surface area, which is an important consideration in the evolution of volatiles from materials such as coal dust (with such volatiles subsequently undergoing homogeneous gas-phase combustion). A smaller particle surface area also means fewer active sites in the case of those powders that undergo heterogeneous gas/solid combustion. In either reaction mode, the result is the previously mentioned preventive effect on explosion parameters such as P_{max}, $(dP/dt)_{max}$, MEC, and inerting level (a *chemical* effect). A dust explosion criterion required by the aforementioned explosion pentagon is fuel/oxidant mixing, i.e., the formation of an airborne dust cloud, often achieved by an aerodynamic disturbance directed at a layered dust deposit. Larger particles are also more difficult to disperse than smaller-size powders (a *physical* effect).

The importance of particle size considerations in dust explosion prevention and mitigation is recognized in various process-safety standards and engineering methodologies. Examples include the following:

1. *Dow fire and explosion index*: The Dow F&EI (Section 10.4.2) incorporates a particle-size effect in the index calculation procedure, as shown in Table 13.2,[31] where particle size is indicated by the actual particle diameter and the mesh or sieve number. Also shown in Table 13.2 is the penalty factor designed to deal with the issue of a potential dust explosion in the process unit under consideration. This dust explosion penalty is clearly seen to decrease as the particle size of the dust increases.

2. *Definition of a dust*: Particle size is a key distinguishing factor in the definition of a dust, as determined by various professional organizations. In addition to the NFPA definition given in Section 13.1, the former U.S. Bureau of Mines developed the following classification for dusts based in part on particle size:[32]

Dust (coal mines): particles <850 μm
Dust (surface industries): particles <425 μm
Float coal dust: particles <75 μm
Mine-size dust: particles <850 μm, with 20% of the particles <75 μm

Table 13.2   Dow Fire and Explosion Index Dust Explosion Penalty

Particle Size (μm)	Tyler Mesh Size	Penalty [a]
175+	60–80	0.25
150–175	80–100	0.5
100–150	100–150	0.75
75–100	150–200	1.25
<75	>200	2.00

Source: Dow Chemical Company, Fire and Explosion Index,
  7th ed. (New York: American Institute of Chemical
  Engineers, 1994).

[a] Use ½ if an inert gas.

3. *Particle size for explosibility testing*: In recognition of the increased hazard posed by fine dust, the American Society for Testing and Materials (ASTM) standard for dust explosion pressure and rate of pressure rise testing makes the following statements:

"Tests may be run on an as-received sample. However, due to the possible accumulation of fines at some location in a processing system, it is recommended that the test sample be at least 95% minus 200 mesh (74 μm)."[33] and

"When a material is tested in the as-received state, it should be recognized that the test results may not represent the most severe dust deflagration possible. Any process change resulting in a higher fraction of fines than normal or drier product than normal may increase the explosion severity."[33]

The following examples have been drawn primarily from the discussion in Reference 26 to provide a wide range of industrial applications demonstrating the need for careful attention to the particle size of dusts being processed in a variety of ways. In some cases, the fineness of the dust posed particular challenges (essentially a negation of the attenuation principle); in others, attenuation was used (or at least attempted) to achieve a degree of risk reduction. In all the examples presented, explosible dust particle size is the predominant factor in assessing the degree of inherent safety.

Returning to the analysis of Frank,[12] the author gives a description of a dust explosion that occurred on May 16, 2002, in a rubber recycling plant operated by Rouse Polymerics International, Inc. (Vicksburg, Mississippi). The rubber dust involved in this incident had a minimum ignition energy (MIE) of 5 mJ, making it highly susceptible to ignition by electrostatic discharge. This low MIE value was attributed to the small particle size of the rubber dust (75–180 μm).[12]

Table 13.3   Magnesium Dust Explosibility Results

Size Fraction (µm)	MEC (g/m³)	MIE (mJ)	MIT (°C)
149–177	900	240	620
0–20	90	4	520

Source: P. R. Amyotte, M. J. Pegg, F. I. Khan, M. Nifuku, and T. Yingxin, "Moderation of Dust Explosions," *Journal of Loss Prevention in the Process Industries* 20, no. 4-6 (2007): 675–687.

A second incident described by Frank[12] is worth noting because of the nature of the bulk material. An explosion occurred at Malden Mills Industries in Methuen, Massachusetts, on December 11, 1995, involving short nylon fibers. These fibers had a high length-to-diameter ratio[12] and were thus not the more-often thought-of spherical or near-spherical dust particles. Flocculent material, therefore, can pose a hazard leading to an explosion risk that must be properly managed. (See the introductory remarks to this chapter.)

In addition to inerting tests, the previously mentioned study by Nifuku et al.[29] (Section 13.5.1) was directed toward an investigation of particle size influence on MEC, MIE, and MIT (minimum ignition temperature) for aluminum and magnesium dusts generated during shredding processes involved in industrial recycling. Both MEC and MIE were observed to increase exponentially with an increase in particle size for the narrow size distributions used; MIT was also found to be directly dependent on particle size.

Examples of the results obtained for magnesium are given in Table 13.3.[26] These and similar results for aluminum led to the conclusion that the recycling shredding process could be made inherently safer by avoiding the production of fine dusts (<74 µm).[29]

A wood-processing facility decided to commission a series of laboratory-scale tests. The data in Table 13.4[26] indicate the inadequacy of explosion prevention and mitigation measures that would occur should these be based solely on the existence of the dust designated as "coarse." In the actual process, although the coarse dust was predominant, pockets of the fine dust were found in a dead space in the process-unit header. Reliance on particle size effects as a line of defense with respect to dust explosions therefore requires a thorough understanding of the dust-handling process during both normal and upset conditions. (See also the earlier comments on particle size concerning the ASTM test methodology described in Reference 33.)

Explosibility data for carbon dust, as would be generated in a carbon-block recycling facility, were also determined by laboratory-scale testing. Some of the results are shown in Table 13.5.[26] Particle size is seen to have a significant influence on explosibility, with only the finest sizes tested resulting in explosions. Carbon-dust

Table 13.4   Wood Dust Explosibility Results

Dust	Moisture (weight %)	Particle Size	P_{max} (bar[g])	K_{St} (bar.m/s)
Coarse	7	50 wt% < 1 mm 0.3 wt % < 75 μm	4.7	9
Fine	9	93 wt% < 1 mm 35 wt% < 125 μm 16 wt% < 75 μm	8.3	130

Source: P. R. Amyotte, M. J. Pegg, F. I. Khan, M. Nifuku, and T. Yingxin, "Moderation of Dust Explosions," *Journal of Loss Prevention in the Process Industries* 20, no. 4-6 (2007): 675–687.

explosions were observed only with an energetic ignition source of 10 kJ. In general, then, carbon is difficult to ignite. The heterogeneous nature of the resulting gas/solid combustion means that only the finest sizes will propagate a flame during an explosion. Similar to the conclusion reached by Nifuku et al.,[29] carbon-block recycling can be made inherently safer by avoiding both the production of fine dusts and massive ignition energies.

Table 13.5   Carbon Explosibility Results

Particle Size (μm)	MEC [5-kJ Ignition Energy]	P_{max} [10-kJ Ignition Energy]	K_{St} [10-kJ Ignition Energy]
300–710 (−24 mesh, +48 mesh)	No explosions from 200 to 2750 g/m³	Not tested	Not tested
8–196[a]	Not tested	No explosions from 250 to 2750 g/m³	No explosions from 250 to 2750 g/m³
4–36[a]	No explosions from 200 to 2250 g/m³	Not tested	Not tested
3–44[a]	Not tested	5.3 bar(g)	15 bar·m/s
4–31[a]	Not tested	5.8 bar(g)	19 bar·m/s

Source: P. R. Amyotte, M. J. Pegg, F. I. Khan, M. Nifuku, and T. Yingxin, "Moderation of Dust Explosions," *Journal of Loss Prevention in the Process Industries* 20, no. 4-6 (2007): 675–687.

[a]   10 wt% < lower limit; 10 wt% > upper limit.

Table 13.6   Proximate Analysis and Particle Size of Datong Float Coal Dust

Moisture (wt%)	Ash (wt%)	Volatiles (wt%)	Fixed Carbon (wt%)	Particle Size (wt%)
0.1	11.6	25.8	62.5	100% < 75 μm; 98% < 45 μm; 78% < 20 μm; 51% < 10 μm

Source: P. R. Amyotte, M. J. Pegg, F. I. Khan, M. Nifuku, and T. Yingxin, "Moderation of Dust Explosions," *Journal of Loss Prevention in the Process Industries* 20, no. 4-6 (2007): 675–687.

Coal mining inevitably leads to the generation of coal dust deposits in working galleries. As mentioned in Section 13.5.1, a primary defense against mine explosions is inerting with rock or stone dust (e.g., limestone or dolomite). The particle size of the coal dust has a profound influence on its explosibility, as illustrated in Figures 13.4[26] and 13.5,[26] which provide laboratory-scale data for float coal dust from the Datong mine in PR China. The coal dust had the physical properties shown in Table 13.6.[26]

As seen in Figures 13.4 and 13.5, the Datong coal dust explosibility data are comparable with those from laboratory-scale testing for a larger-particle-size, higher-volatile-content Columbian coal (Table 13.7),[26] with attainment of explosion pressure and rate of pressure rise maxima at leaner dust concentrations for the Datong sample. The explosion hazard is therefore significantly magnified by the presence of

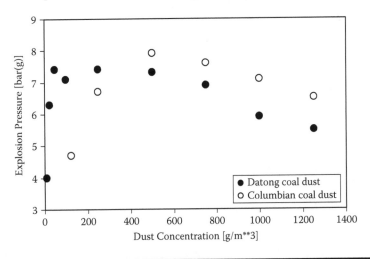

Figure 13.4   Explosion pressure as a function of dust concentration for coal dust. Tests were done using a Siwek 20-L chamber with 5-kJ ignition energy. (From P. R. Amyotte, A. Basu, and F. I. Khan, "Reduction of Dust Explosion Hazard by Fuel Substitution in Power Plants," *Process Safety and Environmental Protection* 81, no. 6 (2003): 457–462. With permission.)

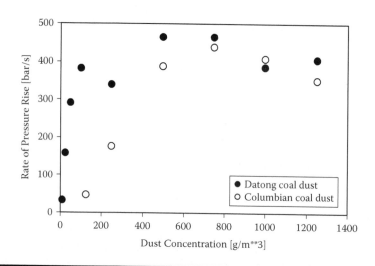

Figure 13.5 **Rate of pressure rise as a function of dust concentration for coal dust. Tests were done using a Siwek 20-L chamber with 5-kJ ignition energy. (From P. R. Amyotte, M. J. Pegg, F. I. Khan, M. Nifuku, and T. Yingxin, "Moderation of Dust Explosions,"** *Journal of Loss Prevention in the Process Industries* **20, no. 4-6 (2007): 675–687. With permission.)**

Table 13.7   Proximate Analysis and Particle Size of Columbian Coal Dust

Moisture (wt%)	Ash (wt%)	Volatiles (wt%)	Fixed Carbon (wt%)	Particle Size (wt%)
0.4	7.7	36.4	55.5	83% < 125 µm; 73% < 75 µm; 53% < 45 µm; 29% < 20 µm

Source: P. R. Amyotte, M. J. Pegg, F. I. Khan, M. Nifuku, and T. Yingxin, "Moderation of Dust Explosions," *Journal of Loss Prevention in the Process Industries* 20, no. 4-6 (2007): 675–687.

very fine coal dust in a mine. Further, the inerting level for the Datong coal dust was determined to be >80 wt% dolomite for a coal dust concentration of 500 g/m³. This means that a mixture of this coal dust with inert dolomite would have to be <20 wt% coal dust for the mixture to be nonexplosible. It is thus imperative that the generation of float coal dust be minimized to the greatest extent possible, and that the hazard be attenuated by the admixture of sufficient rock dust for thermal inhibition.

13.5.3 Avoidance of Hybrid Mixtures

Hybrid mixtures consist of a flammable gas and a combustible dust, each of which may be present in an amount less than its lower flammable limit (LFL) and minimum explosible concentration (MEC), respectively, and still give rise to an explosible mixture. The focus when discussing hybrid mixtures is, in fact, on admixture of a flammable gas in concentrations below the lower flammable limit of the gas itself. If the LFL for the gas is exceeded, one soon has a situation where the worst-case scenario for a primary explosion would be a pure gas explosion.

The influence of the copresence of a flammable gas on the explosibility parameters of a fuel dust alone is well established.[30] Effects include higher values of maximum explosion pressure and maximum rate of pressure rise (and hence K_{St}) as well as lower values of minimum explosible concentration and minimum ignition energy. There is, of course, already a hazard that exists when an explosible dust is present in a quantity above its minimum explosible concentration. With flammable gas admixture, the scenario is now one of magnification of an already-existing hazard, not the creation of a problem that did not already exist in some form already.

Perhaps the most well-known hybrid mixture is the methane/coal dust system often encountered in underground coal mining. There are also examples of hybrid mixture formation in other industries, such as the natural-gas/fly-ash system in fossil-fuel-burning power plants and various hydrocarbon/resin combinations occurring in the production of plastic powders.

A process-safety newsletter gives other examples of the origin of hybrid mixtures in activities other than coal mining.[34] This brief article explains that hybrid mixtures can occur in a diverse range of industries, including fine chemicals, paints and inks, and foodstuffs. General examples are given of flammable gases arising from solvent dried off a powder, trace solvent left on a powder, powder charging into a reactor or mixer, and unintended/unknown side reactions leading to powder decomposition. Three specific cases are also given:

Case 1: "A fluid bed drier fitted with explosion relief completely disintegrated when the powder bed slumped on failure of the air. The residual heat caused the powder to decompose which released a flammable gas. The hybrid mixture exploded, but the explosion vents on the drier had only been designed to cope with a dust explosion. The company concerned had no knowledge that the powder could decompose to release a flammable gas."[34]

Case 2: "An explosion occurred during FIBC (big bag) filling. The presence of solvent increased the sensitivity (minimum ignition energy reduced) of the dust to ignition to the point where electrostatic discharges from the bag ignited the hybrid mixture."[34]

Case 3: "A powder was being added from plastic bags into the manway of a batch reactor. The mixture of solvent vapour from the reactor combined with the dispersed dust cloud to again provide the sensitive hybrid mixture that exploded, seriously injuring the operator."[34]

Avoidance of hybrid mixtures would therefore fit within attenuation as a means of processing materials in less hazardous forms, especially when achieved through deliberate design measures. The previously mentioned production of plastic powders, specifically the manufacture of high-density polyethylene (HDPE), provides a further example.[35] One such HDPE process[36] uses a catalytic fluidized-bed reactor in which oxygen is treated as a contaminant and is therefore excluded from the reactor inventory (which contains hydrocarbon gas, polyethylene resin, and catalyst). Product from the reactor enters a purge unit in which hydrocarbon gas (either free or dissolved in the resin) is removed in a nitrogen environment. The resin then enters a series of storage bins; the first of these units poses a particular challenge for explosion prevention and mitigation because of the potential for residual hydrocarbon gas to be present. Additional risk factors include the generation of fines due to particle attrition and electrostatic discharges due to bulk powder movement.

The process described above employs the expected engineered and procedural measures: safety valves, dust collectors, explosion hatches and doors, water injection system, and observation and monitoring. An additional measure of prevention and mitigation can be gained by determining where in the process fine dust is being produced and examining whether the process operating conditions can be altered accordingly. Purger optimization to minimize the copresence of flammable gas and combustible dust, as well as design issues related to the dust collectors and storage bins, are also important with respect to explosion risk reduction. These points fit well with the concept of attenuation.

A quantitative measure of the importance of avoiding hybrid-mixture formation in the above scenario (and of minimizing particle size reduction) is given by Figure 13.6. The parameter K_{St} is shown for three different sizes of HDPE, for the polyethylene dispersed in air, and also for polyethylene with flammable gas admixed in the dispersing air (2 vol% ethylene and 1 vol% hexane). These gas concentrations are just below their respective lower flammable limits. One can observe a near tripling of K_{St} in moving from the coarsest HDPE sample in air to the finest sample in the presence of 2 vol% ethylene (C_2H_4). Thus, the explosion hazard in this case is attenuated threefold with respect to K_{St} by increasing the particle size and eliminating the copresence of ethylene gas.

13.6 Simplification

The concept of simplification can similarly be applied to several aspects of dust explosion risk reduction.[18] For example, if considered at an early stage, plant design can perhaps be simplified by eliminating long dust-extraction ducts. Other applications of simplified design and presentation of hazard information also exist.

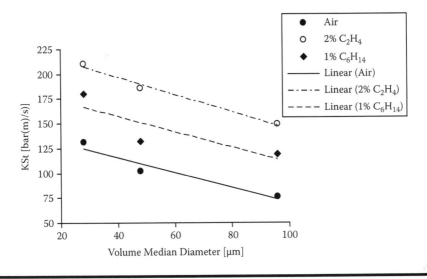

Figure 13.6 **Volume-normalized maximum rate of pressure rise as a function of particle diameter (polyethylene) and copresence of flammable gas (ethylene and hexane). Tests were done using a Siwek 20-L chamber with 10-kJ ignition energy. (Adapted from P. R. Amyotte et al., "Prevention and Mitigation of Polyethylene and Hydrocarbon/Polyethylene Explosions," in** *43rd Annual Loss Prevention Symposium, 2009 Spring National Meeting, American Institute of Chemical Engineers* **[Tampa, FL, 2009]. With permission.)**

13.6.1 Robust Equipment

As discussed in Section 7.2, simplification can be achieved by making process equipment robust enough to withstand process upsets and other undesired events. A more user-friendly and inherently safer facility can therefore be obtained by means of pressure- or shock-resistant design.

Figure 13.7[26] shows a hammer mill used in a wood-processing facility to accomplish size reduction of wood chips. (If ever there was an apt name for a piece of process equipment, "hammer mill" would be it!) Magnets to remove tramp metal (a potential ignition source), explosion vents, and administrative controls restricting access to enclosures housing hammer mills, are all employed in an attempt to reduce the explosion risk to an acceptable level. Yet the likelihood of explosion remains high because of the very nature of the functioning of the device; all that is needed in normal operation to complete the explosion pentagon is an ignition source. Hence, hammer mills are designed and built strong enough to withstand the overpressure resulting from a dust explosion originating inside the unit. Of course, the issue then becomes one of relieving the overpressure and limiting the effects to upstream and downstream equipment (Section 13.7).

Figure 13.7 Photograph of a hammer mill with the cover open. (From P. R. Amyotte, M. J. Pegg, F. I. Khan, M. Nifuku, and T. Yingxin, "Moderation of Dust Explosions," *Journal of Loss Prevention in the Process Industries* **20, no. 4-6 (2007): 675–687. With permission.)**

13.6.2 Material Safety Data Sheets

A key goal of user-friendliness and inherent safety via simplification is the provision of clear, unambiguous information on hazardous materials and how to properly handle them. This is analogous to the need to create "friendly" instructions (Section 9.7). A recent and important development in this regard is the analysis of Material Safety Data Sheets (MSDSs) conducted by the U.S. Chemical Safety Board (CSB).[11] The CSB determined that MSDSs for combustible dusts do not clearly and consistently communicate dust explosion hazards because of their inadequate identification of dust explosibility parameters (K_{St}, MEC, and others listed in Table 13.1) and the hazards that may be expected to arise through material handling and processing.

In its analysis of 140 MSDSs for combustible powders, the CSB used six criteria to assess the quality of information on dust combustion and explosion hazard warnings:[11]

1. Is the hazard stated explicitly and unambiguously?
2. Is the hazard presented in the Hazard Information Section of the MSDS?
3. Is the hazard warning repeated in appropriate sections of the MSDS?
4. Are physical data appropriate to dust explosibility included (e.g., K_{St})?
5. Does the MSDS refer to an appropriate NFPA (National Fire Protection Association) standard for additional information?
6. Does the MSDS include warning against accumulation of dusts and guidance on appropriate methods for removing accumulations?

A similar analysis has been performed for polyethylene.[37] A simple procedure was deliberately chosen to simulate the action a plant worker might take in the absence of both an MSDS and knowledge of the particular manufacturer. An Internet search for "polyethylene MSDS" yielded the expected multiple hits. The

Table 13.8   Review of Polyethylene MSDSs According to CSB Criteria

MSDS No.	Hazard Stated Explicitly?	In Hazard Information Section?	Hazard Warning Repeated?	Dust Explosibility Data?	Reference to NFPA Standard?	Warning against Accumulation?
1	yes	yes	yes	no	yes	yes
2	no	no	no	no	no	no
3	no	no	no	no	no	no
4	yes	yes	yes	no	no	yes
5	yes	yes	yes	no	no	yes
6	yes	yes	yes	no	yes	yes
7	yes	no	no	no	yes	yes
8	yes	yes	yes	no	no	yes
9	no	no	no	no	no	no
10	yes	yes	yes	no	yes	no
11	no	no	no	no	no	no
12	no	no	no	no	no	no

Source: P. Amyotte, R. Domaratzki, M. Lindsay, and F. Khan, "The Role of Material Safety Data Sheets in Dust Explosion Prevention and Mitigation," to be presented at *13th International Symposium on Loss Prevention and Safety Promotion in the Process Industries* (Brugge, Belgium, 2010).

first 12 MSDSs were selected for analysis (12 being a number someone might arbitrarily choose to investigate before deciding that as much information as was available had been retrieved). The results are shown in Table 13.8[37] and are in general agreement with those in Reference 11, especially with respect to physical data on dust explosibility. None of the 140 MSDSs reviewed by the CSB[11] and none of the 12 MSDSs reviewed in Table 13.8 contained any explosibility data for the dust in question. A recent guidance document[38] from the U.S. Occupational Safety and Health Administration (OSHA) attempts to address these concerns.

The CSB conclusion on MSDSs is significant and represents a compelling case for the principle of simplification. One would typically consider the incorporation of MSDSs within, say, safe work practices to be an example of procedural safety. Yet the MSDSs themselves, as just described, must present hazard information in a complete and easy-to-understand manner for maximum effectiveness. As indicated in Section 1.3, the characteristics of a user-friendly plant are sometimes not sharply divided and may well merge into one another.

13.7 Other Examples

As explained in Section 13.1.2, primary dust explosions usually occur inside process equipment because of the fact that the explosible concentration range for dusts is orders of magnitude higher than the dust concentrations permitted by industrial hygiene considerations. Hence there is a strong need for consideration of process-unit segregation and avoidance of knock-on or domino effects (Section 9.1). Location of dust collectors outside building enclosures is an example of domino risk reduction. So is hammer-mill operation outdoors with an adequate no-entry zone surrounding the process. Of course, we are then relying on people to follow the procedure and not enter the exclusion zone during mill operation.

Limitation of effects is important with respect to the prevention of flame-front and blast-wave propagation to downstream equipment. While active engineered devices such as isolation valves are available for this purpose, inherently safer alternatives may also exist. For example, creation of a product choke can help prevent downstream damage by replacement of a portion of the auger in a screw-feed system with a straight section of pipe. This makes use of the physics of bulk powder movement, in which limitation of effects is achieved by means of equipment design (Section 6.1). An alternative would be the use of a baffle plate (a passive-engineered device) to create a screw-conveyor choke.[26]

13.8 Summary of Prevention and Mitigation Measures

Table 13.9 summarizes the typically employed risk-reduction measures for dust explosions according to the categories of the hierarchy of controls.

13.9 Concluding Remarks

The previous sections have given many examples of inherently safer ways to both prevent and mitigate dust explosions. These are not theoretical concepts untried and unproven in industry. They work, and they are available for implementation if considered early in process design. What these measures are not usually given, however, is the name "inherently safer ways to prevent and mitigate dust explosions" or something similar. There are exceptions (as evidenced by the reference citations for this chapter), but perhaps more could be done in this regard. As the field of inherent safety has shown, there can be great value in naming a concept and then consistently using that name in practice.

The same can be said for other process-safety concerns beside dust explosions: fires, gas explosions, mist explosions, toxic releases, corrosion-related incidents, etc. Perhaps it would be beneficial to systematically review the various control measures for these problems and provide guidance on their prioritized use within a

Table 13.9   A Hierarchical View of Various Means of Preventing and Mitigating Dust Explosions

Explosion Prevention		Explosion Mitigation
Preventing Explosible Dust Clouds	*Preventing Ignition Sources*	*Explosion Mitigation*
Process design to prevent undesired generation of dust clouds and particle size reduction and segregation *Inherent Safety*: all principles	Smoldering combustion in dust, dust fires *Procedural Safety*: May also involve aspects of *Inherent Safety or Engineered Safety*	Good housekeeping (dust removal/cleaning) (Mitigation with respect to secondary dust explosions; prevention with respect to primary dust explosions) *Inherent Safety*: Intensification
Keeping dust concentration outside explosible range *Inherent Safety*: Intensification	Other types of open flames (e.g., hot work) *Procedural Safety*: May also involve aspects of *Inherent Safety or Engineered Safety*	Explosion-pressure resistant construction *Inherent Safety*: Simplification
Inerting of dust cloud by adding inert dust *Inherent Safety*: Attenuation	Hot surfaces (electrically or mechanically heated) *Procedural Safety*: May also involve aspects of *Inherent Safety or Engineered Safety*	Explosion isolation (sectioning) *Inherent Safety*: Limitation of effects (product choke) *Inherent Safety*: Avoiding knock-on effects (unit segregation) *Engineered Safety*: Passive (physical barrier) *Engineered Safety*: Active (isolation valve)

(continued)

Table 13.9   A Hierarchical View of Various Means of Preventing and Mitigating Dust Explosions (Continued)

Explosion Prevention		Explosion Mitigation
Preventing Explosible Dust Clouds	*Preventing Ignition Sources*	*Explosion Mitigation*
Intrinsic inerting of dust cloud by combustion gases *Engineered Safety:* Active	Heat from mechanical impact (metal sparks and hot spots) *Procedural Safety:* May also involve aspects of *Inherent Safety or Engineered Safety*	Explosion venting *Engineered Safety:* Passive
Inerting of dust cloud by N_2, CO_2, and rare gases *Engineered Safety:* Active	Electric sparks, arcs and electrostatic discharges *Procedural Safety:* May also involve aspects of *Inherent Safety or Engineered Safety*	Automatic explosion suppression *Engineered Safety:* Active
		Partial inerting of dust cloud by inert gas *Engineered Safety:* Active

Source: Adapted from P. R. Amyotte and R. K. Eckhoff, "Dust Explosion Causation, Prevention and Mitigation: An Overview," *Journal of Chemical Health and Safety* 17, no. 1, January/February 2010: 15–28.

framework that makes explicit use of the vocabulary of inherent safety. This is one way for a field of practice to advance.

On the subject matter of the current chapter, let us conclude by saying that we know small dust particles will explode more violently than large particles of the same material. This is fundamental science. So we should clearly take great care in processing fine particulate materials. But perhaps we can avoid these small particle sizes altogether and eliminate the dust explosion hazard. This would be good engineering based on fundamental science.

References

1. R. K. Eckhoff, *Dust Explosions in the Process Industries*, 3rd ed. (Boston: Gulf Professional Publishing/Elsevier, 2003).
2. T. Abbasi and S. A. Abbasi, "Dust Explosions: Cases, Causes, Consequences, and Control," *Journal of Hazardous Materials* 140, no. 1-2 (2007): 7–44.

3. R. K. Eckhoff, "Understanding Dust Explosions: The Role of Powder Science and Technology," *Journal of Loss Prevention in the Process Industries* 22, no. 1 (2009): 105–116.
4. R. K. Eckhoff, "Dust Explosion Prevention and Mitigation, Status and Developments in Basic Knowledge and in Practical Application," *International Journal of Chemical Engineering* 2009 (2009): 12.
5. P. R. Amyotte and R. K. Eckhoff, "Dust Explosion Causation, Prevention and Mitigation: An Overview," *Journal of Chemical Health and Safety* 17, no. 1, January/February 2010: p. 15–28.
6. NFPA 68, *Standard on Explosion Protection by Deflagration Venting*, 2007 ed. (Quincy, MA: National Fire Protection Association, 2007).
7. N. Piccinini, *Account of a Violent Explosion* (Turin, Italy: Politecnico di Torino, 1996).
8. CSB, *Investigation Report—Dust Explosion—West Pharmaceutical Services, Inc.*, Report No. 2003-07-I-NC (Washington, DC: U.S. Chemical Safety Board, 2004).
9. CSB, *Investigation Report—Combustible Dust Fire and Explosions—CTA Acoustics, Inc.*, Report No. 2003-09-I-KY (Washington, DC: U.S. Chemical Safety Board, 2005).
10. CSB, *Investigation Report—Aluminum Dust Explosion—Hayes Lemmerz International-Huntington, Inc.*, Report No. 2004-01-I-IN (Washington, DC: U.S. Chemical Safety Board, 2005).
11. CSB, *Investigation Report: Combustible Dust Hazard Study*, Report No. 2006-H-1 (Washington, DC: U.S. Chemical Safety Board, 2006).
12. W. L. Frank, "Dust Explosion Prevention and the Critical Importance of Housekeeping," *Process Safety Progress* 23, no. 3 (2004): 175–184.
13. BGIA, "GESTIS-DUST-EX: Database Combustion and Explosion Characteristics of Dusts," http://www.dguv.de/bgia/en/gestis/expl/index.jsp.
14. C. W. Kaufmann, "Agricultural Dust Explosions in Grain Handling Facilities," in *Fuel-Air Explosions*, ed. J. H. S. Lee and C. M. Guirao (Waterloo, ON, Canada: University of Waterloo Press, 1982), 305–347.
15. P. R. Amyotte and F. I. Khan, "An Inherent Safety Framework for Dust Explosion Prevention and Mitigation," *J. Phys. IV France* 12 (2002): Pr7-189–Pr7-196.
16. P. R. Amyotte, F. I. Khan, and A. G. Dastidar, "Dust Explosion Risk Reduction Measures Based on an Inherent Safety Framework," in *Proceedings of Process Plant Safety Symposium, 2003 Spring National Meeting, American Institute of Chemical Engineers* (New Orleans, 2003), 70–90.
17. P. R. Amyotte, F. I. Khan, and A. G. Dastidar, "Reduce Dust Explosions the Inherently Safer Way," *Chemical Engineering Progress* 99, no. 10 (2003): 36–43.
18. P. R. Amyotte, M. J., Pegg, and F. I. Khan, "Application of Inherent Safety Principles to Dust Explosion Prevention and Mitigation," *Process Safety and Environmental Protection* 87, no. 1 (2009): 35–39.
19. C.-S. Kao and Y.-S. Duh, "Accident Investigation of an ABS Plant," *Journal of Loss Prevention in the Process Industries* 15, no. 3 (2002): 223–232.
20. A. A. Pekalski, J. F. Zevenbergen, S. M. Lemkowitz, and H. J. Pasman, "A Review of Explosion Prevention and Protection Systems Suitable as Ultimate Layer of Protection in Chemical Process Installations," *Process Safety and Environmental Protection* 83, no. 1 (2005): 1–17.
21. D. E. Kaelin, "OSHA Combustible Dust National Emphasis Program Gap Analysis," in *Focus 2009* no. 2 (Plainsboro, NJ: Chilworth Technology, June 2009).

22. K. P. Richard, *The Westray Story: A Predictable Path to Disaster*, report of the Westray Mine Public Inquiry (Halifax, Canada: Province of Nova Scotia, 1997).

23. D. Kong, "Analysis of a Dust Explosion Caused by Several Design Errors," *Process Safety Progress* 25, no. 1 (2006): 58–63.

24. K. J. Mintz, M. J. Bray, D. J. Zuliani, P. R. Amyotte, and M. J. Pegg, "Inerting of Fine Metallic Powders," *Journal of Loss Prevention in the Process Industries* 9, no. 1 (1996): 77–80.

25. P. R. Amyotte, A. Basu, and F. I. Khan, "Reduction of Dust Explosion Hazard by Fuel Substitution in Power Plants," *Process Safety and Environmental Protection* 81, no. 6 (2003): 457–462.

26. P. R. Amyotte, M. J. Pegg, F. I. Khan, M. Nifuku, and T. Yingxin, "Moderation of Dust Explosions," *Journal of Loss Prevention in the Process Industries* 20, no. 4-6 (2007): 675–687.

27. P. R. Amyotte, "Solid Inertants and Their Use in Dust Explosion Prevention and Mitigation," *Journal of Loss Prevention in the Process Industries* 19, no. 2-3 (2006): 161–173.

28. P. R. Amyotte, A. Basu, and F. I. Khan, "Dust Explosion Hazard of Pulverized Fuel Carry-Over," *Journal of Hazardous Materials* 122, no. 1-2 (2005): 23–30.

29. M. Nifuku, M. Nifuku, S. Koyanaka, H. Ohya, C. Barre, M. Hatori, S. Fujiwara, S. Horigichi, and I. Sochet, "Ignitability Characteristics of Aluminum and Magnesium Dusts Relating to the Shredding Processes of Industrial Wastes," in *Sixth International Symposium on Hazards, Prevention, and Mitigation of Industrial Explosions* (Halifax, NS, Canada, 2006).

30. K. L. Cashdollar, "Overview of Dust Explosibility Characteristics," *Journal of Loss Prevention in the Process Industries* 13 no. 3-5 (2000): 183–199.

31. Dow Chemical Company, *Fire and Explosion Index*, 7th ed., and *Chemical Exposure Index*, 1st ed. (New York: American Institute of Chemical Engineers, 1994).

32. J. Nagy, *The Explosion Hazard in Mining*, IR 1119 (Washington, DC: U.S. Department of Labor, 1981).

33. ASTM, *ASTM E1226-05: Standard Test Method for Pressure and Rate of Pressure Rise for Combustible Dusts* (West Conshohocken, PA: American Society for Testing and Materials, 2009).

34. W. Chilworth, "When Dust Clouds and Solvent Vapours Mix and Explode," *Process Safety News* no. 3 (Plainsboro, NJ: Chilworth Technology, 2003), 2.

35. P. R. Amyotte, M. Lindsay, R. Domaratzki, N. Marchand, A. Di Benedetto, and P. Russo, "Prevention and Mitigation of Polyethylene and Hydrocarbon/Polyethylene Explosions," in *43rd Annual Loss Prevention Symposium, 2009 Spring National Meeting, American Institute of Chemical Engineers* (Tampa, FL, 2009).

36. K. Y. Choi and H. Ray, "Recent Developments in Transition Metal: Catalyzed Olefins Polymerization—A Survey, I: Ethylene Polymerization," *JMS–REV–Macromolecular Chemistry and Physics* C25 (1985): 1–55.

37. P. Amyotte, R. Domaratzki, M. Lindsay, and F. Khan, "The Role of Material Safety Data Sheets in Dust Explosion Prevention and Mitigation," *13th International Symposium on Loss Prevention and Safety Promotion in the Process Industries* (Brugge, Belgium, 2010).

38. OSHA, *Hazard Communication Guidance for Combustible Dusts*, OSHA 3371-08 (Washington, DC: U.S. Department of Labor, Occupational Safety and Health Administration, 2009).

Chapter 14

Inherent-Safety
Case Studies

> To paraphrase G. Santayana, one learns from history or one is doomed
> to repeat it.
>
> **—Crowl and Louvar**[1]

There really is not a need for a separate chapter in this book to deal with inherent-safety case studies, at least not in the sense of providing more examples. Numerous cases of both good and poor use of inherently safer design principles are interspersed throughout the previous chapters. So the objective of this chapter is a bit different. Here the goal is to provide some thoughts on where to look for case studies that involve inherent-safety considerations. The advice given should be useful when engaged in training efforts (Section 11.2.8).

Here are a few general points to keep in mind while reading this chapter:

1. A good case study does not have to be long and detailed, nor does it have to involve a catastrophic failure.
2. Most of the cases discussed here do, however, involve a failure or shortcoming of some sort. Human beings tend to pay attention when a story is being told and a loss is involved. The opening quote illustrates the importance of learning from case histories and avoiding hazardous situations; the alternative is to ignore the mistakes of others and be involved in potentially life-threatening incidents.[1]
3. There are likely other ways of finding or developing case studies. But an even dozen seemed a nice round number and therefore a good place to stop.

14.1 Everyday Life Experiences

Most readers will agree that life provides ample opportunity to experience the foibles of human nature. Sometimes one needs to look no further than one's own life experiences for an example of inherent safety (e.g., the discussion on stairs at the end of Section 2.1). Recall the traveler's story in Section 8.1.9 concerning the complex entry path at an airport terminal. Here is another one.

The same traveler was vacationing with his family in a major city and they decided to visit a popular tourist attraction—an observation tower located in the downtown area. At almost 200 m in height, the view from the top of the tower was spectacular. Before heading back to ground level, the traveler decided to treat his family to lunch in the rotating restaurant. Alas, there were no seats in the non-smoking section, and with asthmatic children, dining would have to wait until they were back on the ground.

While in the elevator going down, the traveler realized, "Wait a minute; you mean to tell me you can smoke at the top of this tower?" He began to think about potential fires from lit matches and cigarettes as well as limited evacuation routes. Once off the elevator and back on firm ground, he asked to speak to a manager and proceeded to deliver an impassioned lecture on inherent safety. The traveler's children, save one who had been promised a souvenir in the gift shop, abandoned him at this point. His wife watched with an amused look on her face; she had seen this sort of thing before.

The manager patiently listened and then delivered his own lecture on sprinkler systems, automatic fire-suppression devices in the restaurant kitchen, safety procedures for emptying ashtrays, and evacuation plans. "Yes," said the traveler, "but what about all those uncontrolled ignition sources? All you have to do is ban smoking in the tower and then you've eliminated them." The traveler and the manager eventually agreed to disagree and life continued on for both of them.

As a postscript, a smoking ban was instituted for this facility shortly after the incident just described. The decision was most likely made in response to health concerns and as part of a general tightening of smoking regulations, not because of the traveler's safety lecture. The decision may also have been prompted by the August 2000 electrical fire in the Ostankino television tower in Moscow, which killed three people and occurred one week after the traveler's tower incident.

14.2 Newspapers and Magazines

Pick up a newspaper or magazine and you are pretty much guaranteed to find a story, or part of a story, relating to safety. The catch sometimes is to recognize that safety is involved and to analyze the story for inherent-safety aspects. The example of the worker mistakenly placing a hot-water hose in a gas tank (Section 7.4) would fit in this category, as it was taken from a magazine (albeit an occupational health and safety publication). Several examples have been given in this book based on newspaper articles; another follows below.

A newspaper reported that a woman was temporarily paralyzed after the wrong fluid was injected into her spine during a hospital examination procedure known as a myelogram. According to the article,[2] in this procedure a small needle is inserted into the spinal canal, and a contrasting agent is injected so problems in the spine can be seen more easily in an x-ray. The article also mentioned that the wrong contrasting agent was used. The problem arose because there were three different liquids (only one of which was the correct agent), all clear and colorless and stored in similarly labeled bottles. Apparently, a technician would choose the liquid on orders from the doctor; a radiologist was then supposed to check that it was the right liquid before filling the syringe.[2]

So what should an incident investigation report recommend in this case: Address the procedure to ensure that the radiologist check is done properly? Only if the hospital wishes to have repeat occurrences of this and similar incidents. No, the hazard needs to be addressed at its source so that the technician is not set up for failure. Having similar liquids with similar labels stored in the same location is not a design that is free from opportunities for human error (Section 7.4).

Cases such as these, and those drawn from life experiences (Section 14.1), help to personalize inherent safety and infuse the concepts with the familiar. Almost everyone has been in a hospital; everyone has had everyday encounters with safety issues, such as when walking, bicycling, driving, etc. The remainder of the chapter deals with more technical examples.

14.3 Topical Conferences

Many process-safety conferences (such as those identified in the preface to this second edition) routinely feature case-study papers. Some—for example, the Annual Loss Prevention Symposia of the American Institute of Chemical Engineers—have special sessions devoted solely to case studies. Of course not all of these papers deal with inherent safety, but the emphasis is always on the broadly applicable lessons that were learned from process incidents.

A paper of this type was presented by Sanders[3] in one of the Mary Kay O'Connor Process Safety Center Annual Symposia organized at Texas A&M University in the United States. He gives several case histories of designs that lacked inherent safety, including issues related to facility siting, incompatible chemicals, and piping arrangements.[3]

14.4 Technical Papers

The archival journal literature is another rich source of case studies on inherently safer design. Process-safety journals (again, such as those identified in the preface to this second edition) have carried an increasing number of papers on inherent safety over the past decade. Some of these publications are themselves case studies, while others incorporate some form of case-study validation for a developed methodology. The Westray mine explosion has been analyzed from both of these perspectives—as a loss-causation case study to identify management system deficiencies[4] and as a case study to validate the inherent-safety based incident investigation methodology[5] described in Section 11.2.9.

The Westray coal mine explosion occurred in Plymouth, Nova Scotia, Canada, on May 9, 1992, killing 26 miners.[6] An indication of the destructive overpressures generated underground can be seen in Figure 14.1,[7] which shows surface damage at the mine site. The methane levels in the mine were consistently higher than regulations and were caused by inadequate ventilation in the mine. Dust accumulations also exceeded permissible levels due to inadequate cleanup of coal dust (Section 13.3); additionally, there was no crew in charge of rock dusting to inert the coal dust with limestone or dolomite (Section 13.5.1). These and many other factors contributed to the poor work conditions that continually existed in the Westray mine and made it the site of an incident waiting to happen. All of these substandard conditions and practices could be attributed to the lack of concern expressed by management personnel toward safety issues in the mine and an inadequate safety management system. These were the primary root causes of the problems and the explosion at Westray.[4]

With reference to dust explosion causation, prevention, and mitigation as discussed in Chapter 13, the hierarchy of controls (in particular the inherent-safety control level) was essentially ignored at the Westray site. Had there been a commitment to the principles of inherent safety, some of the hazard-avoidance recommendations

Figure 14.1 Damage to portal at No. 1 main, Westray coal mine. (From K. P. Richard, *The Westray Story: A Predictable Path to Disaster, Report of the Westray Mine Public Inquiry* (Halifax, Canada: Province of Nova Scotia, 1997). (With permission.)

Table 14.1   Hazard Avoidance Possibilities Categorized by Inherent-Safety Principles

Principle	Recommendation to Avoid Hazard
Intensification	*Methane*: degasification (extraction of methane prior to mining)
	Coal dust: housekeeping program for continuous removal
	Fuel storage: elimination of underground storage
	Ignition sources: replacement of nonflameproof equipment (i.e., minimization of ignition sources by substitution of equipment)
	Shift length: reduction from 12 hours
Substitution	*Auxiliary ventilation system*: forcing, rather than exhaust, system for adequate airflow to clear methane from the working face of the mine
	Main ventilation fan: alternative design and location so as not to pick up dust and other debris from the coal return conveyor belt (also linked to limitation of effects)
Attenuation	*Coal dust*: purchase and maintenance of an adequate inventory of rock dust; implementation of a program for rock dusting underground
	Coal dust: roadway consolidation (a process to control the formation and dispersal of dust on mine roadways by application of rock dust, moisture-absorbing material, and a binding agent)
Simplification	*Methane*: installation of a reliable, robust, simplified mine air-monitoring system

Source: Adapted from A. Goraya, P. R. Amyotte, and F. I. Khan, "An Inherent Safety-Based Incident Investigation Methodology," *Process Safety Progress* 23, no. 3 (2004): 197–205.

given in Table 14.1 would have been implemented. It is particularly poignant to note that all of the measures listed in Table 14.1 are well known and are commonly practiced in the underground coal mining industry.

Occasionally, the more mainstream chemical engineering journals (i.e., those without a specific safety focus) publish a paper on inherent safety with a case study. One example is the review by Khan and Amyotte[8] in which the Bhopal tragedy is briefly discussed. Adapted excerpts from Reference 8 indicate the lack of attention to inherently safer and user-friendly design from a number of perspectives:

1. *Intensification*: Methyl isocyanate (MIC) was neither a raw material nor a product; it was an intermediate. Although it was convenient to store MIC onsite, it was not essential to do so, as evidenced by the fact that MIC stocks had been reduced by 75% within a year of the incident. It can be readily agreed that such devastating consequences as did occur would not have happened had a lesser amount of MIC been stored (and, in this case, released).

2. *Substitution*: The product, carbaryl, was manufactured by reacting phosgene and methylamine to produce MIC, which was then reacted with alpha-naphthol. The same product can be made from the same raw materials by reacting them in a different order and avoiding the production of MIC. In this alternative process route, phosgene is reacted with alpha-naphthol, and then the intermediate (a less hazardous chloroformate) is reacted with methylamine. The chemistry of these two processes is shown in Figure 14.2. Clearly, application of the substitution principle at the process-route-selection stage could have played a role in averting the incident consequences.

3. *Attenuation*: Attenuation of the storage conditions (temperature) would have been enabled by the refrigeration system had it been operational. Rather than being stored at 0°C or lower as standard procedures required, the MIC was actually at ambient temperature—obviously much closer to its boiling point of 39.1°C. With the contaminant presence leading to an exothermic reaction and elevated MIC temperatures and vapor generation, it is not clear how pressure might have been effectively attenuated.

Methyl Isocyanate Route:

$$CH_3NH_2 \quad + \quad COCl_2 \quad \rightarrow \quad CH_3CNO \quad + \quad 2HCl$$

Methylamine Phosgene Methyl Isocyanate Hydrochloric Acid

$$CH_3CNO \quad + \quad C_{10}H_8O \quad \rightarrow \quad C_{12}H_{11}NO_2$$

Methyl Isocyanate Naphthol Carbaryl

Non-Methyl Isocyanate Route:

$$C_{10}H_8O \quad + \quad COCl_2 \quad \rightarrow \quad C_{11}H_7ClO_2 \quad + \quad HCl$$

Naphthol Phosgene Naphthol Chloroformate Hydrochloric Acid

$$C_{11}H_7ClO_2 \quad + \quad CH_3NH_2 \quad \rightarrow \quad C_{12}H_{11}NO_2 \quad + \quad HCl$$

Naphthol Chloroformate Methylamine Carbaryl Hydrochloric Acid

Figure 14.2 **Alternative route for carbaryl production. (From F. I. Khan and P. R. Amyotte, "How to Make Inherent Safety Practice a Reality," *Canadian Journal of Chemical Engineering* 81, no. 1 (2003): 2–16, 2002]. With permission.)**

4. *Simplification*: One of the important lessons learned from this incident is the need to make operation and control systems simple and to maintain them in good working order. The Bhopal facility had many additional end-of-pipe monitoring and control systems attached to the storage vessels, but their reliability was questionable. This can cause a dual problem: The safeguards may not be available when needed, and they can provide a false sense of security for process operators who may ignore initial warning signs such as a pressure increase. Many of these issues are brought into focus when one analyzes a system through use of the simplification principle.

14.5 Process-Safety Books

Books devoted solely to process safety will usually have several case studies for illustrative purposes. Process-safety theory and examples of application (or lack thereof) seem to go hand in hand.

An example in this category is the text, *Chemical Process Safety: Fundamentals with Applications*, by Crowl and Louvar.[1] Chapter 13 in that book gives various case histories categorized in four sections: static electricity, chemical reactivity, system design, and procedures. Chapter 1 contains a brief section on inherent safety followed by accounts of "four significant disasters": Flixborough (England), Bhopal (India), Seveso (Italy), and Pasadena (Texas). Each of these four major incidents has been discussed in the current book with respect to inherently safer design—some of them several times.

A second example in this category is the three-volume *Lees' Loss Prevention in the Process Industries*, now in its third edition.[9] Chapter 32 in Volume 2 of that work covers inherently safer design, and Volume 3 contains several appendices with comprehensive accounts of major process incidents. These case studies include the four significant disasters covered by Crowl and Louvar[1] as well as others mentioned in the current book (e.g., Piper Alpha and Chernobyl) and more. Internet sources of case history information are also given in Appendix 1 in Volume 3 of that work (see Section 14.12 in the current book).

When reading the case studies mentioned above (or for that matter any case study), keep these questions in mind:

1. Did a lack of application of the principles of inherently safer and user-friendly design play a role in incident causation?
2. Would intensification, substitution, attenuation, etc., have helped to prevent the incident or mitigate the consequences?
3. How effective were the passive and active engineered safety devices with respect to prevention and mitigation?
4. How effective were the procedural safety measures with respect to prevention and mitigation?

14.6 Case Study Books

Some process-safety books are devoted solely to case studies. This is the purpose of a work by Kletz:[10] *What Went Wrong? Case Histories of Process Plant Disasters and How They Could Have Been Avoided.* Several of these incident descriptions and application examples have been referenced in the current book. Not all of the case studies in Reference 10 deal with inherent safety, but many do, including those in Chapter 21 on inherently safer design. And, of course, the vocabulary of inherently safer and user-friendly design is used quite explicitly.

14.7 Other Inherent-Safety Books

There are other books dealing exclusively with the topic of inherent safety (although there do not appear to be many). For example, the recent second edition of *Inherently Safer Chemical Processes: A Life Cycle Approach,*[11] a publication of the Center for Chemical Process Safety of the American Institute of Chemical Engineers, is of interest. Chapter 11 contains several worked examples and case studies.

14.8 Books from Other Industries and Applications

Books from industries other than the process industries, and applications other than strictly inherently safer design, can provide helpful case studies to bridge between personal experiences (Section 14.1 and Section 14.2) and process-engineering applications. Examples include:

1. The book on safety culture by Hopkins[12] contains case studies from the railway and aircraft maintenance industries that have been referenced in Section 5.3 and Section 11.2.1, respectively.
2. Petroski[13] writes on the successes and pitfalls of the design process, and raises the issue of a lack of design standardization and whether this has serious consequences for end users. He gives the example of telephone and calculator keypads having different layouts for their respective push buttons (numbers 1,2,3 on the top row for the telephone and 7,8,9 on the top row for the calculator).[13] Is this a complexity that is problematic? Are we exceeding the ability level of the "operator"? Do we need a more friendly, standardized design? Or would changes at this point be cost prohibitive and even more problematic?
3. Casey[14] gives several interesting "true tales of design, technology and human error." One particularly chilling account relates the problems that occurred with the Therac-25 radiotherapy treatment machine and the delivery of lethal

doses to cancer patients. The problem was rooted in the computer control system[14] (Section 9.6).

4. In his book on engineering safety and reliability, Wong[15] gives an example similar to that in Section 14.2. He describes a case where a patient was administered the wrong antibiotic and died. Warnings were ignored and procedural checks failed to prevent the error.[15] Was there an inherently safer way to address this hazard?

14.9 Training Packages

The training packages identified in Section 11.2.8 both contain several case-study examples. The Institution of Chemical Engineers package[16] incorporates the Westray case study[5] summarized in Section 14.4. The American Institute of Chemical Engineers package[17] has been updated in 2006[18] and expanded to include design conflicts in 2008.[19]

14.10 Trade Literature

Literature from manufacturers and trade associations can be helpful in giving short, practical examples of devices that might be inherently safer or more user friendly. This would not strictly fit with the definition of a case study as an incident description with accompanying lessons learned. Nevertheless, training questions can be posed around such devices.

The Safe Spade described in Section 4.1.1 could be placed in this category. Another example would be the cleaning agent advertised in an equipment supplier's trade publication. This multipurpose product was described as an environmentally friendly cleaner/degreaser that is "biodegradable, non-flammable, non-toxic and non-corrosive."[20] One might also reasonably conclude that this particular cleaner offers some significant inherent-safety benefits.

14.11 Loss Prevention Bulletins

Loss prevention bulletins are published by the Institution of Chemical Engineers in the United Kingdom. They represent a good source of case studies arising from process-incident and near-miss reports. In reading the case descriptions, it is again helpful to keep in mind the four questions listed at the end of Section 14.5.

Two recent examples will illustrate the usefulness of the loss prevention bulletins in this regard. Soderlund et al.[21] give an example of aboveground domino effects caused by failure of underground water pipelines. They recommend the replacement of pipe material to avoid catastrophic failure (Section 4.1.1) and establishing safe distances between underground pipework and process units to avoid knock-on effects (Section 9.1).

In their examination of the role of bunding in the Buncefield incident (Section 3.6), Whitfield and Nicholas[22] observe that many of the bund walls suffered loss of containment at points of penetration by pipework. Leakage occurred due to the lack of a good seal between the pipes and the bund wall (Section 3.1 and Section 6.1). As one of the lessons learned, they comment that, in the case of aboveground storage of flammables, bunds should have fire-resistant structural integrity, joints, and pipework penetrations (Section 4.1.1).[22]

14.12 Chemical Safety Board Reports

As noted on its Web site (http://www.csb.gov/), the U.S. Chemical Safety Board is an independent, nonregulatory federal agency that conducts root-cause investigations of chemical accidents at fixed industrial facilities. The reports of its investigations are available on the CSB Web site for downloading and are often accompanied by video footage and animation of the incident sequence. As with other avenues for case study development, the language of inherent safety is not typically explicitly present in the CSB reports. However, the concepts of inherently safer design are embodied in the reports, and the questions in Section 14.5 will bring them to the surface. (Witness the use of the CSB dust explosion reports throughout Chapter 13.) In short, the U.S. Chemical Safety Board investigation reports are excellent resources for training and other process-safety improvement efforts.

As an example, consider the report of the 2006 methanol tank explosion and fire at a wastewater treatment plant in Daytona Beach, Florida.[23] Two employees were killed and a third severely burned in this incident. The facility processed wastewater with a treatment scheme that used methanol, which was stored in an aboveground tank. Root causes identified by the CSB were related to inadequate hot-work controls and an ineffective hazard communication program with respect to the hazards associated with methanol.[23] Contributing causes were:[23]

1. No program for evaluation of nonroutine tasks
2. Inappropriate material of construction of methanol system components (piping and valves made of polyvinyl chloride instead of steel)
3. Inappropriate material of construction of flame arrester on methanol tank vent (aluminum, which is corroded by methanol)
4. No requirement to maintain flame arrester

One sees the appearance of procedural measures in the root-cause list and in item 4 of the contributing causes. Item 1 in the above list relates to hazard assessment, and items 2 and 3 indicate the need for material substitution in the case of a passive engineered device that failed. Digging deeper into the report, other inherently safer design lessons begin to appear.

The original plant had been modified in 1993 to incorporate a biological nutrient removal (BNR) process in which bacteria converted nitrogen compounds in the wastewater into nitrogen gas. Methanol was used as an organic nutrient for the bacteria and was continuously fed to the process from a 10,000-gallon carbon steel storage tank.[23]

Quoting from the investigation report: [23]

> In 1999, the City of Daytona Beach modified the BNR process to operate without the continuous methanol feed; however, the facility continued to use the methanol system and 10,000-gallon storage tank for sporadic methanol addition. As a result, the facility maintained a large inventory of methanol even though demand was substantially reduced. The methanol storage tank contained between 2,000 and 3,000 gallons when the incident occurred.

At this point, the word *intensification* fairly jumps off the page. Follow this with questions as to whether process alternatives were available that did not involve flammable methanol, and an inherent-safety case study is (unfortunately) born.

14.13 Concluding Remarks

Various ways of locating and developing case studies based on the principles of inherent safety and user-friendly design have been presented in this chapter. There is at least one other option available to those readers who have access to their company incident reports and investigation reports. If you were to take your most recent report of an incident, what would be the answers to the four questions at the end of Section 14.5?

And if you considered the investigation report of your most recent incident, what would be the answer to this question: "What recommendations were made to avoid the hazards and to permanently remove them where possible?"

References

1. D. A. Crowl and J. F. Louvar, *Chemical Process Safety: Fundamentals with Applications*, 2nd ed. (Upper Saddle River, NJ: Prentice Hall, 2002).
2. "Woman Paralysed after Getting Wrong Injection during Exam," *Mail Star* (Halifax, Canada), 2001.
3. R. E. Sanders, "Designs That Lacked Inherent Safety: Case Studies," *Journal of Hazardous Materials* 104, no. 1-3 (2003):149-161.

4. P. R. Amyotte and A. M. Oehmen, "Application of a Loss Causation Model to the Westray Mine Explosion," *Process Safety and Environmental Protection* 80, no. 1 (2002): 55–59.

5. A. Goraya, P. R. Amyotte, and F. I. Khan, "An Inherent Safety-Based Incident Investigation Methodology," *Process Safety Progress* 23, no. 3 (2004): 197–205.

6. P. R. Amyotte and R. K. Eckhoff, "Dust Explosion Causation, Prevention and Mitigation: An Overview," *Journal of Chemical Health and Safety*, 17, no. 1, January/February 2010: 15–28.

7. K. P. Richard, *The Westray Story: A Predictable Path to Disaster, Report of the Westray Mine Public Inquiry* (Halifax, Canada: Province of Nova Scotia, 1997).

8. F. I. Khan and P. R. Amyotte, "How to Make Inherent Safety Practice a Reality," *Canadian Journal of Chemical Engineering* 81, no. 1 (2003): 2–16.

9. S. Mannan, ed., *Lees' Loss Prevention in the Process Industries*, 3rd ed. (Oxford, U.K.: Elsevier Butterworth-Heinemann, 2005).

10. T. Kletz, *What Went Wrong? Case Histories of Process Plant Disasters and How They Could Have Been Avoided*, 5th ed. (Oxford, U.K.: Gulf Professional Publishing, 2009).

11. Center for Chemical Process Safety, *Inherently Safer Chemical Processes: A Life Cycle Approach*, 2nd ed. (Hoboken, NJ: John Wiley & Sons, 2009).

12. A. Hopkins, *Safety, Culture and Risk: The Organizational Causes of Disasters* (Sydney: CCH Australia Ltd., 2005).

13. H. Petroski, *Small Things Considered: Why There Is No Perfect Design* (New York: Alfred A. Knopf, 2003).

14. S. Casey, *Set Phasers on Stun and Other True Tales of Design, Technology, and Human Error*, 2nd ed. (Santa Barbara, CA: Aegean Publishing, 1998).

15. W. Wong, *How Did That Happen? Engineering Safety and Reliability* (London: Professional Engineering Publishing Ltd., 2002).

16. IChemE, *Inherently Safer Process Design*, Training Resource STP001 (Rugby, U.K.: Institution of Chemical Engineers, 2005).

17. SACHE, *Inherently Safer Plants*, Slide Package, Safety and Chemical Engineering Education (New York: Center for Chemical Process Safety, American Institute of Chemical Engineers, 1996).

18. D. C. Hendershot, *Inherently Safer Design*, Safety and Chemical Engineering Education (New York: Center for Chemical Process Safety, American Institute of Chemical Engineers, 2006).

19. D. C. Hendershot and J. Murphy, *Inherently Safer Design Conflicts and Decisions*, Safety and Chemical Engineering Education (New York: Center for Chemical Process Safety, American Institute of Chemical Engineers, 2008).

20. Acklands-Grainger, "Multi-Purpose Cleaner is Environmentally Friendly," *SupplyLink* (Spring 2006).

21. M. Soderlund, P. Bots, P. Eriksson, P. Nilsson, and J. Hartlen, "Evaluating the Domino-Effect of Critical Constructions due to Damage of Underground Water Pipelines," *Loss Prevention Bulletin* 195 (2007): 22–27.

22. A. Whitfield and M. Nicholas, "Bunding at Buncefield: Successes, Failures and Lessons Learned," *Loss Prevention Bulletin* 205 (2009): 19–25.

23. CSB, *Investigation Report—Methanol Tank Explosion and Fire—Bethune Point Wastewater Treatment Plant*, Report No. 2006-03-I-FL (Washington, DC: U.S. Chemical Safety Board, 2007).

Chapter 15

Do We Go Too Far in Removing Risk?

I would give all my fame for a pot of ale and safety.

—**Shakespeare,** *Henry V*

What is safe is distasteful; in rashness there is hope.

—**Tacitus,** *History,* **Book 3**

Earlier chapters have shown that much complication in plant design is the result of adding protective equipment to control risk. I have tried to show that in many cases we can change the design, avoid the risk, and end up with a simpler plant. Another way of simplifying the design is to add less protective equipment and accept a higher degree of risk. Do we go too far in trying to control every conceivable risk, however trivial, however unlikely?

To answer this question, let us first consider risks to output and efficiency and then risks to life and limb.

15.1 Risks to Output and Efficiency

During the 1960s, as a result of changes in technology and a drive for minimum capital cost, many plants had difficult start-ups (Section 10.1.2). By the 1970s, many companies had learned the lesson and tried to foresee and overcome as many problems as possible; many, though not all, very large plants had easy start-ups and soon achieved flowsheet output. For ethylene plants, the time from "oil on" to flowsheet output varied from 2 weeks to over 6 months.

The successful companies found that they could sell the products, and the cost of the extra design features and procedures was soon recovered. At the end of the decade the position changed. Some plants were started up just as successfully, but there was no market for all the product. It would hardly have mattered if they had taken 6 months or even 2 years to achieve flowsheet output. If we think this will recur, there may be less need to add features that will help us achieve an easy start-up. We do not know, when we are designing a plant, what the commercial situation will be when it comes on-stream. Nevertheless, it can be argued that the chemical industry, or parts of it, has been too risk averse.

Certainly, some companies have gone too far in providing expensive and hazardous storage capacity to prevent loss of output because raw materials arrive late, product dispatch is delayed, or one section of a plant is shut down. When Bhopal drew attention to the hazards of storing large quantities of chemicals, many companies found that they could manage with less (Section 3.6).

On other occasions, companies have added alarms and trips to prevent inconvenience to the operators, minor plant upsets, or infrequent and not very hazardous events such as the overflow every few years of a tank containing a nonhazardous material.

15.2 Risks to Life and Limb

Since the early 1970s, the chemical and nuclear industries have increasingly adopted a numerical approach to risk by means of quantitative risk assessment (QRA). Many attempts have been made to quantify risks and set standards or targets. (Recall Section 10.4.3 and the extensive reviews of collectively almost 100 techniques by Tixier et al.[1] and Reniers et al.[2]) Risks above the target at new or existing plants should be reduced; risks below the target should be left alone, at least for the time being. Resources are not unlimited, and so we should deal with the biggest risks first.[3,4]

In a development of simple target setting, there are two targets: an upper level of risk that should never be exceeded and a lower or trivial level. If this level is reached, we need not try to get below it. Between the two levels, we should reduce the risk if it is "reasonably practicable" (to use the U.K. legal phrase) to do so (see Table 10.3, Note 1, and Figure 10.4).

The various targets proposed are all very low. Are they too stringent? Should they be relaxed?

I do not think so. As far as risks to employees are concerned, the targets do not amount to much more than saying, "Do a little better than in the past"—a philosophy with which it is hard to disagree. Also, once companies have accepted targets for safety and risk, it is hard to go back on them. They can hardly say that they propose to kill or injure more people than in the past (see appendix to Chapter 8, item 18).

Of course, there are individual instances in which companies have gone too far in removing hazards, particularly after an accident. In these cases, numerical methods of assessing risk should be used, if possible.

The targets proposed for risks to the public are naturally much lower than those proposed for employees, for people choose to work for a company, but risk is imposed on the public without its permission. (In the United Kingdom, the Health and Safety Executive has proposed that the maximum tolerable risk for members of the public should be one-tenth as large as that for employees. See Table 10.3, Note 1.) The targets also differ in another respect. Very few members of the public have been killed by industrial accidents. In some countries such as the United Kingdom, the chemical industry has killed no one in this way. The targets cannot, therefore, be based on doing better than in the past. Instead, they have been set by fixing a level that is insignificant compared with all the other risks to which we are exposed or by deciding what level of risk the public will tolerate.[5] There is consequently much greater variation in the targets. Those suggested by some government authorities are absurdly low and probably impossible to achieve if the calculations are carried out honestly. The nuclear industry in particular, in many countries, has adopted, or been made to adopt, very low targets, not because there is a technical or economic case for them, but because the industry is trying to overcome the irrational fears of the public.

15.3 The Contribution of the Operator

A specific area in which many people contend that we have gone too far is in estimating the unreliability of the operator. Much protective equipment has been added to our plants to prevent operators from making errors, to warn them that they have made errors, or to guard against the consequences of the errors.

Suppose that, when an alarm sounds, an operator has to go outside, select the right valve out of many, and close it within, say, 10 minutes, or there will be a spillage of hazardous material. Should we rely on the operator, or should we use automatic equipment? Some people vote for automatic equipment because experience, they say, shows that people are unreliable. Others consider that the operators can reasonably be expected to do what we ask.

Both these attitudes are unscientific. We should not ask whether the operator will or will not close the right valve in the required time, but what is the probability that he or she will do so? The answer will depend on the degree of stress and distraction and will include a large measure of judgment (guidance is available[6–8]), but it will be far better than saying the operator always will or never will. The use of performance-influencing factors (e.g., stress), expert judgment, and quantitative analysis was described in Section 11.2.7 for offshore-platform emergency scenarios.[9,10]

Once we have estimated the probability, however roughly, of operator error, we are in a better position to decide whether we should change the design or whether we should accept an occasional error. In some cases, we may wish to draw a fault tree, estimate the risk to life, and compare it with our target or criterion.

People are actually very reliable, but there are many opportunities for error in the course of a day's work (Sections 14.2 and 14.8), and if we handle hazardous materials, we may find that we have to reduce our reliance on the operators. This is better than expecting them to make fewer mistakes than experience shows they will make and then blaming them for acting in a way that could have been foreseen.

If we change the design by adding protective equipment, it may not be more reliable than the operator.[11] Each design should be checked in detail. If we wish to reduce our dependence on the operator, we should do so, if possible, by changing the design rather than by adding protective equipment. Estimates of operator reliability should therefore be made early in design, not left until the end.

The consideration of human error and human factors in design is an area in need of attention by the process industries. According to one recent report, engineers have little understanding of the human component in relation to system development, as evidenced by the high rate of latent design error as a contributing factor to accidents.[12] It does seem that expertise in this field resides largely in other professions (e.g., human performance) and other industries (e.g., air traffic management).[13,14] Different industries can, however, learn from one another; the first step is dialogue and an understanding that there are common problems.[13]

References

1. J. Tixier, G. Dusserre, O. Salvi, and D. Gaston, "Review of 62 Analysis Methodologies of Industrial Plants," *Journal of Loss Prevention in the Process Industries* 15, no. 4 (2002): 291–303.
2. G. L. L. Reniers, B. J. M. Ale, W. Dullaert, and B. Foubert, "Decision Support Systems for Major Accident Prevention in the Chemical Process Industry: A Developers' Survey," *Journal of Loss Prevention in the Process Industries* 19, no. 6 (2006): 604–620.
3. T. A. Kletz, *Hazop and Hazan: Identifying and Assessing Process Industry Hazards*, 4th ed. (Rugby, U.K.: Institution of Chemical Engineers, 1999).
4. S. Mannan, ed., *Lees' Loss Prevention in the Process Industries*, 3rd ed. (Oxford, U.K.: Elsevier Butterworth-Heinemann, 2005), chap. 9.
5. Health and Safety Executive Report, *The Tolerability of Risks from Nuclear Power Stations*, 2nd ed. (Sudbury, U.K.: HSE Books, 1992).
6. T. A. Kletz, *An Engineer's View of Human Error*, 3rd ed. (Rugby, U.K.: Institution of Chemical Engineers, 2001).
7. Center for Chemical Process Safety, *Guidelines for Preventing Human Error in Process Safety* (New York: American Institute of Chemical Engineers, 1994).
8. B. Kirwan, "Human Error Identification Techniques for Risk Assessment of High Risk Systems—Part 1: Review and Evaluation of Techniques," *Applied Ergonomics* 29, no. 3 (1998): 157–177.

9. D. G. DiMattia, F. I. Khan, and P. R. Amyotte, "Determination of Human Error Probabilities for Offshore Platform Musters," *Journal of Loss Prevention in the Process Industries* 18, no. 4-6 (2005): 488–501.

10. F. I. Khan, P. R. Amyotte, and D. G. DiMattia, "HEPI: A New Tool for Human Error Probability Calculation for Offshore Operation," *Safety Science* 44, no. 4 (2006): 313–334.

11. T. A. Kletz, *Hazop and Hazan: Identifying and Assessing Process Industry Hazards*, 4th ed. (Rugby, U.K.: Institution of Chemical Engineers, 1999), sec. 3.6.6.

12. Y. Toft, P. Howard, and D. Jorgensen, "Changing Paradigms for Professional Engineering Practice towards Safe Design: An Australian Perspective," *Safety Science* 41, no. 2-3 (2003): 263–276.

13. F. Drogoul, S. Kinnersly, A. Roelen, and B. Kirwan, "Safety in Design: Can One Industry Learn from Another?" *Safety Science* 45, no. 1-2 (2007): 129–153.

14. A. Hale, B. Kirwan, and U. Kjellen, "Safe by Design: Where Are We Now?" *Safety Science* 45, no. 1-2 (2007): 305–327.

Chapter 16

The History and Future of Inherently Safer and User-Friendly Design

> A man takes a mustard seed and sows it in his field. It is the smallest
> of all seeds, but when it grows up, it is the biggest of all plants. It
> becomes a tree.
>
> **—Matthew 13:31–32, *Good News Bible***

16.1 The Start of the Story

No idea suddenly appears out of a blue sky; every idea has a pedigree. However
far back we look for the origin of an idea, we find someone who had thought of it
before, though perhaps in an incomplete or primitive form, or without realizing
its full implications. One example of inherently safer design from the 1870s was
described in Section 4.2.5. In the early days of the Solvay process for the manufacture of sodium carbonate, a manhole at the top of each batch distillation column
had to be opened and solid lime tipped in every time the column was charged.
Because ammonia vapor could escape, this operation was hazardous. Ludwig Mond
suggested that milk of lime should be pumped into the columns instead. However,
it was not until 100 years later that inherently safer design was recognized as a specific concept and engineers began to consider it systematically.

My first introduction to the idea occurred in 1970, when I was a member of the organizing committee for the symposium on "Loss Prevention in the Process Industries" held in Newcastle, United Kingdom, the following year. (This was the first of the regular triennial European loss-prevention symposia, though it was not numbered, and the 1974 symposium in Holland was called the first.) Among the papers offered for the symposium was Bell's now-classic and often-quoted paper on the manufacture of nitroglycerin described in Section 3.2.3. A small, well-mixed, continuous reactor containing about a kilogram of raw materials and product replaced the traditional batch reactor containing about a ton. In another paper (Section 9.5), Pickles solved a difficult control problem (controlling a temperature to within 1°C), not by adding complex control equipment, but by finding another temperature that showed greater variation and so did not have to be controlled so accurately. The chairman of the committee, my former ICI colleague, T. A. Kantyka, remarked that it was far better to avoid the need for complex safety or control systems, as Bell and Pickles had done, than to install them.

A few times in a lifetime, like a single stone setting off a landslide, a single remark makes one look at a whole range of problems in an entirely different way. Kantyka's remark had this effect on me, though the seed he sowed did not germinate until the explosion at Flixborough four years later.

There were many lessons to be learned from this explosion (Section 3.2.4), but one of the most important was missed by the official inquiry and by most commentators: The leak was big (about 50 tons) and the explosion devastating because there was so much flammable material in the plant (about 400 tons). If the inventory could be reduced, the plant would be safer: What you don't have, can't leak. The inventory was so large because the conversion was low, about 6% per pass, and so most of the raw material, cyclohexane, had to be recovered and recycled; 94% of it got a "free ride" through the plant—in fact many "free rides." If the conversion could be increased, the inventory would be lower.

In a paper on the lessons of Flixborough presented at the 1975 loss prevention symposium,[1] I included a page on inventory reduction. I wrote,

> Can plant inventories be reduced? Generally speaking, in designing processes we have accepted whatever inventory was called for by the design. If the material in the plant is hazardous, we have tried to contain the hazard by good design and by provision of control features such as emergency isolation valves, steam curtains, etc. Rarely, if ever, have we asked if it is possible to reduce the inventory by changes in design.
>
> Prevention of leaks by good design and operation should continue to be the main approach. But where the inventory can be reduced by changes in design without too much increase in cost, this should be done. There may be some payback in reducing the need for other safety equipment such as emergency isolation valves and steam curtains.

I then went on to list some examples of what might be done, mentioning Bell's paper, but giving no details.

Rereading my paper after an interval of many years, it seems very diffident. I did not point out that reducing inventories would make plants smaller and therefore cheaper; I did not use the words "inherently safer" or "intensification." Following my presentation, there was a long discussion,[2] mostly about the details of the explosion, with no reference to inventory reduction.

Reducing the inventory in the Flixborough process is not easy, for reasons I have discussed in Section 3.2.4. One company, which operated similar plants, did start a research program, and a process using a tubular reactor showed much promise. The program was abandoned because there seemed little chance that a new plant would be needed in the foreseeable future. The research team was disappointed, but commercially the decision was right. So far as I am aware, no new plant has been built in the West since Flixborough; all the new capacity has been in Eastern Europe or the Far East.

In accordance with the preface for this second edition, to this point the reader will have interpreted the word "I" to mean the author of the first edition—Trevor Kletz. For the remainder of Section 16.1, it is Trevor's coauthor for the second edition—Paul Amyotte—writing in the first person.

My entry into the world of inherent safety was not so much the result of a single remark but more an event followed by growth over a period of years. I had completed my doctorate in a rather specialized area of dust explosion research, which is an even more specialized field of chemical engineering. As I continued my research, I focused on the science of dust explosions—the physics of turbulence generation in dust clouds and the chemistry of explosion propagation and termination. I recall giving a seminar on dust explosions to our Department of Chemistry at Dalhousie University in 2000. (I thought there was a lot of chemistry in the talk, but I'm not sure the audience did.)

The precipitous event was the decision in the late 1990s by what was then the Department of Chemical Engineering to introduce a mandatory course for senior students in industrial safety and loss management. I had previously developed a graduate course in a similar area, and I was the logical choice to teach the undergraduate course. I knew this new course could not be a course on dust explosions (although that surely had to be the subject of a few lectures!). As I developed the course material, I read much of Trevor's work, and I learned of the training packages referenced in Section 14.9. I then realized that the course, a sort of hybrid of process safety and occupational health and safety, had to include a component on inherent safety.

Perhaps more importantly, what I also realized was that much of my previous work on dust explosions actually incorporated the concepts of inherent safety. I went back through some of my earlier papers and found that many did indeed deal with matters related to inherently safer design. The best example is "Inerting of Fine Metallic Powders";[3] at the time of doing this work, I thought it was about explosion inerting. It was, of course, but it was also about substitution and attenuation. I then decided to focus my research efforts, as well as my teaching efforts, on inherent safety.

My reason in telling this story is simple. How many other researchers and practitioners would find inherent-safety applications in their previous work by having a second look at it after reading this book?

16.2 The Idea Develops

In 1976, ICI held an all-day internal company symposium on intrinsically safer plants (as they were then called), and this did much to publicize the concept in the company. The next year, I (back to Trevor) was invited to give the annual Jubilee lecture to the Society of Chemical Industry in London. I chose as my title, "What You Don't Have, Can't Leak" (which some people thought rather flippant). This was the first paper devoted entirely to inherently safer (IS) design. I described the nitroglycerin process and also discussed the constraints that hindered the development of IS designs.[4]

During the following years, I continued to advocate IS designs in papers and articles, at conferences and courses, and in lectures to learned societies. I noticed a curious fact. If I was asked to talk on Hazop (hazard and operability study), QRA (quantitative risk assessment), human error, or any other subject and I was not available, the organizers found someone else. If I was asked to talk on inherently safer design and I was not available, the subject was left out of the program. (This is no longer true; many very able chemical engineers are now involved in the subject.) My early lectures on inherently safer design produced fewer questions than those on other topics.

In 1979, a company held an internal symposium on simplification (see the appendix to Chapter 8). During a discussion of the problem described in Section 4.1.3, someone said

> As long as we argue about whether we need two, four, six or eight trips we'll never get anywhere.
>
> We'll always prove we need eight, and [a company that had had several explosions on similar plants] will use two and go on blowing their plants up. The answer is to use a different design for that unit operation which eliminates the problem. Any arguments about how many trips we need will be sterile.

Later that year, I held a series of morning discussions for groups of 12–20 managers and engineers. One of the tasks given to the groups was the simplification of the plant shown in Figure 8.8. Few were able to simplify it, and some made it more complex (Section 8.3.2). In the same year, I published a paper on simplification[5] that described many of the examples now included in Chapters 7 and 8.

In 1984, after my retirement from ICI, I collected many examples of inherently safer and user-friendly design into a short book called, rather long-windedly, *Cheaper, Safer Plants or Wealth and Safety at Work*,[6] on which this book is based. The title was chosen to emphasize my belief that inherently safer plants are usually cheaper than traditional designs.

A few weeks after the book was published, the Bhopal tragedy occurred. As discussed in Section 3.6, the material that leaked and killed over 2000 people was an intermediate, not a product or raw material, and although it was convenient to store it, it was not essential to do so—a point overlooked by many commentators. Following Bhopal, many companies reduced their stocks of hazardous intermediates, and inherently safer designs were in the news. In 1987, the newly formed Center for Chemical Process Safety held its first symposium, and the program included four papers on these designs.[7]

In my lectures, I emphasized that inherently safer designs need a change in the design process (Section 10.1.1). Members of the audience often said, "You are talking to the wrong people. We can't change the design process. You should talk to our senior managers." The same comments would hold today.

I made a video lecture[8] (shown at the 1987 Loss Prevention Symposium) in the hope that safety advisers might show it to their seniors. It has sold well, but I wonder how many presidents and vice-presidents have seen it.

Today there is much more interest in inherently safer design; many people are writing and lecturing on the subject, and many books devote a chapter or more to it (see Further Reading at the end of this chapter). In 1996, the Center for Chemical Process Safety devoted much of a conference to the subject.[9] As described in Section 14.3, inherent-safety papers and sessions have now become fixtures in the programs of various topical conferences. Nevertheless, as also described in Section 10.1.2, reports have concluded that, although safety advisers are now familiar with the concept, there is still a general lack of awareness in design organizations and among senior managers. Common sense (i.e., inherent safety) has become common knowledge in some areas, but not all, and thus has not yet become common practice.

16.3 Extensions of the Idea

The concept of inherently safer design has been extended in several ways. Simplification has already been mentioned. At an American Institute of Chemical Engineers meeting in 1988, I described user-friendly plants, those in which human error (or equipment failure) does not have serious effects on safety (and output and efficiency).[10] Others have called them "forgiving." The most effective way of making plants user friendly is by inherently safer design, but there are also other ways, as described in Chapter 9.

Although I regard inherently safer design as part—the most important part—of user-friendly design, other writers give inherently safer design a wider definition and regard all the items discussed in this book as routes to inherently safer design (see Section 1.1, Item 4). In *Lees' Loss Prevention in the Process Industries*,[11] Table 32.1 provides a comparison of these approaches and the associated terminology.

It is interesting to see the recent proposal by Edwards[12] on the naming of two potentially new inherently safer design principles: hybridization (or transformation) and stabilization (or ensuring dynamic stability).

Passive design, discussed in Section 9.11, is a sort of halfway house to inherently safer design. It is the closest neighbor to inherent safety in the hierarchy of controls (Section 1.3).

As already stated in Chapter 1, the concepts of inherently safer design apply also to protection of the environment, industrial hygiene, and the disposal of waste. The twenty-first century challenges of sustainable development will require the adaptation and evolution of existing design principles and methods as process engineers are faced with developing new processes, not just the optimization of old ones.[13] Inherently safety design will be a key contributor to success in meeting these challenges as well as those related to the continuing search for alternative energy sources.[14]

16.4 Conclusions

The constraints that have hindered the growth of inherently safer designs and the actions needed to overcome them have been discussed in Sections 10.1 and 10.2. Regulatory requirements in some parts of the world may indeed drive the process of further use of inherently safer design.[15] Additionally, we should always remember the value of wide public communication of the benefits of the inherently safer approach.[16]

It has been my privilege to have played a part in nurturing several seeds in the field of loss prevention. Two of them, Hazop (hazard and operability studies) and Hazan (the latter now better known as quantitative risk assessment, or QRA), have been adopted rapidly—far more so than we considered possible 35 years ago. Within that time, crude calculations have been replaced by sophisticated computer programs. Inherently safer design has grown more slowly, but I remain convinced that it is an oak tree that grows slowly but will live a long time.

But there are hurdles to be overcome. Inherently safer design requires a basic change in approach. Instead of assuming that we can keep large quantities of hazardous materials and other hazards under control, we have to try and remove them. Changes in belief and the corresponding actions do not come easily. Many of today's engineers were trained as students (and in industry) to identify hazards—during design or by bitter experience—and then control them. Later, they heard about a new approach, the inherently safer way, but the old perspective is habitual and deeply ingrained; the new one is seen as a novelty to be tried when we have time and money to spare or when a problem requires exceptional treatment. We will not get very far until we learn to abandon the old approach, look upon the inherently safer way as the normal, and teach this to our students. We should do this early in the educational process.[17] Otherwise, we will perpetuate old ways of thinking.

A single datum does not prove a theory, and one example does not make a case; but the following story does offer hope. And while hope is not a strategy, it does provide a glimpse of what is possible.

A True Story (Except for the Names): August 2009

The professor (John) looked forward to his meeting with the former student (Luke). He had taught Luke a course in industrial safety about five years ago and had followed his career with interest. They kept in touch by phone and e-mail, and Luke would, from time to time, send the professor a case study or a safety bulletin—nothing proprietary of course, but enough to keep John in touch with "the real world." (How the professor disliked that term—as if the world he lived in were not real enough!) Luke was doing very well in industry and was taking on positions of greater responsibility within a major, multinational oil and gas company.

As they chatted over coffee, Luke told the professor about a big initiative underway within his company to promote safety culture and to demonstrate management's commitment to safety within the company worldwide. "Fascinating," thought John. "Here's another good example I can use in the safety course starting next month."

Next came the quiz from John: "So Luke, do you have any examples of inherently safer design in your company?" Luke proceeded to tell John about the training seminar he had given recently on inherent safety and how he had used some of his old course material on substitution and the other fundamental principles. He told him about a recent publicly reported process incident in his company that affected the way they now segregated plant sections to avoid domino effects. And he mentioned how he had recently been talking to a contractor and had challenged the individual on the number of flanges in a proposed design. "You know," said Luke, "we just had to minimize those flanges. There were too many of them and each one was a potential source of leakage."

The professor smiled. "Minimize the number of flanges," he thought. "I think I've read that somewhere recently."

A Final Thought

To end, let me quote from two papers written in very different contexts (but human nature is a common factor).

A game isn't won and lost when the king is finally cornered. The game's sealed when a player gives up having any strategy at all. When his soldiers ... move one piece at a time, that's when you've lost.[18]

There is a willingness to listen and to try to understand. But when it comes to practical matters, actually changing what we do, there is a deal of hesitancy—and even obstinacy. We are happy to listen but we don't want to change. We don't dare to take a step into the unknown....[19]

The next step is unknown no longer, as I hope this book has shown. Now is the time for continued and concerted action.

References

1. T. A. Kletz, "The Flixborough Cyclohexane Disaster," *Loss Prev.* 9 (1976): 106–110. Reprinted, with minor changes, as "Preventing Catastrophic Accidents," *Chem. Eng. (U.S.)* 83, no. 8 (1976): 124–128.

2. "Discussion," *Loss Prev.* 9 (1975): 111–113.

3. K. J. Mintz, M. J. Bray, D. J. Zuliani, P. R. Amyotte, and M. J. Pegg, "Inerting of Fine Metallic Powders," *Journal of Loss Prevention in the Process Industries* 9, no. 1 (1996): 77–80.

4. T. A. Kletz, "What You Don't Have Can't Leak," *Chem. Ind.* 6 (6 May 1978): 287–292.

5. T. A. Kletz, "Is There a Simpler Solution?" *Chem. Eng. (U.K.)* 342 (1979): 161–164.

6. T. A. Kletz, *Cheaper, Safer Plants or Wealth and Safety at Work*, 2nd ed. (Rugby, U.K.: Institution of Chemical Engineers, 1985).

7. D. E. Wade, D. C. Hendershot, R. Caputo, and S. E. Dale, in *Proceedings of the International Symposium on Preventing Major Chemical Accidents*, ed. J. L. Woodward (New York: American Institute of Chemical Engineers, 1987).

8. T. A. Kletz, *Inherent Safety*, Safety Training Package No. 9 (Rugby, U.K.: Institution of Chemical Engineers, 1987).

9. Center for Chemical Process Safety, *International Conference and Workshop on Process Safety Management and Inherently Safer Processes* (New York: American Institute of Chemical Engineers, 1996).

10. T. A. Kletz, "Friendly Plants," *Chem. Eng. Prog.* 85, no. 7 (1989): 18–26.

11. S. Mannan, ed., *Lees' Loss Prevention in the Process Industries*, 3rd ed. (Oxford, U.K.: Elsevier Butterworth-Heinemann, 2005), chap. 32.

12. V. H. Edwards, "Develop and Design Inherently Safer Process Plants," in *Global Congress on Process Safety, 2009 Spring National Meeting, American Institute of Chemical Engineers* (Tampa, FL, 2009).

13. G. Gwehenberger and M. Narodoslawsky, "Sustainable Processes: The Challenge of the 21st Century for Chemical Engineering," *Process Safety and Environmental Protection* 86, no. 5 (2008): 321–327.

14. M. S. Mannan, Y. Wang, C. Zhang, and H. H. West, "Application of Inherently Safer Design Principles in Biodiesel Production Processes," in *Institution of Chemical Engineers Symposium Series No. 151* (2006), Rugby, U.K.

15. D. C. Hendershot, "Inherently Safer Design: Back in the Spotlight?" *Journal of Chemical Health and Safety* 16, no. 2 (2009): 33–34.
16. M. J. M. Jongen, A. Dijkman, S. Zwanikken, G. I. J. M. Zwetsloot, and J. Gort, "Spreading the Word of the Concept 'Inherent Safety' in a General Industrial Setting in the Dutch Province of Zeeland," *2007 Spring National Meeting, American Institute of Chemical Engineers* (Houston, TX, 2007).
17. M. Papadaki, "Inherent Safety, Ethics and Human Error," *Journal of Hazardous Materials* 150, no. 3 (2008): 826–830.
18. K. Ishiguro, *A Pale View of the Hills* (New York: Putnam, 1982), 129.
19. M. Hilton, "How Far Dare We Go?" *The World* (June 1995): 222. The author is discussing the problems of interreligious dialogue.

Further Reading

Center for Chemical Process Safety. *International Conference and Workshop on Process Safety Management and Inherently Safer Processes*. New York: American Institute of Chemical Engineers, 1996.

Center for Chemical Process Safety. *Inherently Safer Chemical Processes: A Life Cycle Approach*. 2nd ed. Hoboken, NJ: John Wiley & Sons, 2009.

Crowl, D. A., and Louvar, J. F. Chap. 1 in *Chemical Process Safety: Fundamentals with Applications*. 2nd ed. Upper Saddle River, NJ: Prentice Hall PTR, 2002.

Englund, S. M. Chap. 2 in *Prevention and Control of Accidental Releases of Hazardous Gases*, edited by V. M. Ethenakis. New York: Van Nostrand Reinhold, 1993.

Green, D. W., and Perry, R. H. Chap. 23, pp. 23-38–23-39, in *Perry's Chemical Engineers' Handbook*. 8th ed. New York: McGraw-Hill, 2008.

Hendershot, D. C. Chap. 2 in *Guidelines for Engineering Design for Process Safety*, edited by the Center for Chemical Process Safety. New York: American Institute of Chemical Engineers, 1993.

Kletz, T. Chap. 21 in *What Went Wrong? Case Histories of Process Plant Disasters and How They Could Have Been Avoided*. 5th ed. Oxford, U.K.: Gulf Professional Publishing, 2009.

Mannan, S., editor. Chap. 32 in *Lees' Loss Prevention in the Process Industries*. 3rd ed. Oxford, U.K.: Elsevier Butterworth-Heinemann, 2005.

Appendix: An Atlas of Safety Thinking

In his *Atlas of Management Thinking*,[1] Edward de Bono says that simple pictures can be more powerful than words for conveying ideas. His book is a collection of what he calls "nonverbal sense images for management situations." "The drawings," he says, "do not have to be accurate and descriptive, but they do have to be simple enough to lodge in the memory. They should not be examined in detail in the way a diagram is examined, because they are not diagrams. They are intended to convey the 'flavour' of the situation described."

In the following, I have tried to express some safety and loss prevention ideas in similar simple drawings in the hope that the ideas expressed may stick in people's memories better than when these concepts have been expressed in words.

A.1 Intensify

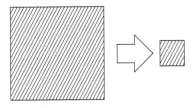

Adapted by permission of *The Institute of Chemical Engineers*.

The best way of preventing large leaks of hazardous materials is not to have so much hazardous material about; to use so little that it does not matter if it all leaks out or all explodes. By using our skill as chemical engineers, it is possible to achieve the same throughout with a lower inventory. For example,

1. Nitroglycerin used to be made in batch reactors containing about 1 tonne. Now it is made in small, continuous reactors containing a few hundred grams.
2. Reaction volumes are often large, not because the rate of reaction is slow, but because the mixing is poor. Various methods of improving mixing have been devised.
3. The Higee system, in which distillation is carried out in a rapidly rotating packed bed, can yield a thousandfold reduction in inventory.

Intensification results not only in a safer plant, but in a cheaper one—cheaper for two reasons. Because we do not have so much hazardous material present, we do not need such big vessels, pipes, structures, foundations, and so on. And because there is less hazard, we do not need to add so much protective equipment such as trips, alarms, emergency valves, and fire protection.

A.2   Substitute

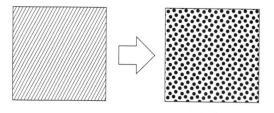

If we cannot reduce the inventory of hazardous material, or use it under less hazardous conditions, perhaps we can use a safer material instead. For example,

1. Water under pressure can be used as a heat-transfer medium instead of flammable oils.
2. Fluorinated hydrocarbons have been used as refrigerants instead of propylene, ethylene, or ammonia.
3. New routes can sometimes be found through which hazardous raw materials or intermediates are avoided.

Substitution results in a cheaper plant as well as a safer one because we do not need to add so much protective equipment.

A.3   Attenuate

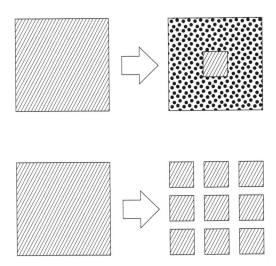

If we cannot reduce the inventory of hazardous materials, perhaps we can use the hazardous materials under less hazardous conditions, that is, at lower temperatures or pressures, as vapors or dissolved in safe solvents. For example,

1. Large quantities of anhydrous ammonia are now stored refrigerated at low temperature and not under pressure at ambient temperature. If a leak occurs, less vapor will be produced.
2. A polypropylene process uses gaseous propylene instead of liquid propylene dissolved in a flammable solvent.
3. In a continuous method for the production of nitro-aromatics, excess acid is used to dilute the aromatics and make a reaction runaway impossible.

Attenuation results in a cheaper plant as well as a safer one because we do not need to add so much protective equipment.

A.4 Simplify

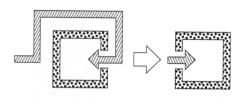

Modern plants are very complicated, and this makes them expensive and provides too many opportunities for error. The following are several reasons for the complexity:

1. The need to keep hazards under control so that intensification, substitution, and attenuation will give us simpler plants as well as cheaper, safer ones.
2. A slavish following of rules, codes, and accepted practice when they are not appropriate.
3. A desire for flexibility.
4. A reluctance to spend money on simplification, though we spend it willingly on complication.
5. Most important of all, a failure to recognize ways of simplifying plants until too late in design. If we could recognize hazards early, we could often avoid them by a change in design instead of controlling them by adding protective equipment. So the final message is ...

A.5 Change Early

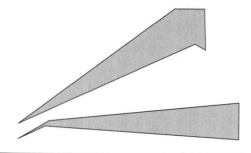

Here are some examples of ways in which plants have been simplified by carrying out Hazop-type studies early in design, on the flowsheet, before detailed engineering design starts and long before a conventional Hazop (hazard and operability) study is carried out:

1. By using stronger vessels, we can sometimes avoid the need for large relief valves and the associated flare systems. The need for the change must be recognized early in design before the vessels are ordered.
2. By using grades of steel suitable for low temperature, we may be able to avoid complex control and trip systems designed to prevent the equipment from getting too cold. The need for the change must be recognized early in design before the equipment is ordered.
3. By introducing intensification, substitution, or attenuation changes early in design, we can make our plants simpler and safer.

Reference

1. E. de Bono, introduction to *Atlas of Management Thinking* (Aldershot, U.K.: Maurice Temple Smith; London: Penguin Books, 1983).

Index

An environmentally friendly book printed and bound in England by www.printondemand-worldwide.com

PEFC Certified

This product is
from sustainably
managed forests
and controlled
sources

www.pefc.org

This book is made entirely of sustainable materials; FSC paper for the cover and PEFC paper for the text pages.

#0125 - 170314 - C0 - 234/156/20 [22] - CB